Springer-Verlag London Ltd.

Rajkumar Roy, Mario Köppen, Seppo Ovaska,
Takeshi Furuhashi and Frank Hoffman (Eds)

Soft Computing and Industry

Recent Applications

With 401 Figures

 Springer

Rajkumar Roy, PhD
Enterprise Integration, Cranfield University, Cranfield, Bedford, MK43 0AL, UK

Mario Köppen, Diplom-Phys
Department of Pattern Recognition, Fraunhofer IPK-Berlin, Pascalstrasse 8-9, 10587 Berlin, Germany

Seppo Ovaska, PhD
Helsinki University of Technology, Institute of Intelligent Electronics, Otakaari 5 A, FIN-02150 Espoo, Finland

Takeshi Furuhashi, PhD
Department of Information Engineering, Mie University, 1515 Kamihama-cho, Tsu 514-8507, Japan

Frank Hoffmann,
NADA/CVAP, Royal Institute of Technology, 10044 Stockholm, Sweden

British Library Cataloguing in Publication Data
Soft computing and industry : recent applications
    Industries - Data processing - Congresses 2.Soft
    computing - Industrial applications - Congresses
    3.Intelligent control systems - Congresses
    I.Roy, R. (Rajkumar), 1966- II.On-line World Conference on
    Soft Computing in Engineering Design and Manufacture (6th : 2001)
    670.2'8563
    ISBN 978-1-4471-1101-6   ISBN 978-1-4471-0123-9 (eBook)
    DOI 10.1007/978-1-4471-0123-9

Library of Congress Cataloging-in-Publication Data
A catalog record for this book is available from the Library of Congress.

ISBN 978-1-4471-1101-6

http://www.springer.co.uk

© Springer-Verlag London 2002
Originally published by Springer-Verlag London Limited in 2002

Typesetting: Camera ready by editor
Printed and bound by Athenæum Press Ltd., Gateshead, Tyne & Wear
69/3830-543210   SPIN 10834744

and suppression calculated by the fast fuzzy neural network [17], and k denotes the dispassion factor to ensure the global stability of the immune network. The described system is illustrated in Figure 1.

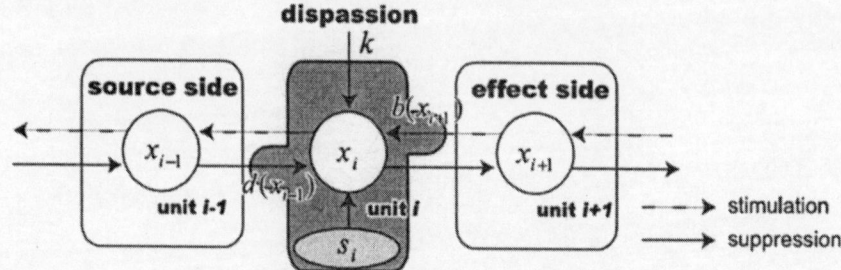

Figure 1. Concentration of antibody

In this paper, feedback systems are decomposed into decision tree structure which has only the forward passes with branches using fuzzy decision tree concept [6] based on knowledge obtained by simulations, as shown in Figure 2, which is automatically achieved by our developed fast fuzzy neural network with general parameter (GP) learning [17].

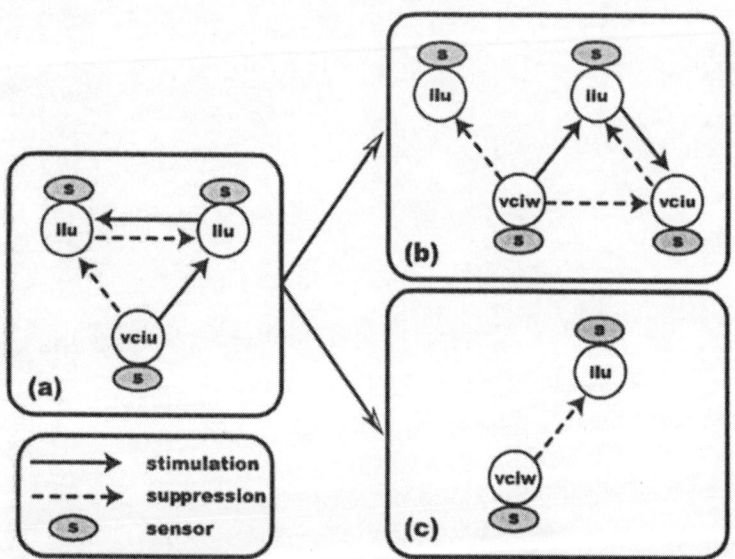

Figure 2. Decomposition of systems into tree structure

2.2 Simulation Results

The UPS system used in our research is shown in Figure 3. For simplicity, we consider four sensors, that is, an inverter line current (iiu), a load line current (ilu), a load u-phase voltage ($vciu$), and a load w-phase voltage ($vciw$).

8

self-tuned using evolutionary computation [18]. The result of sensor fault detection is shown in Figure 6.

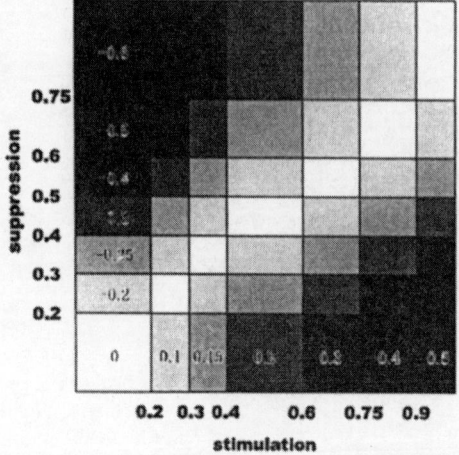

Figure 5. Ratio of stimulation and suppression

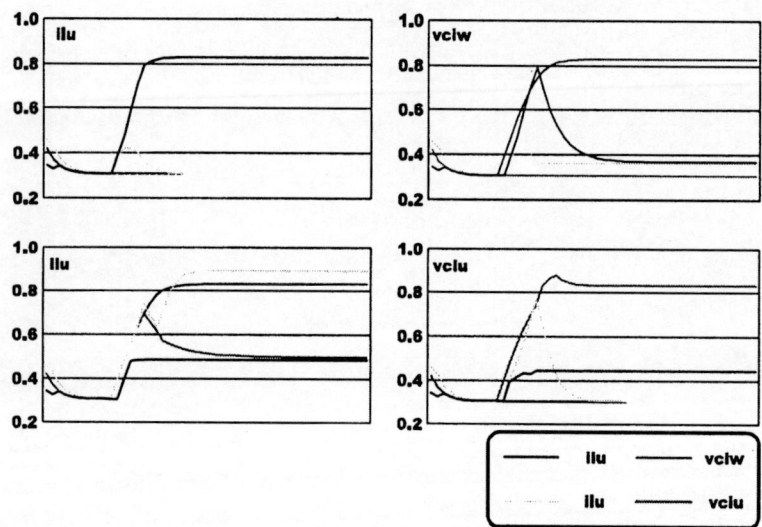

Figure 6. Result of sensor failure detection

3 Surviving robot in a changeable environment [19]

3.1 A surviving robot in a changeable environment using decision making by immune networks

An artificial decision making robot 'immunoid' by interactions among antibodies in artificial immune networks is considered. In this simulated environment, there are following three kinds of objects: (1) predator, (2) obstacles, and (3) food. It is assumed that pre-specified quantity of initial energy is given to the immunoid at

other languages as well. In order to link the MATLAB and JAVA programs, a path, describing where on the computer the JAVA files are stored, has to be added to the class-path file of MATLAB. The input values are received in MATLAB and sent to the JAVA program, where the calculations are performed. The calculated values are then returned to MATLAB, where a graphical representation of the results is produced. The above-mentioned software architecture is pictorially illustrated in Figure 4.

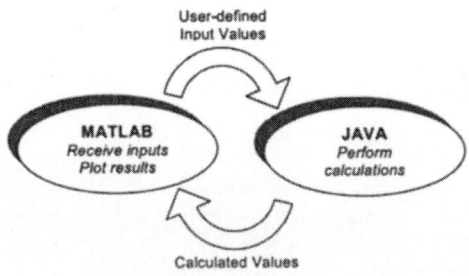

Figure 4. Software architecture

6.2   Case study 1: Design of gas turbine blade cooling system

A preliminary model of the gas turbine blade cooling system was developed in collaboration with Rolls Royce plc. (Bristol, UK) and Plymouth Engineering Design Centre [9]. This model is developed considering one dimensional, single pass coolant flow. This represents a computationally inexpensive mathematical model of the blade cooling system. The model includes a film cooling mechanism and involves twelve design variables. This Turbine Blade COoling system Model (TBCOM) also uses several constants known as design parameters. The values of the constants were set by the design experts from Rolls Royce plc., but may not represent the current practice in the company. TBCOM also includes three nonlinear constraints [9].

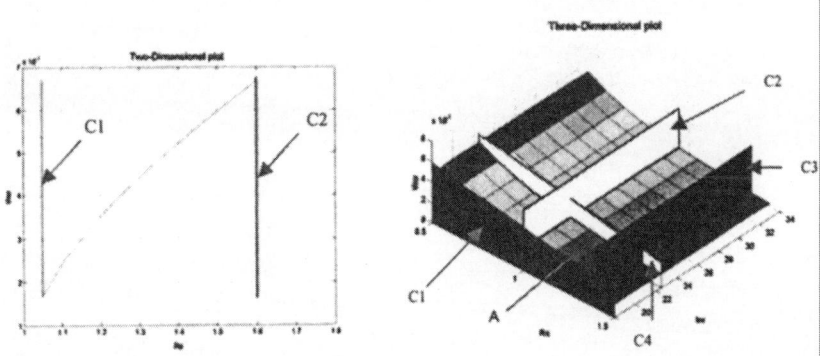

Figure 5. One variable against output   **Figure 6.** Two variables against output

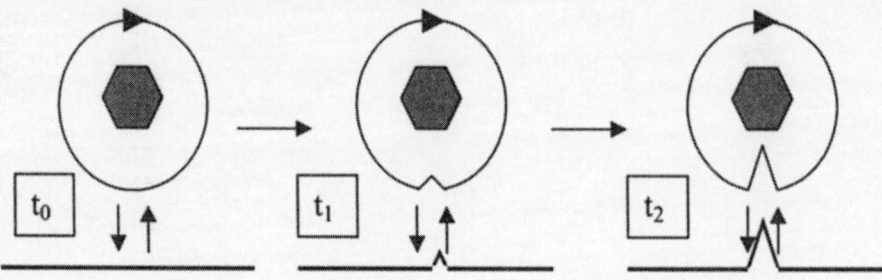

Figure 1. Structural coupling. The diagram depicts the basis of the mechanism of Structural Coupling. The autopoietic system, represented by a circle and defined by its structure and its organization (hatched area), initially confronts a medium without organized "objects" (at t_0). As recurrent interactions (represented by the arrows) between the medium and the system are stabilized, at t_1, an "object" (represented by the triangle) begins to be configured. The "objects" is made of two complementary parts. One part exists in the medium and the other exists as a change in the autopoietic system structure. Finally, at t_2, the "object" is totally configured. At this point the change in structure, but not in organization as the hatched area remains unchanged, could be very important. From the point of view of computing the important aspect is the existence of "objects" defined by spatio-temporal correlations, thus the change in the autopoietic system structure also contains spatio-temporal corrrelations. Thus the study of the temporal changes in the systems's structure can predict the changes in the environment. This scheme of prediction works because the (spatio-temporal) objects in the environment are congruent with the structure of the autopoietic system. The congruence is a consequence of structural coupling.

4.2 Anticipation

An anticipatory system is a system that computes its future state, $X(t+1)$, not only by taking into account its present state $X(t)$ and the perturbation just received $P(t)$, but by taking into account a prediction (or model) of its future state, $\hat{X}(t+1)$. Thus, according to the notion of anticipation, the future of any system can be computed according to one of the following three models:

lesson when we have to deal with a ticket agent whose kids are sick and car is in the shop and vacation just got canceled.

Figure 1. Education Yesterday – Probably Yours, Certainly Mine!

Panels 1.e and 1.f, both involve *books*. I cannot overemphasize how strongly I feel about good reading skills, learned early. Just as running is exercise for the body, reading is exercise for the mind. Read this paragraph from Poe's *Descent into the Maelström* [3]:

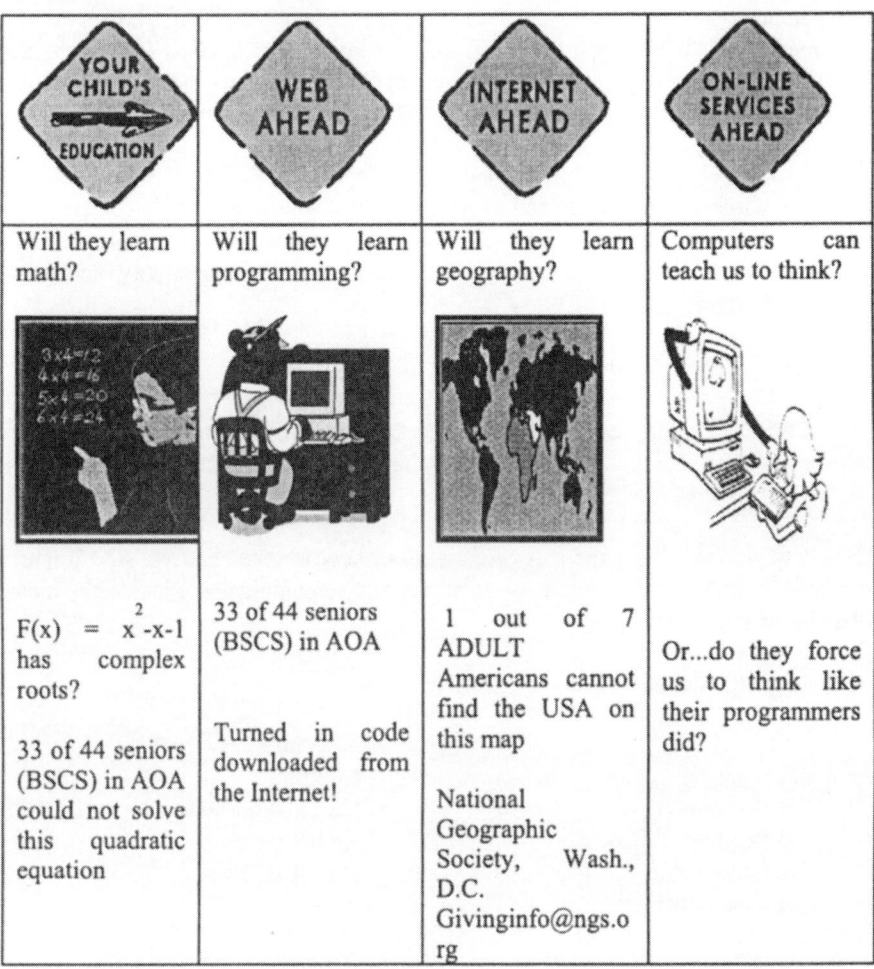

Will they learn math?	Will they learn programming?	Will they learn geography?	Computers can teach us to think?
$F(x) = x^2 - x - 1$ has complex roots? 33 of 44 seniors (BSCS) in AOA could not solve this quadratic equation	33 of 44 seniors (BSCS) in AOA Turned in code downloaded from the Internet!	1 out of 7 ADULT Americans cannot find the USA on this map National Geographic Society, Wash., D.C. Givinginfo@ngs.org	Or...do they force us to think like their programmers did?

Figure 2. Where is computerized education taking us?

The last panel in Figure 2 challenges you to think about the real utility of automated courses and intelligent tutoring systems. There are lots of studies around nowadays touting *the oh-so-much-better* skills of students who take courses from computers, as opposed to the old, human instructor way. I bet you can find hundreds of such studies as references *to National Science Foundation* (NSF) proposals on intelligent tutoring systems. Well, what do you think? *Can* students learn more about, say, data structures, from a computer than they can from you? If your measure of learning is "knowledge" about data structures, gauged by performance on examinations, it's probably true.

What's wrong with this? Well, students who learn from computers learn exactly what the computer is prepared to teach them. This scheme encourages them to think incrementally – "if I can just knock off this question, the computer will skip me along to that big, red box". Kind of like Pokemon, or PacMan isn't it? I see lots of problems with distance learning and automated tutoring. Here are some authors who agree with me:

"A rising number of educators and experts warn that software not only controls how a computer performs its increasingly sophisticated workload, but that it also channels the way people think, write and work in ways that stifle creativity and ground breaking thought"

Nesbitt and Barnett, Wall Street Jo. June 3, 2000

"Software isn't benign; the way it allows consumers to perform tasks and the choices it allows a computer user to make affect how that person thinks, acts, and works"
IBID

"Students who use the same software are less likely to take chances. And important ingredients in the creative process - creativity and serendipity - are lost to the dictates of a software template."

Al Young, UC Berkeley, as quoted in IBID

No Practice	-	No Skill
No Questions	-	No Thinking
No Reading	-	No Imagination
No Imagination	-	No Creativity
No Creativity	-	No Science

Figure 3. The input-output system of computer learning: a crisp rule base for Y2K

Figure 4. Education Tomorrow – Your Children, My Grandchildren?

Figure 4 gives you an idea of my view of the future. How future? Oh, 50 years or so. As you look over the panels in this Figure, do a little pattern matching with yourself, or your children, or my grandchildren. How many panels in Figure 4 do your kids already match? Give yourself one point for each match in Figure 4. If your score is greater than x, award yourself a new mother board. What is x? You decide.

96

References

[1] Rand, Ayn (1957). Atlas Shrugged, republished by Mass Market Paperbacks, NY (1996).

[2] Bezdek, J. C. (2000). Quo Vadis Computational Intelligence? Perseus and the Gorgon: A Cautionary Tale, in Quo Vadis Computational Intelligence?, eds. P. Sincak and J. Vascak, Physica-Verlag, Heidelberg, V-XIV.

[3] Poe, E. A. (1809-1849). A Descent into the Maelstrom, in *The Collected Tales and Poems of Edgar Allan Poe*, Random House, NY, 1992.

[4] Feynman, Richard P. (1996). *Lectures on Computation*, 1984-1986, eds. Tony Hey and Robin W. Allen, Perseus Publ., Cambridge, 15-16.

"Imagination is more important than knowledge"

for resources. A position of the holon in the list is calculated with the scheduling strategy for the resource imposed by the system operator.

To optimise the scheduling a master scheduler is being developed. It will act as a auxiliary holon that will be able to generate the outline of the production plan for the nearest future. Such a plan will be a guideline for the order holons

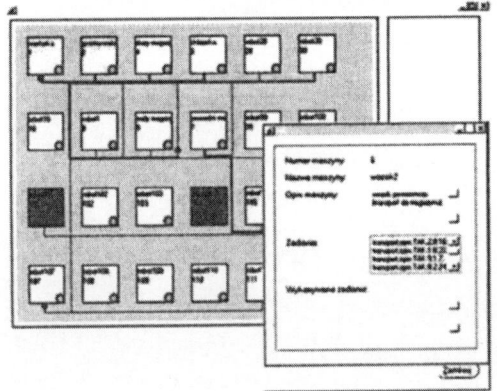

Fig. 2. Map generated by the resource holon

when to request a resource. The introduction of the master plan would however require to define the preferred execution path for each order.

Some agents were equipped with a user interface to make possible the presentation of the system's work to the human. Fig. 2 presents an example map of the resources defined inside the resource database. The map is generated automatically by the resource agent and does not reflect the physical arrangement of the actual manufacturing system. All the machines and storage areas are represented there as a code-coloured rectangles and the transportation paths as lines connecting the rectangles. The availability of a resource (excluding the transport utilities) is indicated with the green dot in the right lower corner of each rectangle. The availability of the transport is indicated by the existence of the line and by the position of the dot (trolley) on the line. Clicking on a resource pops up an information window with the database records on this resource.

References

1. Cechowicz R. (1998) Komputerowy system szeregowania zadan w wielomaszynowym srodowisku wytwarzania: PhD Thesis, Technical University of Lublin, Lublin, Poland
2. Iimura J. (1993): Unskilled Worker-Oriented Manufact., Human Intelligence-Based Manufacturing, Advanced Manufact. Series, Springer Verlag
3. Koestler A. (1969) The Ghost in the Machine, Arkana Books, London
4. Minski, M (1985) The Society of Mind, Heinemann
5. Van Brussel, H., J. Wyns, P. Valckenaers, L. Bongaerts, P. Peeters, (1998) Reference Architecture for Holonic Manufacturing Systems: PROSA, Computers In Industry, Special Issue on Intelligent Manufacturing Systems, Vol. 37, No. 3, pp. 255 - 276
6. Waldrop, M. (1992) Complexity, The emerging science at the edge of order and chaos, Viking, Penguin group

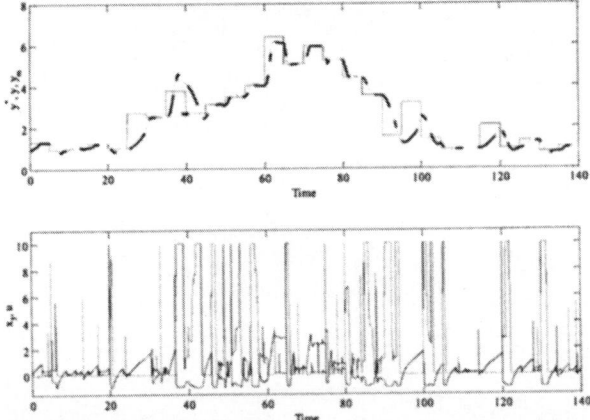

Fig. 3. Evaluation of the performance of hybrid model.
The solid line is the system's output, the dotted line is the prediction of the model (on top). The good prediction performance is also reflected by the perfect estimation of the steady-state characteristic of the process (see Fig. 4).

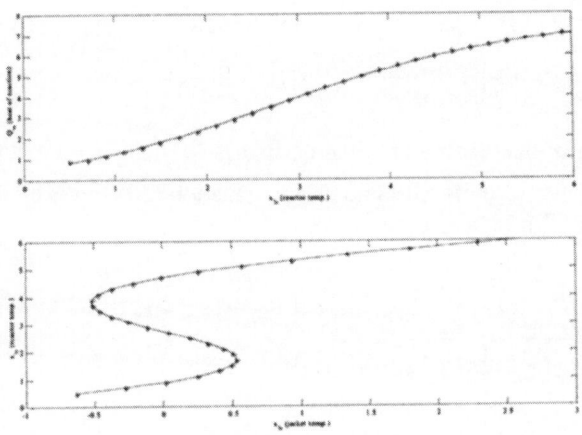

Fig. 4. Prediction of the steady-state behavior of the process.

5.3 Control Scheme

For the temperature control of the chemical reactors, it is advantageous to use cascade-control scheme, where the slave-controller is responsible for the control the jacket temperature based on the set-point given by the master controller as it is shown in Fig. 5.

120

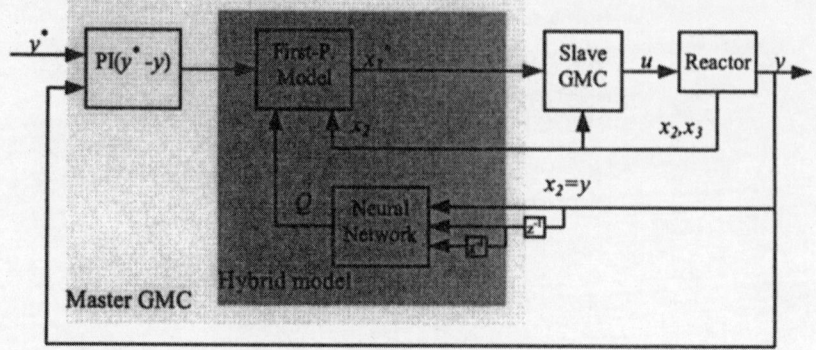

Fig. 5. Closed-loop structure of CSTR with cascade control

The slave controller is also a GMC controller, whose desired closed-loop trajectory is given by equation (3):

$$(\dot{x}_3)^* = k_{1s}(x_3^* - x_3) + k_{2s} \int_0^t (x_3^* - x_3) d\tau \tag{20}$$

With the use of (18) the control input can be analytically expressed by:

$$u = \left((\dot{x}_3)^* - \frac{\delta}{\delta_1 \delta_2}(x_2 - x_3) \right) \frac{\delta_1}{(x_{3f} - x_3)} \tag{21}$$

The input of this controller is the desired jacket temperature, x_3^* given by the master controller (see Fig. 5). Similarly to equation (20) and (21), the control rule of the master GMC controller is

$$x_3^* = \left(\underbrace{(y)^* - NN(x_2(k), x_2(k-1), x_2(k-2))}_{(y)_{FP}} - q(x_{2f} - x_2) \right) \frac{1}{\delta} x_2 \tag{22}$$

where $(\dot{y})^* = k_{1s}(y^* - y) + k_{2s} \int_0^t (y^* - y) dt$

The results were compared to the case when the concentration was measured and the control relevant model was perfect (Fig. 6a, 6b)

matical models for the particular application (according to the type and complexity of the plant or system) and the correct architecture of the neural networks for identification and control. Initial training data can then be used to obtain the initial weights for the networks. The intelligent control system will then be ready for use on-line in the real plant or dynamical system. We have implemented a prototype intelligent control system, with the new hybrid approach for control, using the MATLAB© programming language.

4 Simulation Results for the Inverted Pendulum

To give an idea of the performance of our neuro-fuzzy approach for adaptive model-based control of non-linear plants, we show below simulation results obtained for an inverted pendulum. The desired trajectory for the link was selected to be

$$q_d = t\sin(2.0t) \tag{12}$$

and the simulation was carried out with the initial values: $q(0) = 0.1$ $q'_1(0) = 0$

We used three-layer neural networks (with 15 hidden neurons) with the Levenberg-Marquardt algorithm and hyperbolic tangent sigmoidal functions as the activation functions for the neurons. We show in Figure 2 the function approximation achieved with the neural network for control after 9 epochs of training with a variable learning rate. The identification achieved by the neural network can be considered very good because the error has been decreased to the order of 10^{-4}. We show in Figure 3 the curve relating the sum of squared errors SSE against the

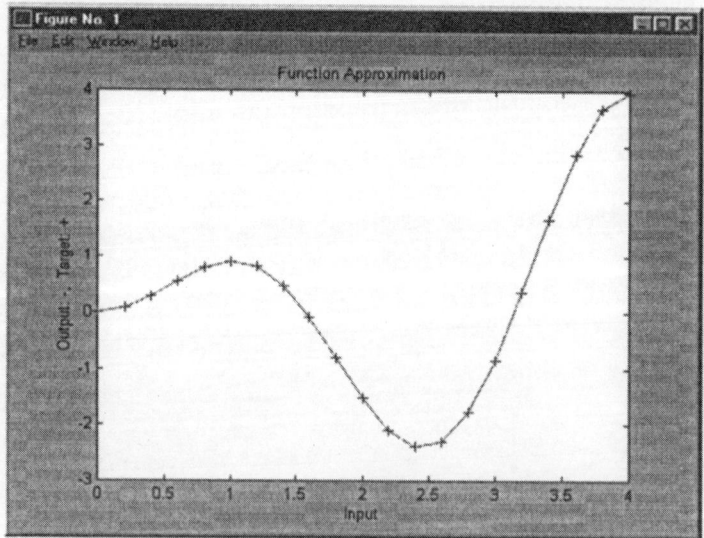

Fig. 2. Function approximation after 9 epochs

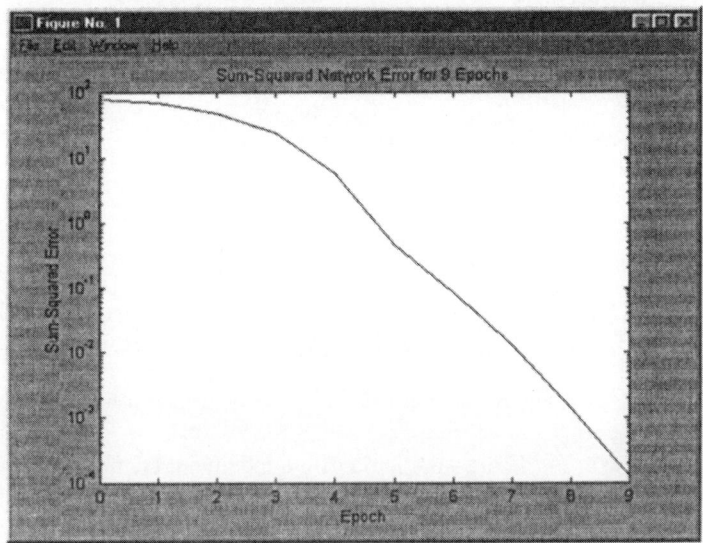

Fig. 3. Curve relating error with number of epochs

number of epochs of neural network training. We can see in this figure how the SSE diminishes rapidly from being of the order of 10^2 to smaller value of the order of 10^{-4}. Still, we can obtain a better approximation by using more hidden neurons or more layers. In any case, we can see clearly how the neural networks learns to control the robotic system, because it is able to follow the arbitrary desired trajectory.

We show in Figure 4 the fuzzy rule base for controlling the inverted pendulum. The fuzzy system was implemented in the fuzzy logic toolbox of MATLAB. We show in Figure 5 the non-linear surface for modelling the non-linear plant. Finally, we show in Figure 6 the membership functions for the angular velocity variable.

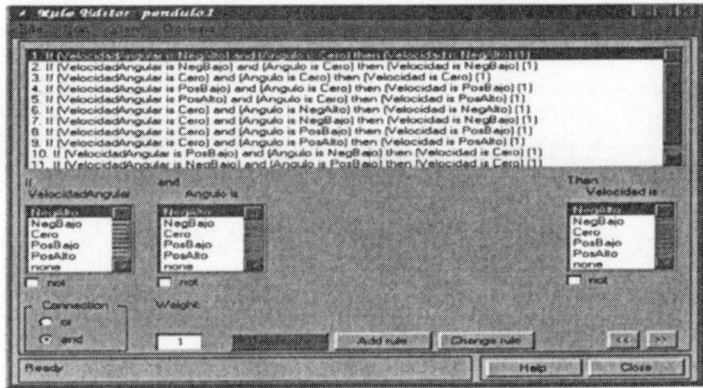

Fig. 4. Fuzzy rule base for control of the plant

130

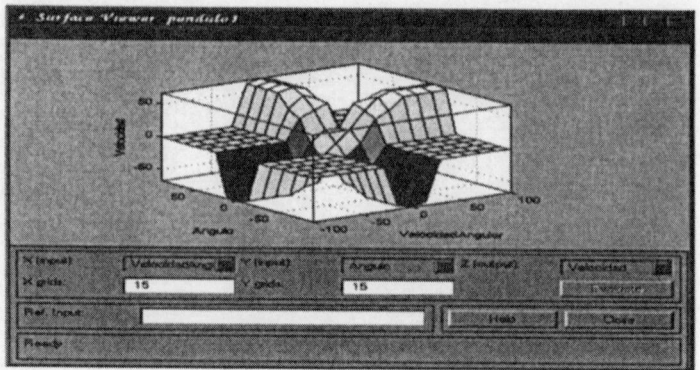

Fig. 5. Non-linear surface for modelling the plant

We also show in Fig. 7 the simulation of the intelligent control system for the inverted pendulum. We implemented the simulation in the SIMULINK toolbox of the MATLAB programming language.

Summarizing, we can say using a set of fuzzy rules for controlling the non-linear plant. However, still we can improve the performance of intelligent system for control by using a hybrid approach combining the advantages of neural networks and fuzzy logic. The neuro-fuzzy approach can be used to optimize the parameters of the fuzzy system (like in the ANFIS method). Another approach is to use neural networks for control and fuzzy logic for modelling the non-linear plant. In this case, we obtained the best results with the last approach.

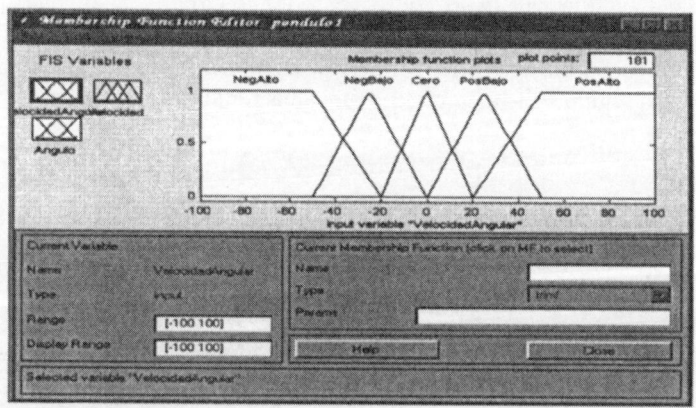

Fig. 6. Membership functions for the angular velocity

- a separator which electrically isolates the positive and negative electrodes.

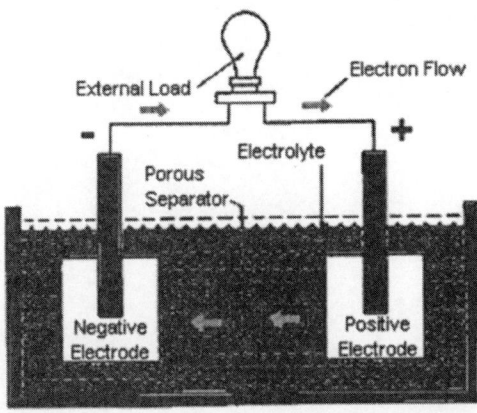

Fig. 1. Components of a battery cell

In some designs, physical distance between the electrodes provides the electrical isolation and the separator is not needed. In addition to the critical elements listed above, cells intended for commercial batteries normally require a variety of packaging and current collection apparatus to be complete.

2.2 How a Cell Works

When a battery or cell is inserted into a circuit, it completes a loop, which allows charge to flow uniformly around the circuit. In the external part of the circuit, the charge flow is electrons resulting in electrical current. Within the cell, the charge flows in the form of ions that are transported from one electrode to the other. As mentioned above, the positive electrode receives electrons from the external circuit on discharge. These electrons then react with the active materials of the positive electrode in "reduction" reactions that continue the flow of charge through the electrolyte to the negative electrode. At the negative electrode, "oxidation" reactions between the active materials of the negative electrode and the charge flowing through the electrolyte results in surplus electrons that can be donated to the external circuit. It is important to remember that the system is closed. For every electron generated in an oxidation reaction at the negative electrode, there is an electron consumed in a reduction reaction at the positive. As the process continues, the active materials become depleted and the reactions slow down until the battery is no longer capable of supplying electrons. At this point the battery is discharged.

2.2.1 Recharging

The world of batteries divides into two major classes: primary and secondary batteries. Primary batteries such as the common flashlight battery are used once and replaced. The chemical reactions that supply current in them are irreversible. Secondary batteries (for example, car batteries) can be recharged and reused. They use reversible chemical reactions. By reversing the flow of electricity i.e. putting

137

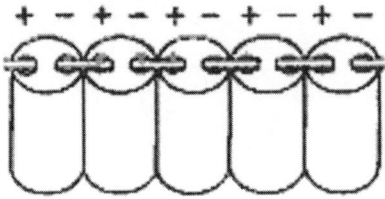

Fig. 2. Series Connected Batteries

The other way to connect cells within a battery is to connect the negative terminal from one cell to the negative of the next cell and to connect the positive terminal to the positive terminal. When this is done throughout the battery, the result is the parallel-connected battery shown in Figure 3. Here the capacities of the individual cells add to make the battery capacity but the battery voltage remains as the voltage of the individual cell. Series-connected batteries are far more common than parallel-connected. Usually it is easier to get added capacity by just using a larger cell rather than a parallel-connected battery. All of the battery connections may be made internally so that it is difficult to determine the number of cells by external examination. However, knowing the voltage of the basic cell, it is easy to determine the number of cells by dividing the cell voltage into the battery voltage. Cells used for batteries should always be identical. Mixing cells of different chemistry or different size may be hazardous and should be avoided.

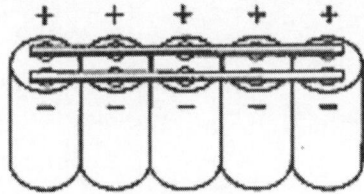

Fig. 3. Parallel Connected Batteries

While there are various choices for a rechargeable system, lead-acid batteries are still the workhorses. They represent about 60% of all batteries sold worldwide. Lead-acid batteries are usually more economical and have a high tolerance for abuse. The fact that all of the batteries used for starting, lighting, and ignition (SLI) service on automobiles and trucks are lead-acid indicates their ability to withstand varied forms of maltreatment. Lead-acid batteries also provide motive power for everything from forklifts to submarines. Lead-acid batteries are also mainstays of the backup systems that provide power when the electrical grid fails. Now, development of the sealed-lead battery has allowed lead technology to be used in applications such as electronics that need a clean power source. All lead batteries work on the same set of reactions and use the same active materials. At the positive electrode, lead dioxide (PbO_2) is converted to lead sulfate ($PbSO_4$) and at the negative electrode, sponge metallic lead (Pb) is also converted to lead sulfate ($PbSO_4$). The electrolyte is a dilute mixture of sulfuric acid that provides the sulfate ion for the discharge reactions.

membership functions were tuned manually until they give the best values for the problem.

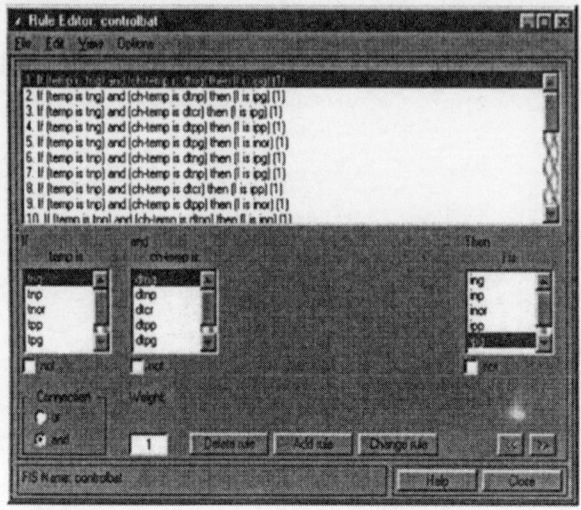

Fig. 5. Fuzzy rule base for controlling the process.

5. Neuro-Fuzzy Method for Control

Since it is difficult to tune a particular inference system to model a complex dynamical system [2] it is convenient to use adaptive fuzzy inference systems. Adaptive neuro-fuzzy inference systems (ANFIS) can be used to adapt the membership functions and consequents of the rule base according to historical data of the problem [6]. In this case, we can use the data from Table 2 and apply the ANFIS methodology to find the best fuzzy system for our problem. We used the fuzzy logic toolbox of MATLAB to apply the ANFIS methodology to our problem with 5 membership functions and first order Sugeno functions in the consequents. We show in Figure 6 the non-linear surface for control.

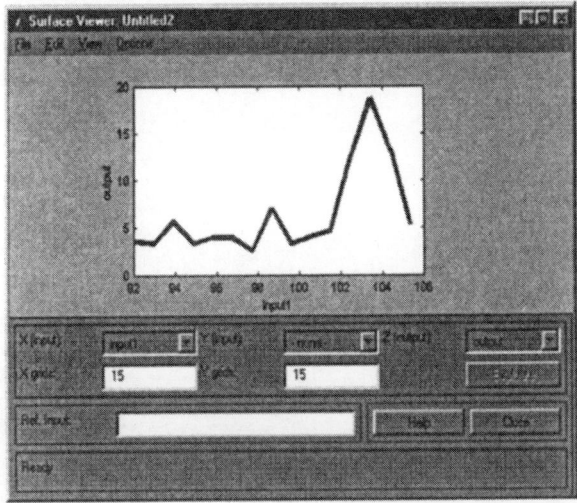

Fig. 6. ANFIS surface for the process

6. Neuro-Fuzzy-Genetic Control

In this case, neural networks are used for modelling the electrochemical process, fuzzy logic for controlling the electrical current and genetic algorithms for adapting the membership functions of the fuzzy system [2, 9]. A multilayer feedforward neural network was used for modelling the electrochemical process. We used the data form Table 2 and the Levenberg-Marquardt learning algorithm to train the neural network. We used a three layer neural network with 15 nodes in the hidden layer. The results of training for 2000 epochs are as follows. The sum of squared errors was reduced from about 200 initially to 11.25 at the end, which is a very good approximation in this case. The fuzzy rule base was implemented in the Fuzzy Logic Toolbox of MATLAB. In this case, 25 fuzzy rules were used because there were 5 linguistic terms for each input variable.

7. Experimental Results

The three hybrid control systems were compared, by simulating the formation (loading) of a 6 Volts battery. This particular battery is manually loaded (in the plant) by applying 2 amperes for 50 hours under manufacturer's specifications. We show in Table 3 the experimental results.

152

3 Experimental Results

We implemented the specific genetic algorithm in the MATLAB programming language to validate and measure the efficiency of our approach. We used as range for the coefficients of the filter the interval [-3,3] and the parameters of the genetic algorithm as described before. We have to mention that no elitism was used in the genetic algorithm, as previous results did not show that this operator improved the efficiency or accuracy of the results at all. Experiment 1 was done using 500 individuals in the population with 100 generations and with random noise as training signal. The parameters of the genetic algorithm are %T=0.30 δ=16, and 80 samples were used to calculate the ASE. The best fitness found was 68.08. We show in Figure 4 the unit impulse test signal, in which the error from sample 37 and forward is between the allowed limits. We also show in Figure 5 the unit step test signal. We show in Figure 6 the performance of the genetic algorithm for this case.

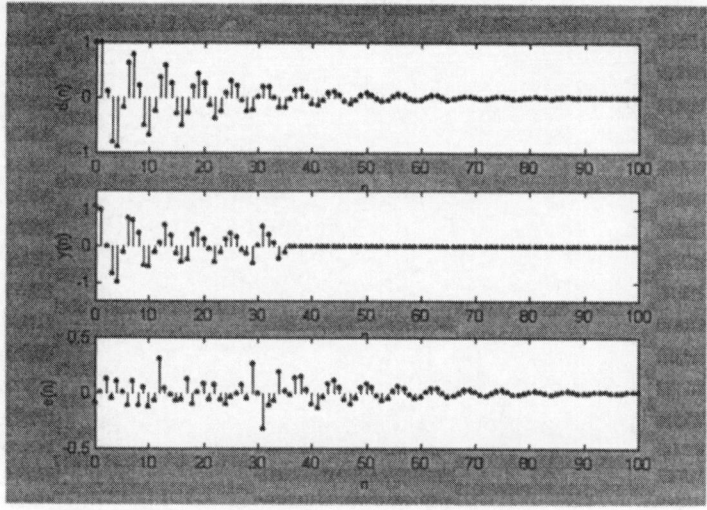

Fig. 4. . Unit impulse test signal (Experiment 1)

153

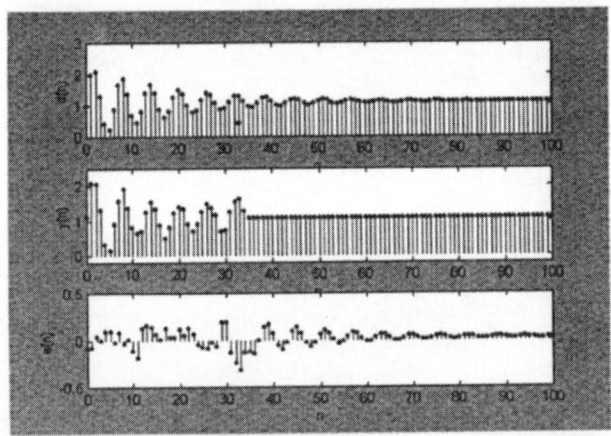

Fig. 5. Unit step test signal (Experiment 1)

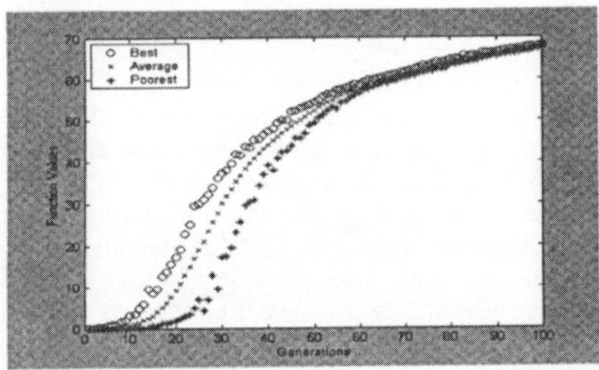

Fig. 6. Results of experiment 1 with a genetic algorithm (%T=0.30, δ=16, best fitness =68.08 using 80 samples to calculate ASE)

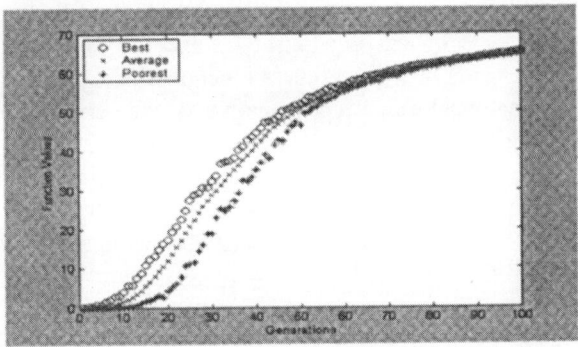

Fig. 7. Results of experiment 2 with the genetic algorithm (%T=0.20, δ=16, best fitness=65.53 using 80 samples to calculate ASE)

154

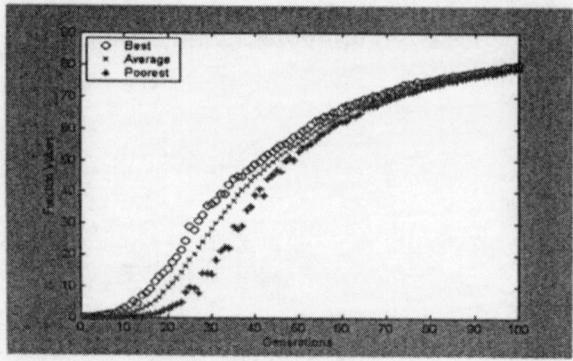

Fig. 8. Results of experiment 2 with a genetic algorithm (%T=0.30, δ=16, best fitness=80.21 using 60 samples calculate ASE)

4 Conclusions

We described in this paper a genetic algorithm to optimize the coefficients of an adaptive filter. The genetic algorithm can be used at least to find a small region, where a "hill climbing" method can be applied to improve the results. For example, we can use LMS as hill climbing, and we can have a very robust method because the GA can use its random nature to find feasible regions and then the LMS can fine tune the results, which is the principal advantage of this method. This could be a good combination because real world signals are complex, multimodal, and with discontinuities, which makes them very difficult for traditional search methods.

The BGA performs well for the problem of optimizing the finite impulse adaptive filter. The BGA is slower than LMS, however it is more robust. Consequently the BGA can be used, for the moment, for off-line applications to digital filter optimization. Regarding the training signals, the best results were achieved when using random noise. We also used as training signal for the filter the unit impulse signal, in this case the BGA was able to fit the signal but the coefficients were not good. Another conclusion is that the fitness function, defined as 1/ASE, requires a lot of samples to really achieve a representative average error. In this case, for a filter of 36 coefficients we needed 80 samples to achieved a good average error (this can be seen in the Figures of Experiment 1). We made a lot of tests and the behavior of the algorithm was consistent with results shown in Figures 5, 6, 7, and 8. There was no significant change in the behavior of the algorithm when δ was changed from 16 to 20.

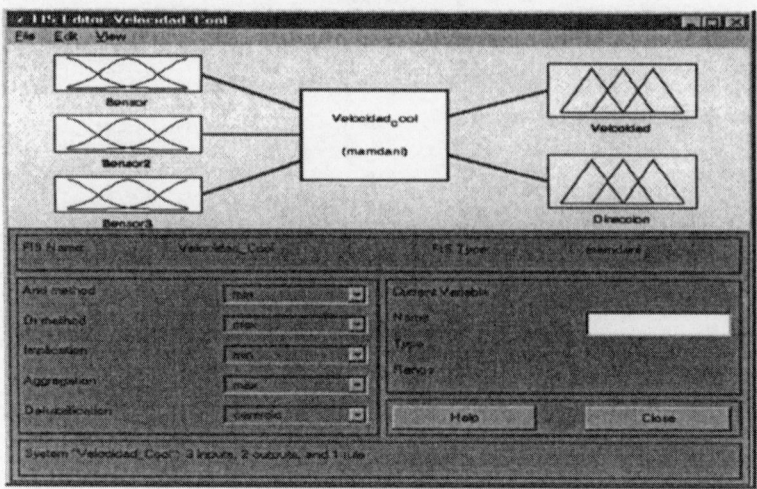

Fig. 1. General architecture of the fuzzy system for control

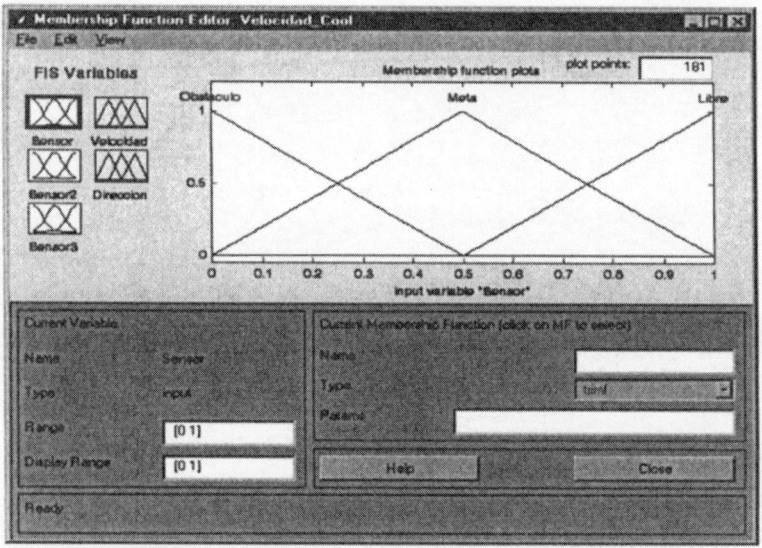

Fig. 2. Triangular membership functions for one of the input variables

We show in Fig. 3 the implementation of the fuzzy rule base for control in the MATLAB programming language. We also show in Fig. 4 the non-linear surface for controlling the robot. Finally, we show in Fig. 5 the rule viewer of MATLAB, where the fuzzy rules áre used for specific values of the input variables.

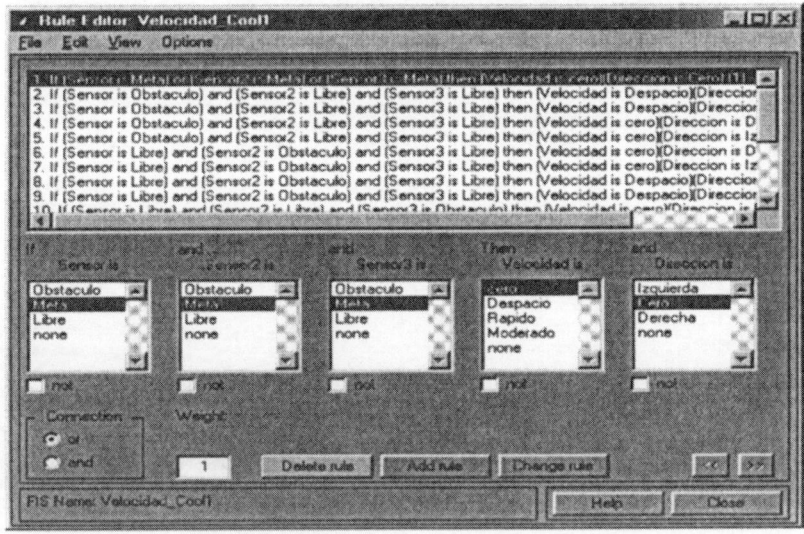

Fig. 3. Fuzzy rule base for controlling the robot

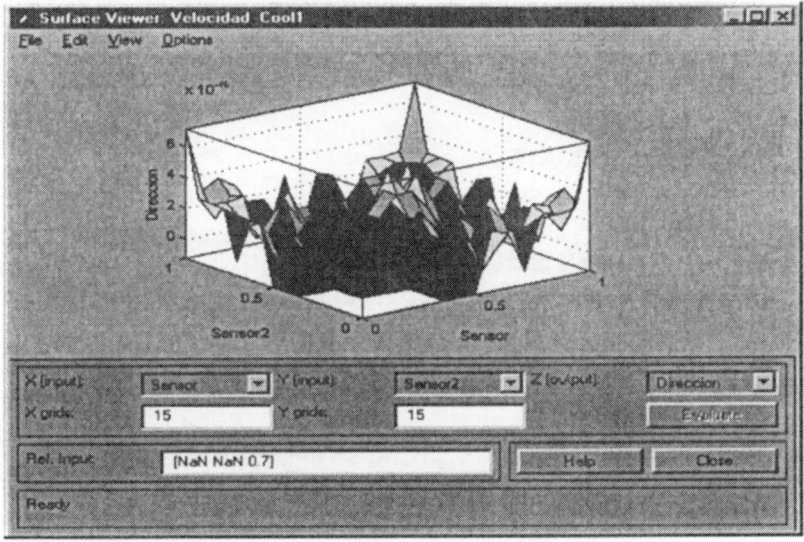

Fig. 4. Non-linear surface for controlling the robot

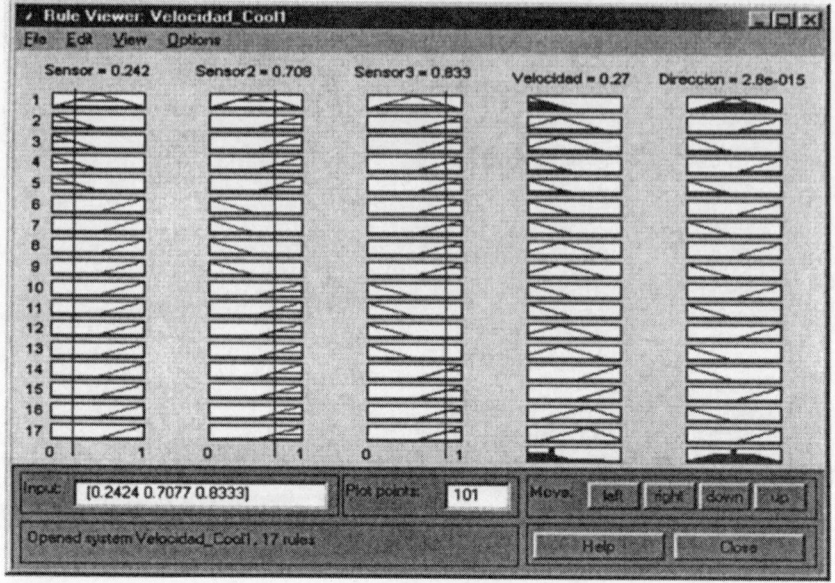

Fig. 5. Use of the rule viewer of MATLAB for specific values of the input variables

The fuzzy system for control was optimized using a specific genetic algorithm. We encoded the parameters of the fuzzy system an used a breeder genetic algorithm to find the best values possible for the fuzzy controller.

5 Simulation Results

To give an idea of the performance of our neuro-fuzzy-genetic approach for adaptive model-based control of robotic systems, we show below simulation results obtained for a simple autonomous robot. As mentioned before the robot has three sensors and the readings from these sensors are used as inputs in the fuzzy controller. We also use a neural network model of the robot as a reference model in the adaptive control. The breeder genetic algorithm is used to optimize the fuzzy controller.

To test our hybrid approach we have used a simulation program that was developed in MATLAB. With this computer program we can simulate the behavior of the autonomous robot for different conditions. We show simulation results only for two cases, to give an idea of the performance of the hybrid approach.

We show in Fig. 6 one simulation result of the robot avoiding an obstacle. We also show in Fig. 7 another simulation of the robot in which it is clear that the robot avoids hitting the obstacle completely. We have many simulation results similar to these ones, suggesting that the hybrid approach is a good alternative for this problem. Still, we need to consider controlling the real physical robot with this approach and we plan to do just that in the near future.

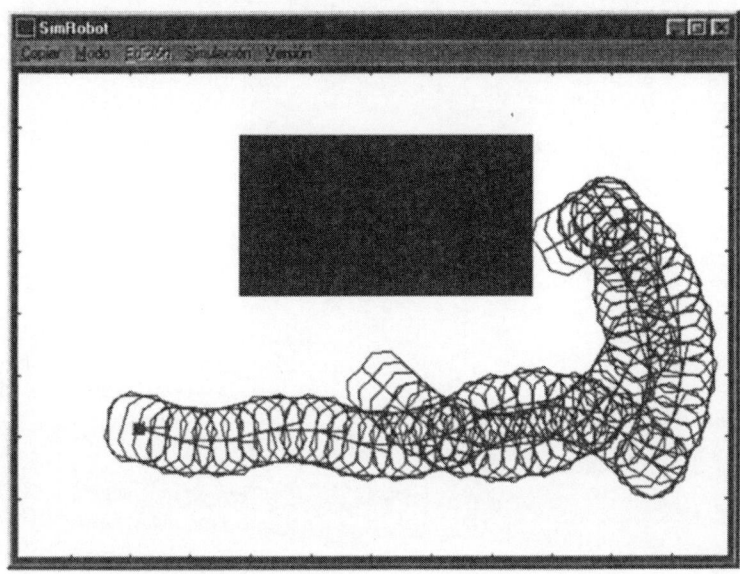

Fig. 6. Simulation of the robot avoiding an obstacle (case1)

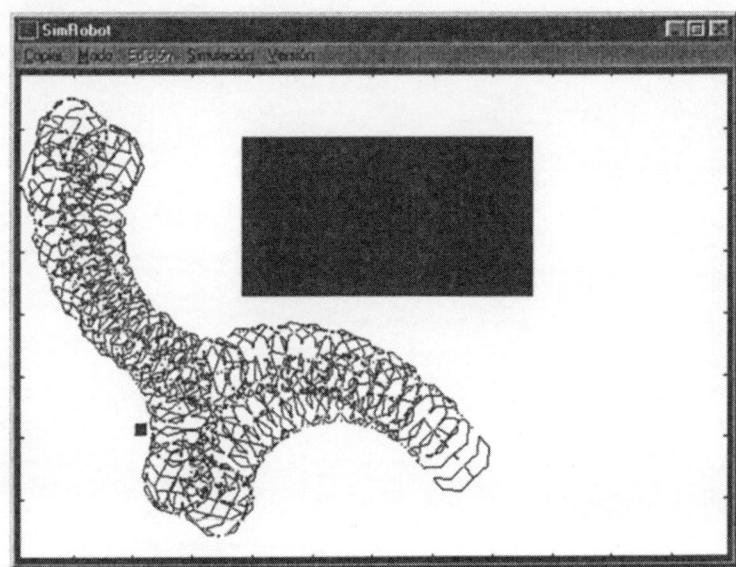

Fig. 7. Simulation of the robot avoiding the obstacle (case 2)

analyzed in section 3. The *GPC* design based on multiple models is studied in section 4. Finally, conclusions are given in section 5.

2 Fan-and-plate experimental process

The fan-and-plate experiment is composed of a fan driven by a DC motor, a 50 cm long air duct with funneling characteristic and having on its left extremity a small rectangular plate (see figures 1 and 2). The 24 volts DC motor is driven by an actuator circuit whose input that is compatible with the *D/A* converter output. The angular deflection of the plate is measured by a photoconductive cell and connected to the measurement circuit. The practical prototype presents a non-minimum phase, dead time, resonant and turbulent disturbance behavior. Therefore, it can used as tangible evidence of the usefulness of advanced control techniques in hard situations.

The evaluation of the nonlinear characteristic of the process is based on Thomson *et al.* [11]. So, from a normal operating level $[u_o(k), y_o(k)]$, apply a step change in input signal $(u_o(k)+\Delta u_1)$ to the process and record the output signal $y_1(k)$. Return the plant to its normal operating level and apply a second input signal $(u_o(k)+\Delta u_2)$ to the process, where the step change Δu_2 is ρ times larger that Δu_1, and record the output signal $y_2(k)$, i.e.

$$\rho(k) = \frac{\Delta u_2}{\Delta u_1} \tag{1}$$

Remove the nominal operating level y_o from each step response and compute

$$\delta(k) = \frac{y_2(k)-y_0(k)}{y_1(k)-y_0(k)} \tag{2}$$

For linear systems the following condition must be guaranteed: $\rho(k)=\delta(k)$, otherwise, the process is nonlinear. Table 1 presents the experiments, starting from the nominal operation level u_0=3.00 volts and y_0=1.25 volts. Table 1 shows ρ=2$\neq\delta$=1.73. So, the fan-and-plate process presents a nonlinear behavior.

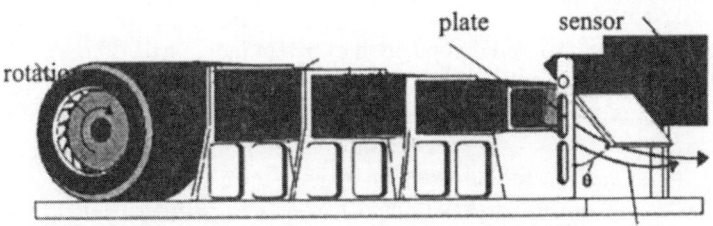

Fig. 1. Physical setup of the fan-and-plate

182

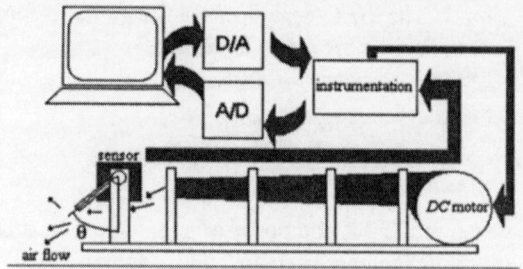

Fig. 2. Control diagram of the fan-and-plate process

Table 1. Nonlinear test for the fan-and-plate process.

Control, $u(k)$	Output, $y(k)$
$u_0(k) = 3.00$ V	$y_0(k) = 1.25$ V
$u_1(k) = 3.50$ V	$y_1(k) = 3.12$ V
$u_2(k) = 4.00$ V	$y_2(k) = 4.50$ V

3 Identification of multiple models based on evolutionary programming

The evolutionary programming technique is employed for identification of multiple models of an experimental case study. The evolutionary programming is different from conventional optimization methods. It does not need to differentiate a cost function and constraints. It is developed based on Darwinian evolution theory that describes a viewpoint of "survival of the fittest" in the life history. The evolutionary programming uses probability transition rules to select individuals in a population to reproduce new generations. An individual competes with some other individuals in the old generation and the mutated old generation. The winners with the same number as the individuals in the old generation form the next generation. The evolutionary programming is carried out mainly with three operations: mutation, competition and reproduction.

One disadvantage of evolutionary programming is its slow convergence to a good near optimum for some function optimization problems. In this paper, fast evolutionary programming [13], which uses a Cauchy instead of Gaussian mutation operator as the primary search operator, is implemented.

General scheme of the evolutionary programming for optimization is summarized as follows:
(i) *initialization*: An initial population of parent individuals is created randomly from a feasible range in each dimension. Typically, the distribution of initial trails is uniform;
(ii) *evaluation of fitness of parent individuals*: In this step is evaluated the error score for each parent individual in terms of the objective function;
(iii) *creating of offspring*: Equal number of offspring are generated by adding a Cauchy random variable with zero mean and pre-selected standard deviation to each component of individuals. Therefore, individuals including parents and offspring exist in the competing pool;

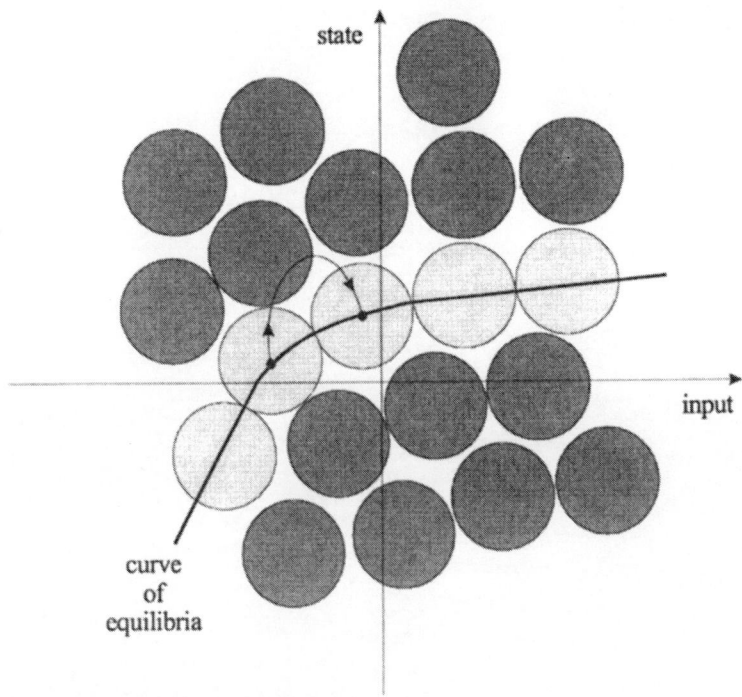

Fig. 1. Validity of conventional multi-model systems

The velocity-based linearisation framework brings together a family of velocity-based linearisations and a non-linear system. Each operating point of the non-linear system, including operating points far from equilibrium, has a counterpart in the velocity-based linearisation family, which describes the dynamic characteristics in the vicinity of that operating point. In contrast to conventional series-expansion linearisation around an equilibrium operating point, the velocity-based linearisation family indicates the plant dynamics not only in the vicinity of a single equilibrium operating point, but also during transitions between equilibrium operating points and when operating far from equilibrium.

In contrast to conventional approaches, velocity-based linearisation analysis offers several advantages. The family of velocity-based linearisations can be pieced together to approximate the solution to a non-linear system, so, according to [5], stability, as well as the transient behaviour of the non-linear system, can be investigated.

A blended local controller system in velocity-based form is

$$\dot{x}_c = w_c$$

$$\dot{w}_c = \left(\sum_i \phi^i_{x_c} \mu_i(\rho) \right) w_c + \left(\sum_i \phi^i_e \mu_i(\rho) \right) \dot{e}$$

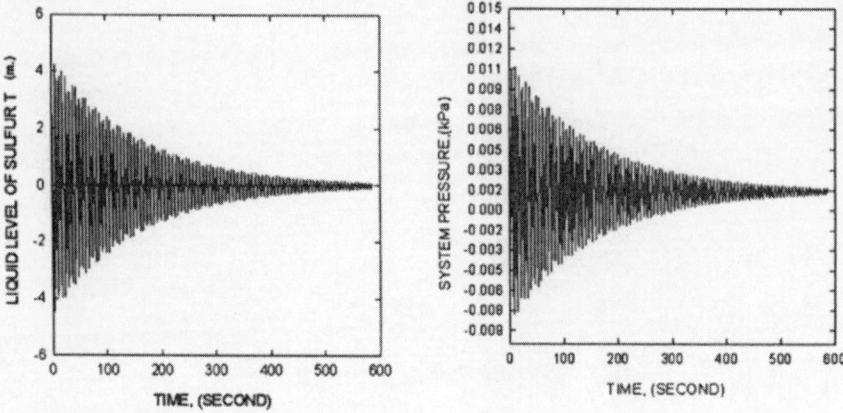

Fig.4.3: Set point tracking for the liquid level control of sulphur tank via the impulse
Disturbance in Fig1, in the case where $y_{fuzzy}=0$ in control.

In the known system case the simulation has been performed using the proposed Fuzzy linear programming algorithm given in chapter 3. The parameter of the model, A has formed in Table 4.1.

$$Y_k^* = A_0^* + A_1^* e_{k-1} + A_2^* e_{k-2} + A_3^* e_{k-3} + A_4^* \Delta U_k + A_5^* \Delta U_{k-1} \qquad (4.2)$$

where y_k: Liquid level of sulphur tank (m.).
ΔU_k: The Change of controller output as the system pressure (kPa).
e_k: $y_k - y_{sett}$

Table 4.1. Fuzzy Parameters A^* (H=0.5) for Eq. 4.2

Fuzzy Para me- ters	A_0^*	A_1^*	A_2^*	A_3^*	A_4^*	A_5^*
center α_i	0.32619	1.10474	0.00	0.00303	0.0	0.0002076
width c_i	11.45476	3.2599	0.00	1.4509	0.0012892	0.00047621
Stan- dard devia -tion	3.7×10^{-2}					

Simulation results are given in Figs4.1-4.3. to obtain the simulation results, both nominal and estimated closed-loop systems have been excited by reference signals , which are unity square waves espectively.

$$\text{If } e(t) \text{ is } A_i \text{ and } de(t) \text{ is } B_i \text{ then } A'_{m,i} \text{ is } C_i; \qquad\qquad i=1...n \qquad (8)$$

where $A_{m,i}$ is the gain margin for the i rule, A_i, B_i and C_i are fuzzy sets on the corresponding supporting sets. The block diagram and the membership functions of fuzzy sets for $e(t)$ and $de(t)$ are shown in Fig. 1b and Fig. 1c, respectively.

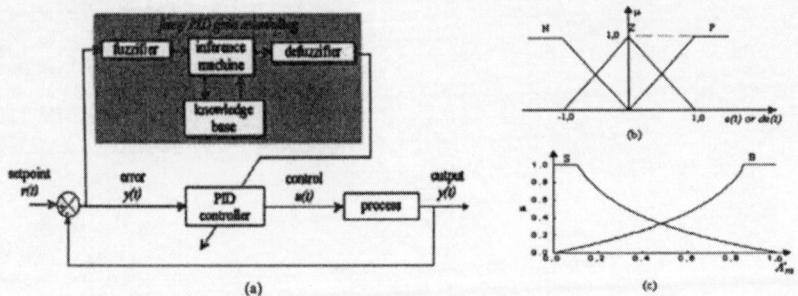

(a)   (b)

                                                     (c)

Fig. 1 (a) Fuzzy *PID* gain scheduling controller diagram; (b) and (c) membership functions.

Fuzzy rule base sets are obtained from operator's expertise by using the step response of the process. Fig. 2a and 2b show a example of a desired response and a fuzzy rules base.

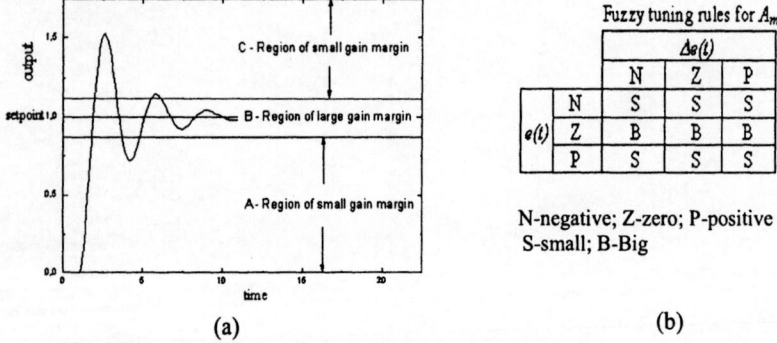

Fuzzy tuning rules for A_m

		$\Delta e(t)$		
		N	Z	P
	N	S	S	S
$e(t)$	Z	B	B	B
	P	S	S	S

N-negative; Z-zero; P-positive
S-small; B-Big

(a)   (b)

Fig. 2 (a) Margin gain regions with process step response and (b) fuzzy rule base sets.

The fuzzy set C_i may be either Big or Small and is characterized by logarithmic membership functions, where the grade of the membership μ and the variable A'_m have the following relation:

$$\mu_G = -\frac{1}{\eta}\ln(A'_m) \quad ou \quad A'_m = e^{-\eta\mu_G} \qquad (9)$$

$$\mu_P(A'_m) = -\frac{1}{\eta}\ln(1-A'_m) \quad ou \quad A'_m(\mu_P) = 1-e^{\eta\mu_P} \qquad (10)$$

The truth value of the ith rule in Eq. (8) μ_i is obtained by the product of the membership function values in the antecedent part of the rule [12].

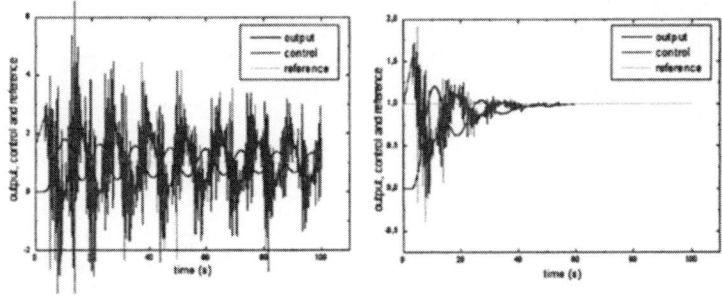

Fig. 6 Output, control and setpoint for (a) *PID* controller and (b) *FPID* controller, considering an error of 50% with variance of 0.01 in the time-delay.

To conclude performance tests, the *PID* and *FPID* control approaches are assessed in a heating tunnel process implemented in the Department of Automation and Systems at the University of Santa Catarina, Fig. 7

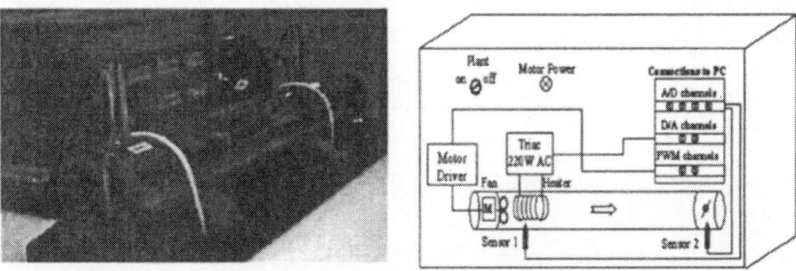

Fig. 7 Heating tunnel plant.

The heating tunnel control system, shown in, is composed of a fan, a *DC* motor, a 50 cm long air duct with uniform transverse area, having on its right extremity an electrical heating resistance. The electrical heating part is driven by a power actuator circuit whose input is compatible with a *D/A* card. The temperature is measured by two sensors placed on the duct extremity. The hot air inside the duct is spread by the fan and the control problem is to regulate the temperature inside the duct (controlled variable) by actuating on the current through the electrical resistance (manipulated variable). Additional information of the process is available in http://www.lcmi.ufsc.br/lcp/.

This practical prototype presents dead-time, resonant and turbulent characteristics, so that it can be used as a tangible real proof of the useful of advanced control techniques in difficult situations.

In order to show the applicability and effectiveness of the proposed *FPID* in a control of a real time heating tunnel plant, experiment have been carried out in servo behavior. Step responses of the *FPID* and the *PID* controller for different setpoints are given in Fig. 8. Although both *FPID* and *PID* controller give good control, *FPID* results show superior performance with minimum overshoot, rise time and control variance.

Example 1.

An example of a fuzzy non-hierarchical clustering method:
Let us choose $n = 10$, $m = 2$. Let the elements of fuzzy objects be in form: $fO=(C_x, C_y, S_x^1, S_x^2, S_y^1, S_y^2)$ and each element of object have membership function in shape of two-sided Gaussian curve with membership function:

$$\mu_1^x(x) = e^{\frac{-(x-C_x)^2}{2S_x^1}} \quad \text{for } x \le C_x \quad \text{and} \quad \mu_1^x(x) = e^{\frac{-(x-C_x)^2}{2S_x^2}} \quad \text{for } x > C_x,$$

$$\mu_2^x(y) = e^{\frac{-(y-C_y)^2}{2S_y^1}} \quad \text{for } y \le C_y \quad \text{and} \quad \mu_2^x(y) = e^{\frac{-(y-C_y)^2}{2S_y^2}} \quad \text{for } y > C_y,$$

where
$fO_1 = (10, 19, 1.7, 0.9, 2.9, 0.3)$, $fO_2 = (20, 23, 1.3, 1.7, 2.1, 1.8)$,
$fO_3 = (22, 16, 1.4, 0.9, 0.2, 2.7)$, $fO_4 = (10, 12, 2.5, 0.3, 1.1, 2.1)$,
$fO_5 = (20, 11, 1.9, 0.9, 2.4, 0.4)$, $fO_6 = (16, 25, 0.7, 1.8, 2.4, 0.3)$,
$fO_7 = (14, 10, 0.1, 2.7, 1.3, 1.7)$, $fO_8 = (7, 9, 0.7, 1.2, 0.4, 2.5)$,
$fO_9 = (15, 20, 0.5, 2.8, 0.3, 2.1)$, $fO_{10} = (23, 8, 2.9, 0.3, 2.3, 0.5)$.

If we think of fuzzy objects in the form $fO = (R \times R, \min\{\mu_1^x(x), \mu_2^x(y)\})$, then the fuzzy objects can be depicted in R^3 as shown Fig. 1.

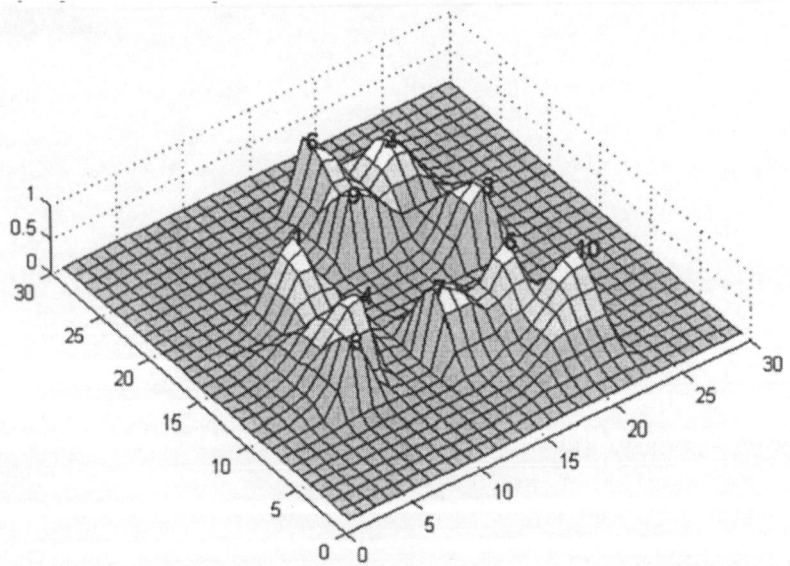

Fig. 1.

We will cluster these objects using non-hierarchical clustering methods for objective functional method with the functional $J_{fw}(U,v) = \sum_{j=1}^{n} \sum_{i=1}^{c} u_{ij}(d_{ij})^2$, where $d_{ij} =$ $fd_E(fO_j, fv_i)$ is extension of Euclidean metric, fv_i is fuzzy centroid of cluster S_i. Let us choose $c = 3$ and initial clusters $S_1 = \{fO_1, fO_2, fO_3, fO_4, fO_5\}$, $S_2 = \{fO_6, fO_7, fO_8\}$, $S_3 = \{fO_9, fO_{10}\}$. We will get the clusters $S_1 = \{fO_2, fO_3, fO_6, fO_9\}$, $S_2 = \{fO_5,$

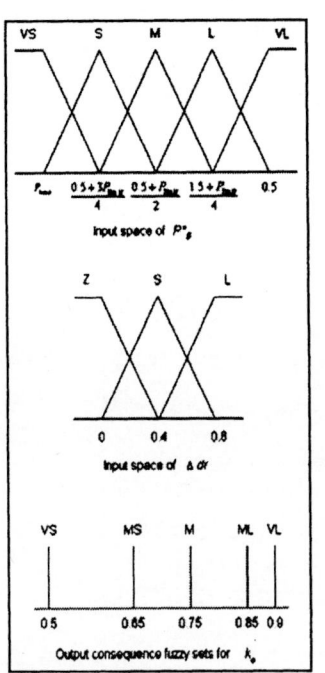 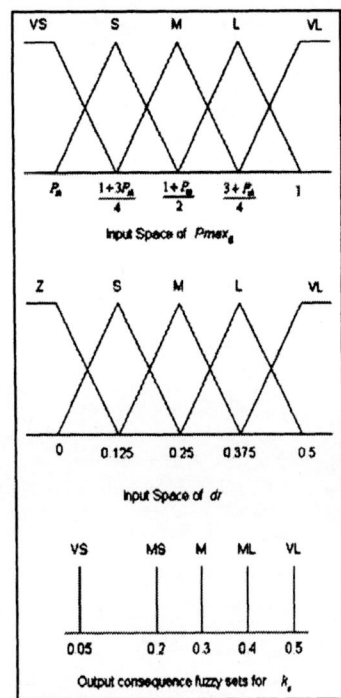

Fig. 3a **Fig. 3b**

Fig. 3. The membership functions and the output consequence fuzzy sets for tuning a) k_O and b) k_ε .

		P^*_B				
		VS	S	M	L	VL
	Z	VL	ML	ML	M	M
Δdr	S	ML	M	M	S	VS
	L	M	M	M	S	VS

Fig. 4a

		$Pmax_B$				
		VS	S	M	L	VL
	Z	ML	M	MS	VS	VS
	S	VL	ML	M	MS	VS
dr	M	VL	VL	ML	M	MS
	L	VL	VL	VL	ML	M
	VL	VL	VL	VL	VL	ML

Fig. 4b

Fig. 4. a) Fuzzy rule table corresponding to k_O . b) Fuzzy rule table corresponding to k_ε .

In order to clarify the results more easily, the simulation study only considers one-dimensional environment for the evaluation of our algorithm with respect to traditional algorithms [2, 7]. For the first simulation, we assumed that some sensor measurements were corrupted by noise. The sensor was located at the origin and the actual position of the obstacle was 1m away from the sensor. Moreover, we assumed that the first 8 sonar measurements were the actual range i.e. $\{r_i = 1\ m : i = 1,2,...,8\}$ and the ninth and tenth were erroneous data and the value was 1.8 m i.e. $\{r_i = 1.8\ m : i = 9,10\}$, which was due to specular reflection. Then, the subsequent measurements of the sensor was equal to 1 m again i.e. $\{r_i = 1\ m : i = 11,12,13...\}$. The profiles of the occupancy probabilities corresponding to 2nd, 5th, 8th, 9th, 10th, 13th readings for different algorithms are shown in Fig. 5. From this simulation, we can observe that the traditional algorithm provides good estimate of the environment from 1st to 8th readings. However, after two noisy readings (9th and 10th), the map occupancy probability corresponding to 1 m away from the sensor is suddenly reduced to zero and can not recover even after correct measurements in next readings. However, our algorithm can resolve the above problem as is shown in Figure 5b.

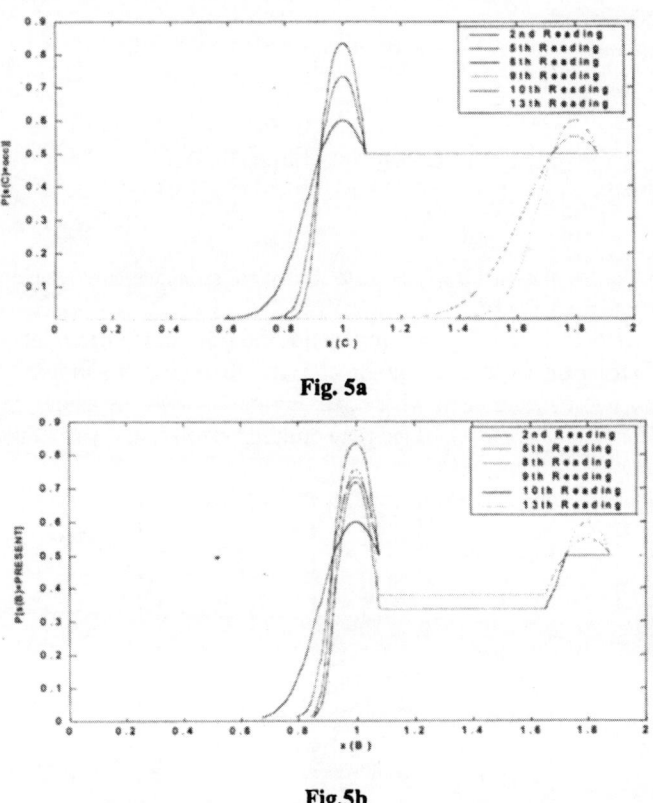

Fig. 5a

Fig.5b

Fig. 5. Result for simple 1 dimensional example a) traditional Bayesian grid based map building, b) Proposed Bayesian state estimation technique.

In the second simulation, we assumed that an obstacle was moving. The motion profile of the object is shown as follows:

278

Position 1: Obstacle is stayed 1 m away from the sensor. i.e. $\{r_i = 1\ m : i = 1,2,...,8\}$

Position 2: Obstacle is moved to 1.6 m away from the sensor. i.e. $\{r_i = 1.6\ m : i = 9,10,...,16\}$

Position 3: Obstacle is moved to 0.5 m away from the sensor. i.e. $\{r_i = 0.5\ m : i = 17,18,19,...\}$

The profiles of occupancy probabilities corresponding to 7th, 8th, 15th, 16th, 23rd, 24th readings for our proposed algorithm are shown in Figure 6. From this figure, we can observe that the proposed algorithm can provide good estimate of the position obstacle is moved from Position 1 to Position 2 and from Position 2 to Position 3. Therefore, it can be concluded that this algorithm is also suitable for dynamic environments.

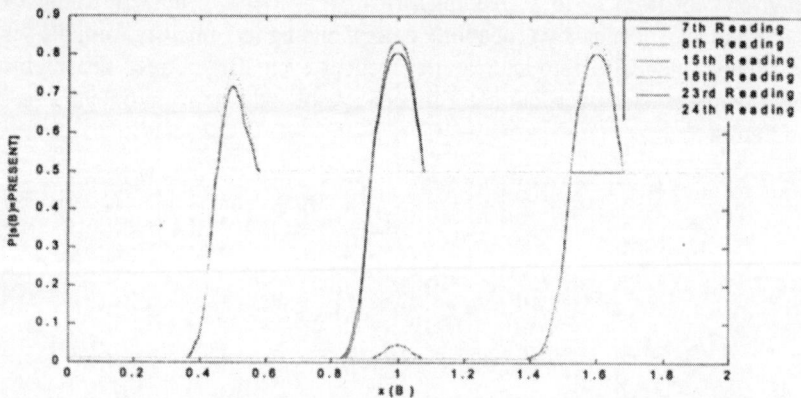

Fig. 6. Result of the proposed Bayesian state estimation technique for moving objects

In the experimental study, the corridor outside the Control Research laboratory was modeled. During the experiment in the corridor, there was one moving object and the doors of two of the offices were open. Figure 7 shows the actual image of the corridor. The on-line segment-based map building result is shown in Figure 8.

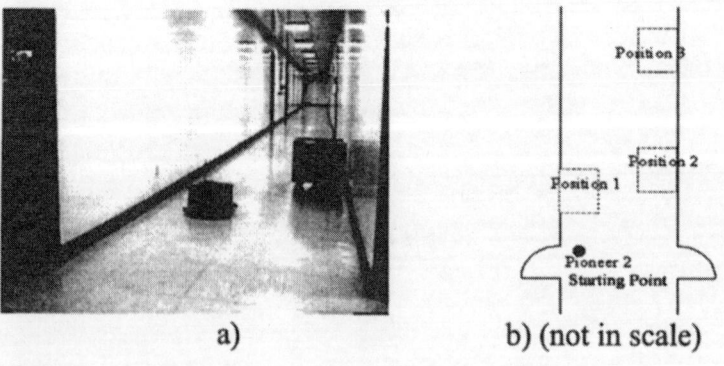

a) b) (not in scale)

Fig. 7. Photo of experimental site (corridor).

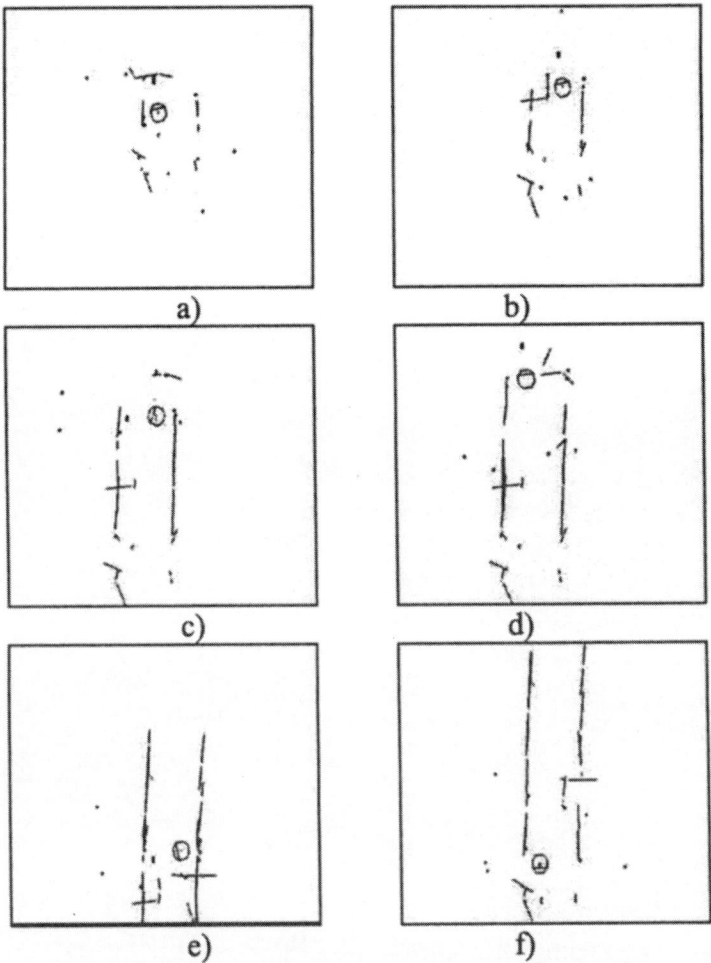

Fig. 8. Online mapping during exploration with Pioneer II.

In this experiment, the dynamic object moved in three different positions (position 1 to position 2 to position 3) which is shown in Figure 7b. In Figure 8a and Figure 8b, the object is placed at position 1. After that, we move the object form position 1 to position 2 as shown in Figures 8c and 8d respectively. Finally, we moved the object form position 2 to position 3. The Pioneer 2 was made to perform a "U turn" at position 2. In Figure 8e, we found that the object in position 2 was disappearing when we compared the map with Figure 8d. In Figure 8f, the Pioneer 2 returned to the staring point. It could be observed that the object in position 1 was also disappearing. From this experiment, it was concluded that the proposed on-line segment-based map-building algorithm could build a map in a dynamic environment.

5. Conclusions

A probability state estimation technique to estimate and update the state of extracted segment has been developed and successfully implemented on EAFC with NC segment based map building algorithm. The development of the algorithm

approach, centers (training points) are added until some preset performance is reached or all the training points have been used.

4   Experimental Results

In this section, we first give an example with a couple of artificial data sets to illustrate the proposed methods and then we use the severe storm cells data set to examine the performance of the proposed methods. The approaches examined are an SVM with the RBF kernel (SVM), a radial basis function network with the centers found using fuzzy clustering (RBF1), and a radial basis function networks with the centers found using the orthogonal least squares approach (RBF2).

For all the experiments the spread is set to $\gamma=1/\sigma$ (with σ being equal to the mean of the standard deviations per dimension in the training data).

For the SVM the experiments were conducted for $\gamma=1/\sigma$ and $C=10$.

For the RBF1 the experiments were conducted for $\gamma=1/\sigma$.

For the RBF2, the performance criterion was set to reduce the mean square error during training up to a value $G=5\%$ for $\gamma=1/\sigma$.

4.1   Artificial Data Sets

The data sets for these experiments are depicted in Fig. 3 and Fig. 4. The vectors were distributed according to a normal density function with unit covariance matrices. The mean vectors (m_1 for class 1 and m_2 for class 2) were:

- For data (a): $m_1=(2.0, 2.0)$ and $m_2=(-2.0, -2.0)$. See Fig. 3.
- For data (b): $m_1=(1.0, 1.0)$ and $m_2=(-1.0, -1.0)$. See Fig. 4.

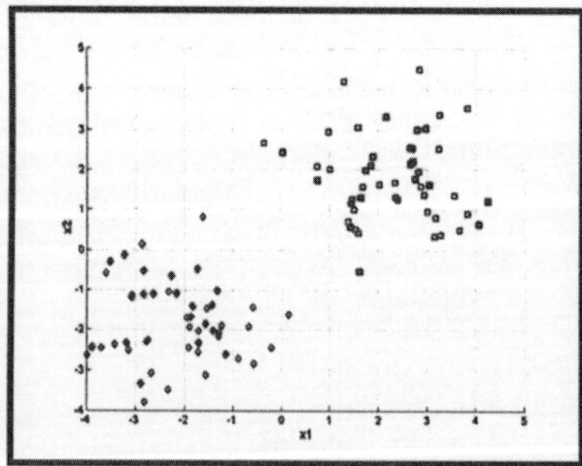

Fig. 3. Synthetic data (a). Squares and diamonds represent class 1 and class 2, respectively. The testing points are the bold patterns.

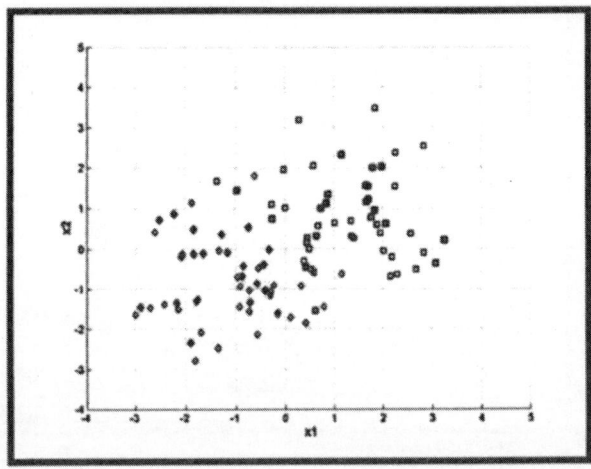

Fig. 4. Synthetic data (b). Squares and diamonds represent class 1 and class 2, respectively. The testing points are the bold patterns.

Each class had 50 elements, giving a data set of 100 elements, 60 of which were used for training and 40 for testing.

To obtain reliable results, a rotation method was employed by randomly re-sampling training and testing data and repeating the experiments ten times. The results were averaged over the ten iterations.

The results for data (a) are summarized in Fig. 5 and the results for data (b) are summarized in Fig. 6. From these results, it can be seen that for data (a) the SVM and the RBF1 had a perfect accuracy (100%) while the performance of the RBF2 was slightly lower (99.5% in testing). In the case of data (b), the best classifier was the SVM, with an accuracy of 90.5% in testing, followed by the RBF1, with an accuracy of 89.75% in testing, and the RBF2, with an accuracy of 88% in testing.

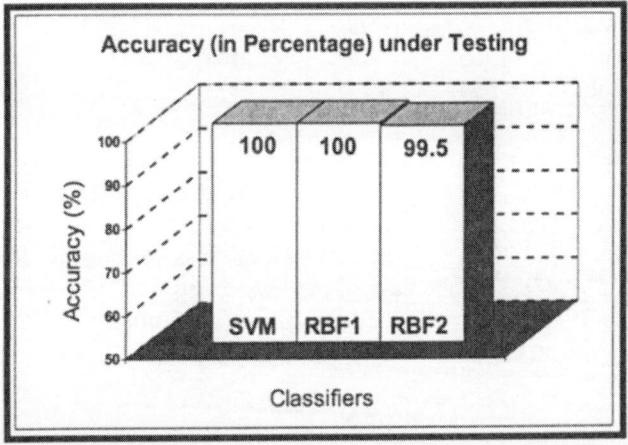

Fig. 5. Classification results obtained for data (a).

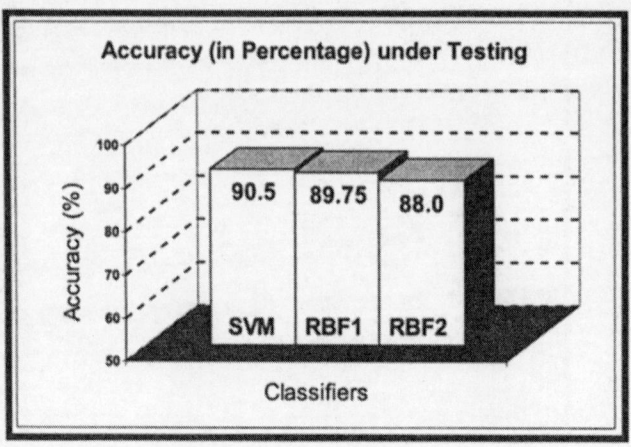

Fig. 6. Classification results obtained for data (b).

4.2 The Severe Storm Cell Data Set

In the experiments, a portion of the data set (60%) was used as a training (learning) set and the remaining 40% was used as a testing set. To obtain reliable results, a rotation method was employed, repeating the experiments ten times. Subsequently, the training data for each one of the ten experiments were normalized to attain a zero mean and one standard deviation. The values for the mean and the standard deviation were then used to normalize the testing data for each experiment. The pattern distribution is depicted in Table 3.

Table 3. Pattern distribution for the storm cell classification problem.

Class	Name	Patterns	Training Patterns	Testing Patterns
1	Hail	166	99	67
2	Tornado	265	159	106
	Totals	431	258	173

The results averaged over ten iterations are summarized in Fig. 7 and Fig. 8. From these results, it can be seen that the approach that showed the best performance for the training data was the SVM (with an accuracy of 96.24%), followed by the RBF2 (with an accuracy of 88.45%) and the RBF1 (with an accuracy of 75.46%). In the case of the testing set, the SVM approach (with an accuracy of 79.13%) outperformed the RBF1 (with an accuracy of 70.29%) and the RBF2 (with an accuracy of 75.9%) approaches.

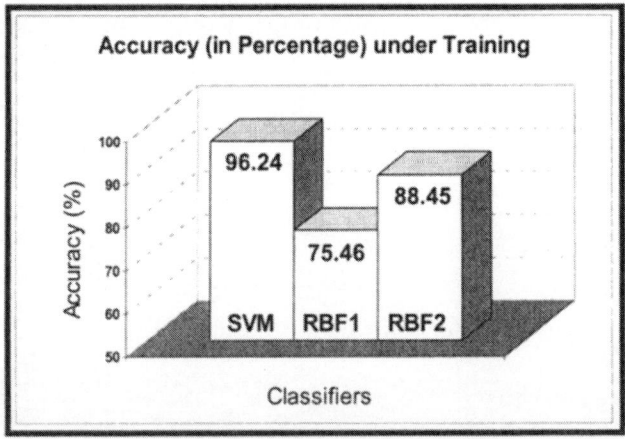

Fig. 7. Training accuracy for the storm cell data.

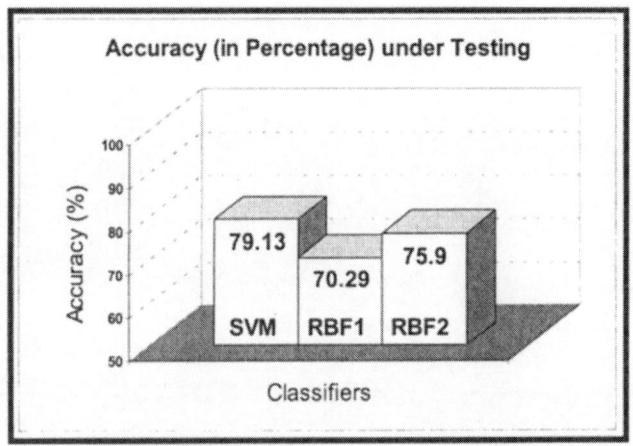

Fig. 8. Testing accuracy for the storm cell data.

Table 4 shows the performance (accuracy, in percentage, on the test set) of the following methods for solving the storm cell classification task: a multilayer perceptron used by Alexiuk et al [1], a linear discriminant analysis used by Li et al. [8], and our support vector machine-based classifier. Direct comparisons, however, could not be done due to the following reasons:

1. The experiments were conducted for different data sets (99 hail cells and 25 tornado cells in the case of Alexiuk et al. and 172 hail cells and 163 tornado cells in the case of Li et al.).

2. The pre-processing strategies were different (in the case of Alexiuk et al., they used principal component analysis, fuzzy interquartile encoding, and genetic algorithms. In the case of Li et al., they used genetic algorithms for optimizing the derived products).

sides of the line at all workstations due to space restrictions caused by the machinery and logistics shelves placement.

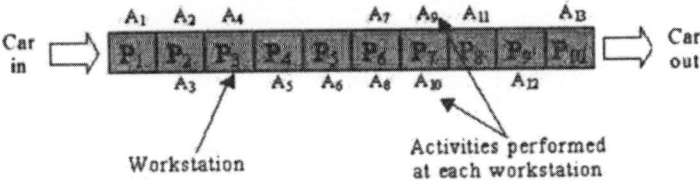

Fig. 1. Representation of the production line under analysis

Each workstation has a fixed length (about 5m) and the vehicle moves on softly, without stopping along the process. The total working time at a given workstation is represented by its longest activity, since both sides start at the same time. For example, the total working time of, say, P_2 workstation is given by the longest duration of activities A_2 and A_3.

The total working time of this line section is given by the sum of all workstations' working times. In the real line, not yet optimized, the current time is 21min 22s (or 21.37min).

4 Solution using genetic algorithms

Variables encoding

Figure 2 shows the chromosome encoding adopted in this work. The chromosome is composed by 13 integer numbers, one for each activity to be performed in the line, all between 1 and 10. The content of each "gene" in the chromosome identifies one of the 10 workstations, and the position it occupies maps the activities performed at that workstation. For instance, in Fig. 1, activities A_2 and A_3 are both performed at workstation P_2, activity A_6 at workstation P_5, and so on.

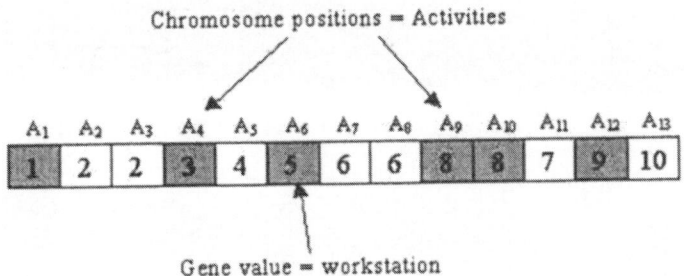

Fig. 2. Chromosome encoding

Although this is a combinatorial problem, the special crossover operators proposed by Goldberg [3], such as PMX, OX and CX, do not apply since the precedence order between the activities does not follow directly their numbering.

ated with the line layout represented by such individual. This worksheet also analyses the chromosome mapping and produces a sketch of the line layout. Part of the worksheet is shown in Fig. 4.

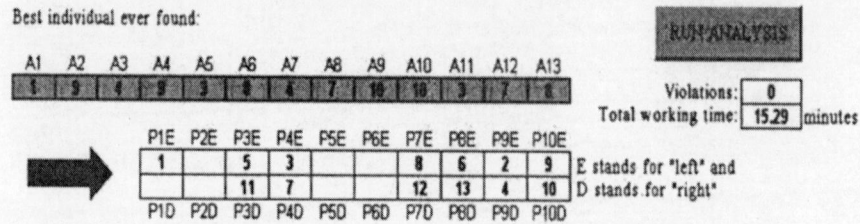

Fig. 4. Worksheet for analysis of the best individual

After running the GA several times, the best individual ever found was the one represented in Fig. 5. This individual represents a solution for the assembly line layout, which is sketched in Fig. 6. The best solution found leads to a total working time of 15.29min, which represents a 28.5% reduction of the current line configuration time (21.37min).

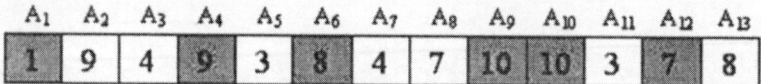

Fig. 5. Best individual ever found

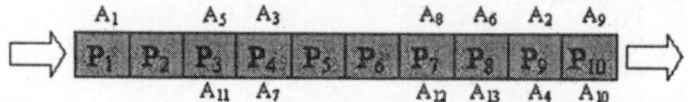

Fig. 6. Best solution ever found (line layout)

6 Conclusions

We have shown with this work that GAs can be very useful to find better alternatives for the layout of an assembly line. If such analysis had been used during the facility design phase, it could have suggested improvements in the layout of the whole production line, optimizing the overall working time.

Considering the dimensionality of the search space and the combinatorial nature of the problem, GAs were found to be very efficient and fast. A comparison with another classical methodologies for the line-balancing problem shall be done in the future to investigate how competitive a GA is.

For the sake of simplicity, the present work did not consider logistics issues, such as parts shelves disposal. It was not analyzed, as well, the costs due to the layout changes in the current line. Notwithstanding, the solution found is real and shows

1. Subtract the host's detail coefficients $f_{k,l}(m,n)$ from those of noisy marked image's $\tilde{f}_{k,l}(m,n)$ only from the locations pointed at by $r_{k,l}(p,q)$. Store the residual marked image for further processing in next step.

2. Inverse transform the extracted watermark coefficients and residual watermarked images' coefficients for 1st level and l level respectively. The extracted watermark is denoted by $\hat{w}_l(p,q)$. Feed the residual marked image to the next level of watermark extraction.

Step IV: Estimate the watermark by averaging the extracted watermarks $\hat{w}_l(p,q)$ and normalize it for binary values.

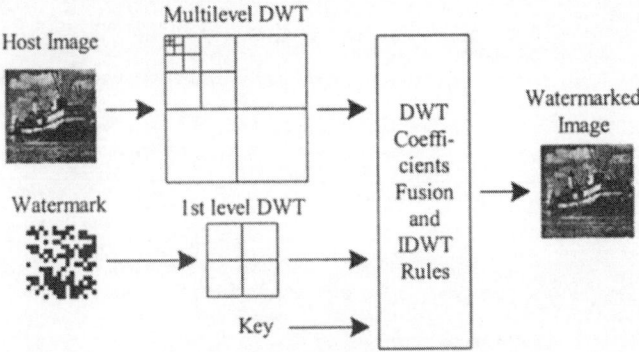

Fig. 1. Proposed watermarking method.

3   Correlation coefficient and signal to noise ratio

In order to find out similarity between embedded and extracted watermarks we calculate correlation coefficients at different signal to noise ratio (SNR) values.

The correlation coefficient, ρ, used for similarity measurement, and SNR are defined as

$$\rho(w,\hat{w}) = \frac{\sum_{i=1}^{N} w_i \hat{w}_i}{\sqrt{\sum_{i=1}^{N} w_i^2}\sqrt{\sum_{i=1}^{N} \hat{w}_i^2}} \tag{1}$$

$$SNR(w,\hat{w}) = 10\log_{10}\frac{\sum_{i=1}^{N} w_i^2}{\sum_{i=1}^{N}(w_i - \hat{w}_i)^2} \tag{2}$$

where N is the number of pixels in watermark, and w and $\hat{w}$ are the original and extracted watermarks, respectively.

Table 2. Wavelet decomposed watermark components.

l	$w_{1,l}(p,q)$	$w_{2,l}(p,q)$	$w_{3,l}(p,q)$
1	8×8	8×8	8×8

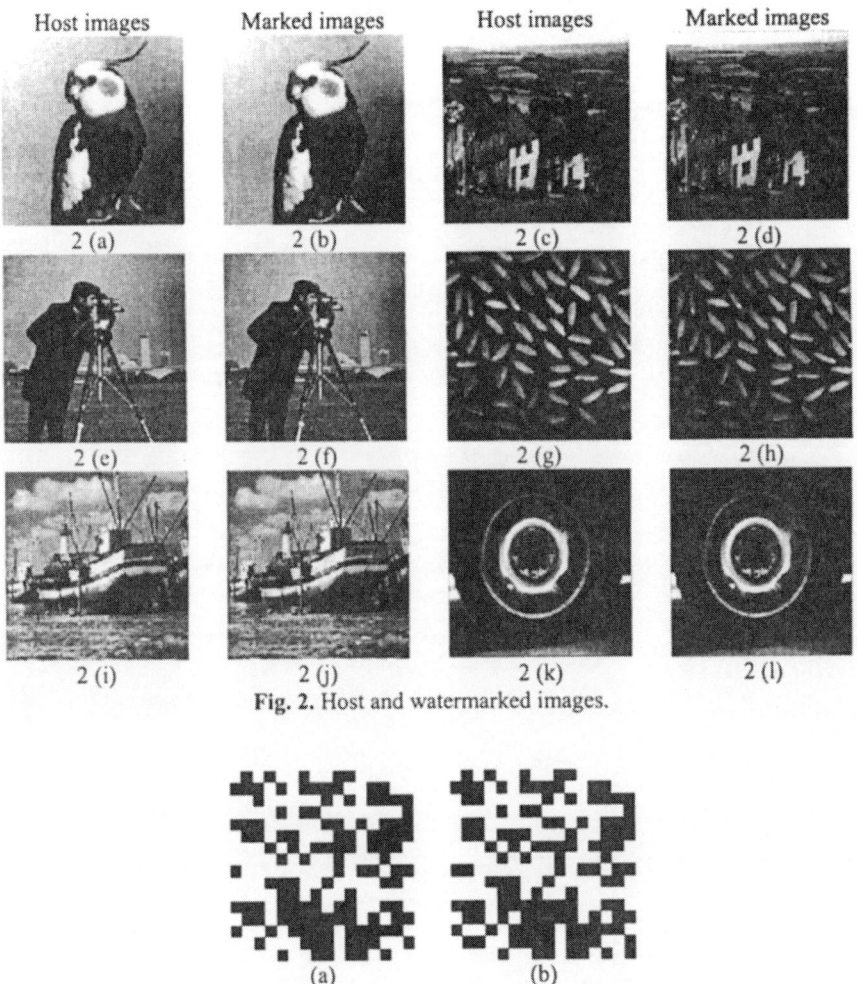

Host images Marked images Host images Marked images

2 (a) 2 (b) 2 (c) 2 (d)

2 (e) 2 (f) 2 (g) 2 (h)

2 (i) 2 (j) 2 (k) 2 (l)

Fig. 2. Host and watermarked images.

(a) (b)

Fig. 3. (a) Original and (b) extracted watermarks (SNR=10.88 dB; ρ =0.99).

4.2 Effect of salt and pepper noise

The random salt and pepper noise of various densities was added to the marked images. Figure 5 exhibits the correlation coefficient, ρ, patterns at different SNR values for the extracted watermarks from all *salt and pepper* noised images. For the given range of noise the ρ started increasing from 0.5 and reached gradually to the maximum value 1. It crossed the desired criteria of 0.75 at about 24 dB SNR. Hence, the presented method satisfied the robustness criteria set for salt and pepper noise of the used density values.

3. Collect appropriate input-output data from the mammogram images.
4. Apply the identification algorithm described in literature [6] to evaluate the model relation matrix (using N reference sets for each coordinates give a N×N×N×N×N relational matrix).

4 Results and Discussion

Our experiments are based on mammogram images taken from a set digitized by Nico Karssemeijer of University Hospital Nijmegen, The Netherlands. The database is publicly available on the internet, http://marathon.csee.usf.edu/Mammography/OtherResources.html#NIJMEGEN.
The database consist of 40 mammograms of 21 patients. For each mammogram image there is ground truth file, in which the locations and the sizes of microcalcification clusters are marked.

We tested the fuzzy feedback decision system on Nijmegen database having 50-micron resolution. If there is a positive detection of a microcalcifcation, this block is marked as a suspicious square region as shown in Figure 3 and Figure 4.

This procedure is repeated all over the image. The obtained results show that the fuzzy dynamic thresholding algorithm proposed here presents a good adaptation capability to varying microcalcification of the each image. It has been designed to be implemented for real time visual processing in outdoor scenerious.

The fuzzy relational matrix algorithm is an adequate and intuitive model to represent the quantitative pattern recognition process accomplished automatically to get a control decision rather than the fuzzy rule based model.

The algorithm has been implemented firstly off-line in FORTRAN and the by considering the on-line adaptation MATLAB® code and C++ Languages. As a continuation of this research, an approach for using the standard relational matrix[5] which explicitly seek to make the residual signals sensitive to faults are under investigation.

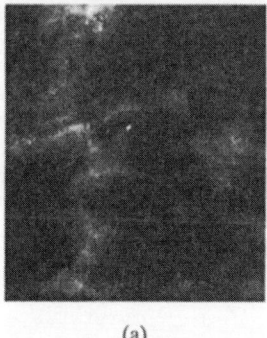   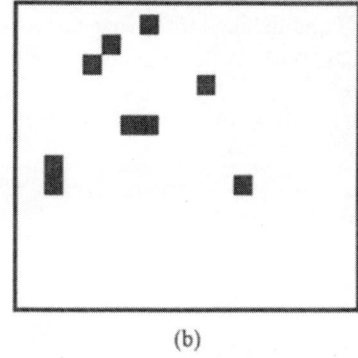

(a)   (b)

Figure 3. Detection scheme output: (a) part of original mammogram image containing a lot of microcalcifications; (b) blocks with microcalcifications as indicated by the detection sheme

378

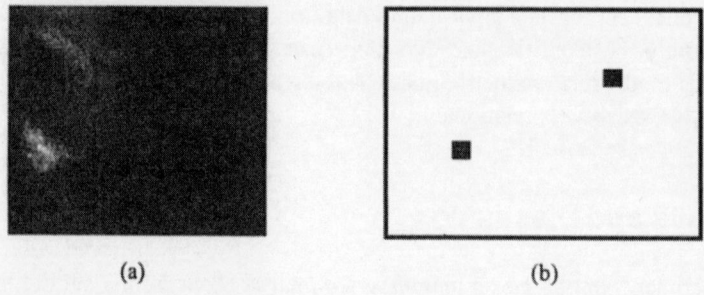

(a)   (b)

Figure 4. Detection scheme output: (a) part of original mammogram image containing two microcalcifications; (b) blocks with microcalcifications as indicated by the detection scheme

References

[1] Gurcan M.N. (1999) Computer aided diagnosis in radiology, Bilkent University, Ph.D. thesis, 125p.

[2] Wang T.C., Karayiannis N.B. (1998) Detection of microcalcifications in digital mammograms using wavelets, IEEE Trans. On Medical Imaging, 17, 4, pp.498-509

[3] Petrick N., Chan H.P. (1996) Sahiner B, Wei D., An adaptive density

[4] weighted contrast enhancement filter for mammographic breast mass detection, IEEE Trans. On Medical Imaging, 15, 1, 59-61

[5] Kupinski M.A., Giger M.L. (1998) Automated seeded lesion segmentation on digital mammograms, IEEE Trans. On Medical Imaging, 17, 4, 510-517

[6] Zeybek, Z., (1997) Modeling and control of chemical processes with fuzzy and neural network methods, Ankara University, PhD Thesis.

[7] Graham,B.P, Newell, R.B. (1989) Fuzzy adaptive control of a first order process, Fuzzy Sets Syst. 31, 47-65.

[8] Krishnapuram R. (1998) Edge detection, in handbook of Fuzzy computation, IOP publishing LTD, Chapters F7.2, F7.3, F7.4 and F7.8

[9] Sugeno M., An introductory survey of Fuzzy control, Information science 36, pp.59-83, 1985.

[10]Zadeh, L.A. (1965), "Fuzzy sets", Information and Control, Vol. 8, pp. 338-53.

Fig. 1. Sample frame from an overtake monitor system simulation.

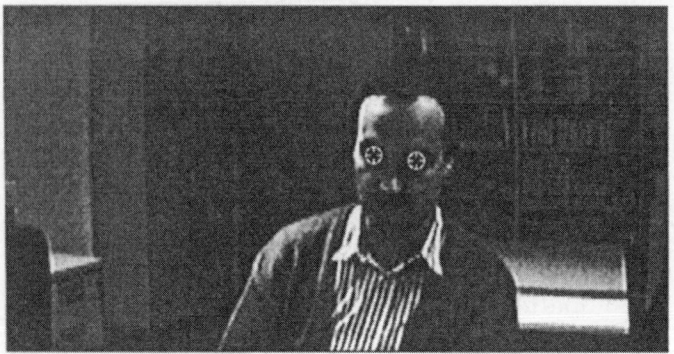

Fig. 2. Sample frame from an eye-tracker simulation system.

consumption, cost, and design time, must be met in addition to a feasible and well performing system model. Thus, an opportunistic and parsimonious design style must be adopted to achieve overall satisfactory and viable embedded system implementations. For instance, in vision systems design, the low-level feature computation is the most time-consuming and decisive processing step. Here, proper methods and architecture provided, invariance properties with regard to illumination, translation, scale, and 2D-rotation are typically achieved. Following decision making by soft-computing methods bases on the achieved, hopefully, compact and discriminant feature space. Therefore, with regard to real-time, power, area, and cost constraints the efficient and par-

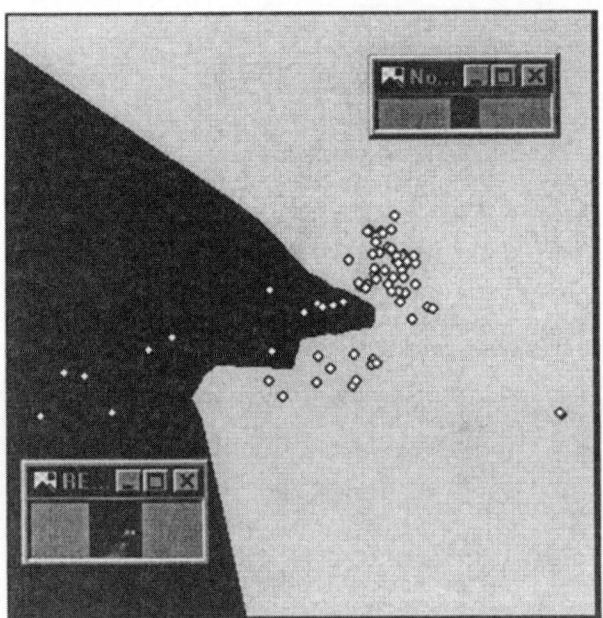

Fig. 5. Resulting feature space for Gabor jet data.

a recognition rate of only 83.6% can be observed for LAC, which is not competitive to the Gabor jet approach inspite of the computational savings.

Finally, the ELAC method was applied to the eye-shape classification problem. Figures 8 and 9 show projections of the resulting feature spaces by Sammons mapping for the respective training data and test data sets. Automatic feature selection chose only features 2, 4, 5, 6, 7, 8, 9, and 12 from the activated features 10 to 21 and the reject feature for this application and the specific training data base. Only one classification error and a recognition rate of 98.4% can now be observed for ELAC, which is competitive to the Gabor jet approach considering the closeness of this method to failure for one pattern and the implied significant computational savings. Carrying in mind, that a Gabor filter mask of 17×17 requires 289 high precision multiplications and additions for real and imaginary response computation, prospective savings of ELAC must now be assessed. Narrowing the view to a single window content, 578 local multiplications seem to be required per applied mask in a 17×17 window. A single Gabor filter requires the same amount of multiplications, but with higher precision and infeasible wiring and weight storage requirements for a parallel microelectronic implementation. In contrast, ELAC requires no convolution parameters but only the pixel values of the adjacent eight local pixels. Further, as the search for an eye-shape is carried out in a scanning mode by a moving window, an even more significant advantage is achieved by the ELAC approach. For each new pixel position

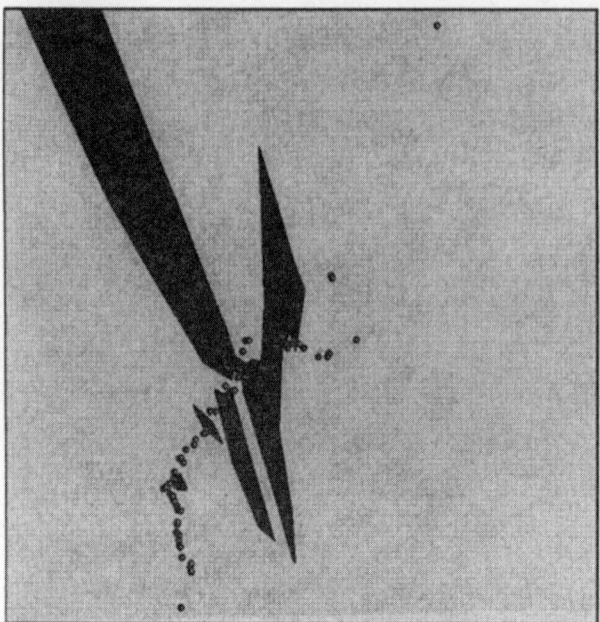

Fig. 6. Resulting feature space for LAC training data.

of the advancing window, the complete Gabor jet has to be computed with the effort given above. In contrast, for ELAC at all pixel positions the best matching kernel is computed once and only once in parallel and its index is stored in digital latches. Then, the moving window computation is simplified to local statistics computation of the winning kernels. Even this can be reduced, exploiting neighborhood relations of adjacent window positions in statistics or histogram computation, i.e., moving the window one pixel to the right requires removing the exposed left column values from the histogram and adding the newly enclosed right column to the histogram. Clearly, no signal processing has to occur in the matrix in the meanwhile. Tremendous computational savings and read-out activity reduction are achieved in this architecture, and all the major computations are digital with very modest resolution requirements.

5 Hardware Implementation Issues

Though most of the presented method and the achieved results could be well exploited by digital architectures and a corresponding microelectronic system design activity, the real benefit is given by the combination of an image sensor with the ELAC approach. Image acquisition, HDR properties and respective illumination invariance are thus combined with massively parallel feature computation. This is elucidated by the proposed vision chip architec-

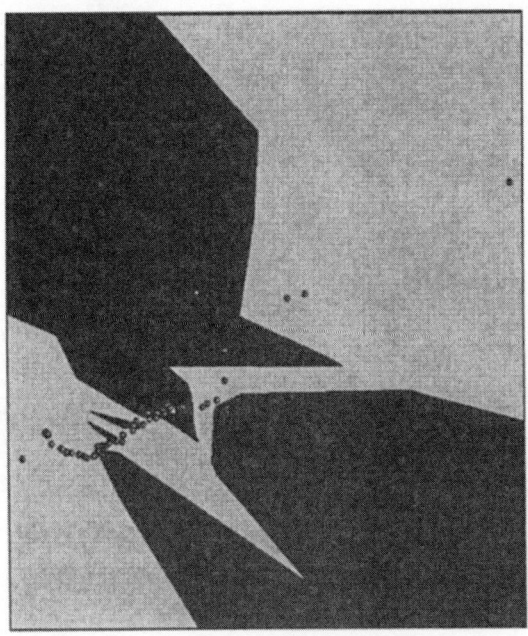

Fig. 7. Resulting feature space for LAC test data with ten classification errors

ture given in Fig. 10. Analog computation is minimized to local autocorrelation computation by five one-quadrant multipliers/dividers, e.g., employing translinear loops, and a comparator for the WTA based on the sequentially computed mask responses as well as the rejection mechanism. For the latter, a standard 2D smoothing network is also required for local normalization computation. The second analog value for comparison is recomputed according to the stored digital value employing five instead of three multipliers. Thus no analog memory is required. Due to the employed feature computation, analog computation is constrained to the pixel cell and no explicit A/D-conversion is required. A sensor is simply assembled from cascadable pixel cells with digital read-out interface, i.e., neclecting some analog bias and control input signals, the pixels can be treated like random access memory cells. Only, address decoders and digital histogram computation resources are required. These are robust, simple to design, and quite fast. Currently, the pixel cell is under design in its first implementation within the scope of a students' semester project, employing 0.8μm CMOS technology and subthreshold circuits. In this design, the masks that proved to be most meaningful by AFS for application-specific image data can be selectively activated. Thus, computational effort and corresponding real-time and power-consumption properties can be optimized in a learning loop, employing the QuickCog system as explained in [8]. A redesign in a 0.6μm 3M CMOS technology and submission of a first small test sample is envisioned in March 2002. Extensions of the

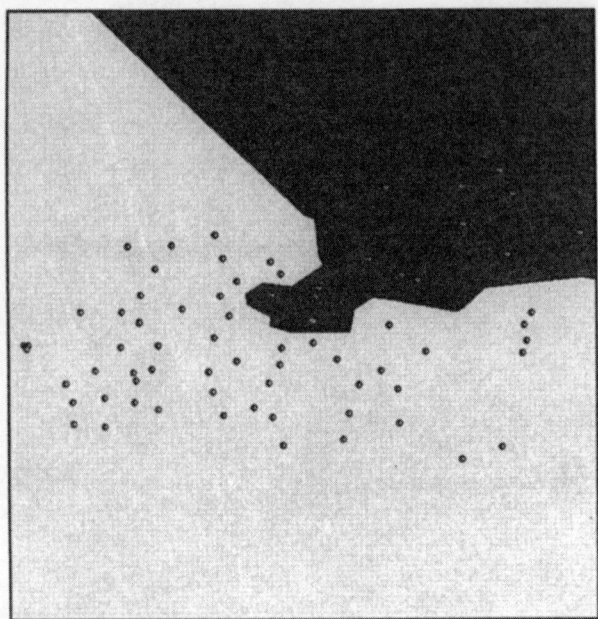

Fig. 8. Resulting feature space for ELAC training data.

approach to texture and especially color texture analysis and classification is a promising and technologically straightforward step, e.g., for high speed in-line visual inspection.

6 Conclusion

In this paper, we have presented an opportunistic and parsimonious feature computation method derived from standard local autocorrelation. We have demonstrated the validity of the approach in the context of our holistic modelling and design framework for application-specific vision and cognition systems. Currently, a parallel VLSI implementation of the ELAC method in subthreshold CMOS is pursued, targeting on the integration of image sensor and feature computation in a cascadable pixel cell for massively parallel computation (see Fig. 10).

In future work, the implementation of a complete feature computing image sensor is aspired. From earlier work, classifier cells, e.g., the applied RNN classifier, are already available for further integration. The long term objective is the SoC integration of a low-power eye-tracker system. The described ELAC approach is also promising for embedded automotive applications and real-time texture segmentation and classification. Especially the texture and color texture analysis option will be regarded in the next steps of the work.

Fig. 9. Resulting feature space for ELAC test data with one classification error

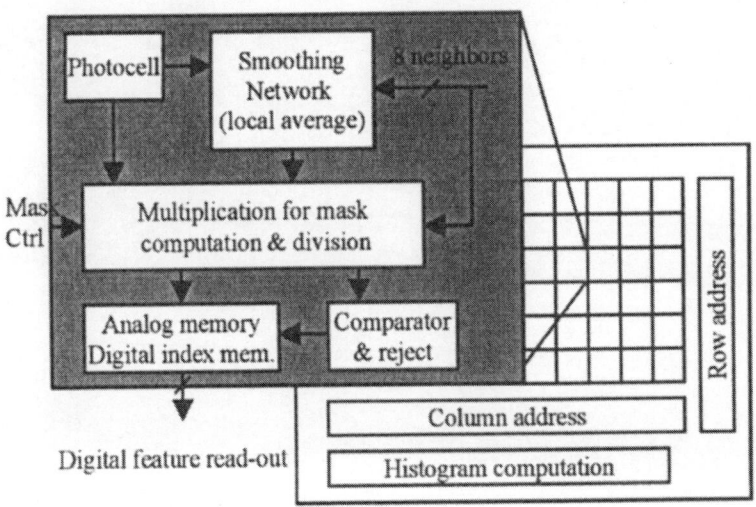

Fig. 10. Proposed low-power pixel cell and vision chip architecture for ELAC

In concurrent work, LOC is subject to the same considerations and investigations. Results will be reported in an upcoming publication including a comparison to ELAC with regard to implementation effort, power requirements, and achievable discriminance.

2.2 An Example

The image in figure 1 is the input image and the pulse images are displayed in figure 2. The signature for this image is shown in figure 3.

Fig. 1. An Original Image

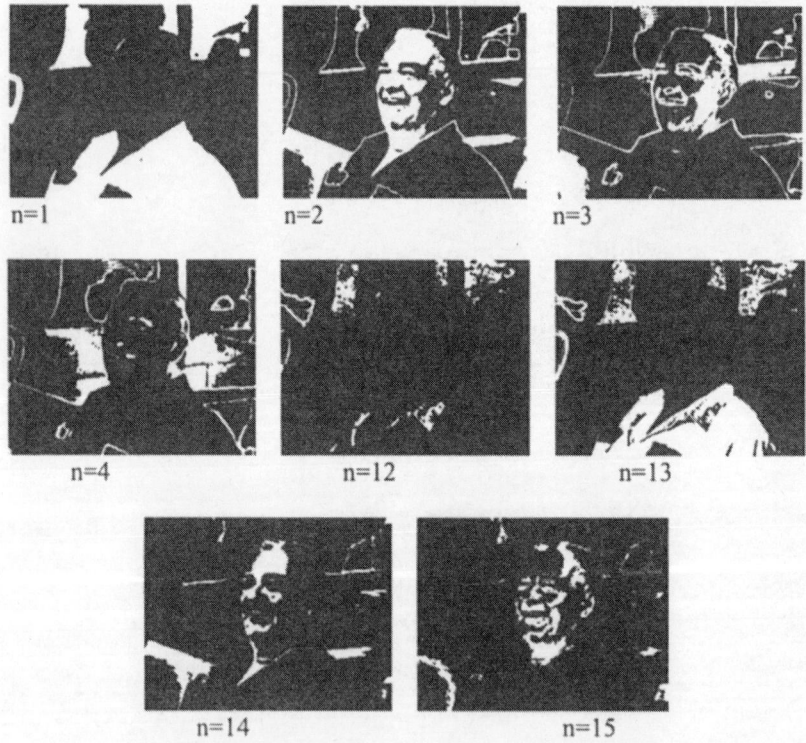

Fig. 2. Pulse Images

Fig. 4. Foveation points overlaid on the image in figure 1

3.3 Texture Classification

Texture is extracted from many different pulse images. As can be seen in figure 2 the pixels on the face pulse in subsequent iterations. However, these pulsations are desynchronized when compared to the first cycle. This desynchronization is due to the texture of the input segment.

A vector or each pixel in which the elements are extracted from the set of pulse images defines the texture for that pixel. In this example, pixels from selected areas (shown in yellow boxes in figure 5) are used for training. Each region is defined by the texture within those boxes. So there were nine different training categories.

The texture vector or each pixel (including non training regions) was compared to the average and standard deviation of the vectors for each training region. The color-coded image in figure 6 displays the categorization of each pixel. As can be seen classification by texture as defined by the pulse images can be performed.

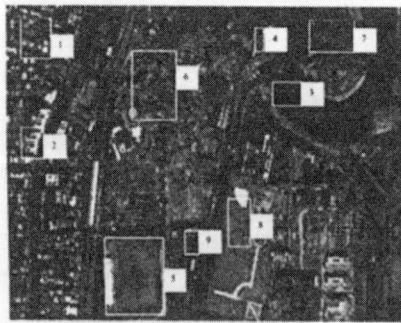 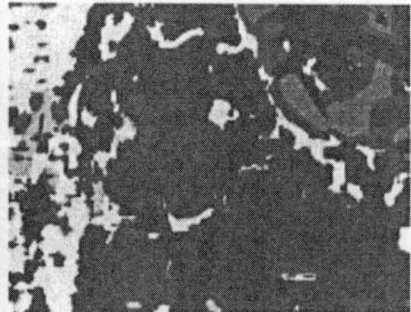

Fig. 5. An original image with training pixels for nine classifications enclosed in yellow boxes

Fig. 6. A color coded representation of classification of each pixel

 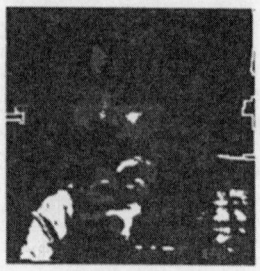

Fig. 7. An original three channel image and one of the pulse images

Fig. 8. The correlation surface of the pulse image in figure 8 with a fractional power filter. The spike in the lower part of the correlation surface indicates that the target was found and its location.

3.5 Identification by Signatures

In order to test the classification via image signatures a database of 1000 images from random web pages was created. Some web pages contained several images concerning a particular target, thus, the database consisted of random images as well as sets of images with strong similarities. Each image was converted to gray scale and the signature was computed for each. So far, there has not been much of an advantage seen in using the signature of color images since the similarity match was concerned with shape.

The measurement of similarity between two signatures Gp and Gq is performed by the subtraction of the normalized signatures. The normalization process eliminates the effects of image scale. Thus, a similarity measure is computed by,

$$a = 1.0 - \sum_n \left| \left(\left\| G_p[n] \right\| - \left\| G_q[n] \right\| \right) \right| \tag{16}$$

The signatures of each in the database are compared to find the best matches. Since there are 1000 images in the database there are 499,000 unique pairings (excluding self-pairings). The top scores are listed and the images were manually compared.

In an earlier study six objects from the VTK (Visualization Toolkit) [15] software package were used. One (the cow) was selected as the target and the others were selected as non-targets. Signatures for every thirty degrees of rotation for all three axes of all objects were computed. The average signatures of each non-target was also computed.

The optimal view is defined as the view that is most distinguishable from the other targets. Thus, the signature of a target view that was most different from the averages was found. It was expected that this signature would represent a view that displayed most of the characteristics unique to the cow (horns, head, four legs, udder, etc.) The image in figure 9 displays the view selected by this method and it can be seen that these attributes are distinctly visible in this view.

Fig. 9. The optimal view when compared to five other targets

4 Summary

Pulse image processing is a relatively new field that uses models of the mammalian visual cortex as the foundation for extracting important image information. The pulse images present information in several fashions. The first presentation is that of pulse images – a set of binary-element images – that display segments inherent in the original image. These segments are usually solid with sharp edges thus making it easier to classify segments than it is to do so directly from the original image.

The second presentation was that of fused pulse images making the system capable of isolating 3D segments in a set of complex-valued 2D pulse images. This facilitates the identification of 3D objects using standard 2D detection methods.

The third presentation was that of image signatures. Image signatures are short representations of the image shapes which commonly translate into the representation of the image content. These signatures can be easily compared to find images containing similar content, to estimate motion, or to find an optimal viewing angle.

Pulse image processing provides a powerful took for extracting important information for a variety of applications.

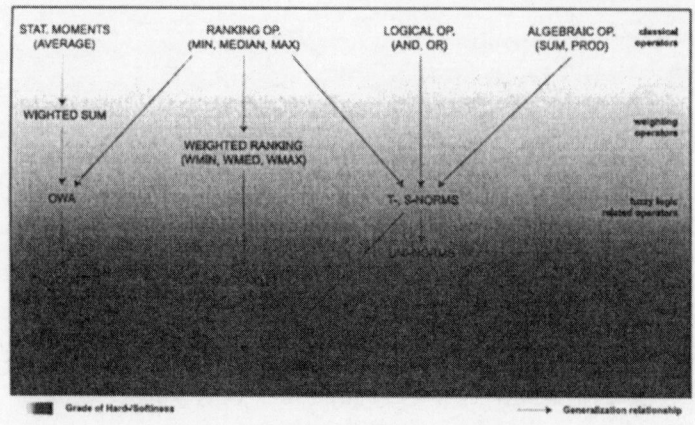

Fig. 1. Relational map of different fusion operators. The grade of softness increase in the vertical axis from the top to the bottom and is a result of successive generalizations. In the horizontal axis the operators are grouped upon its flavor, which defines different families of operators. In the vertical axis, the operators are grouped upon different theoretical frameworks in operation research.

ity, partial aggregation and reinforcement [41]. Finally fuzzy integrals reflect in the fusion result all this mentioned information: the value delivered by the different sources, their *a priori* importance, and their ranking. Therefore fuzzy integrals are the only fusion operators in this theoretical framework taking into consideration the importance of the individual information sources and that of their possible coalitions [10]. Actually the weighting scheme in this operators succeed in form of so-called fuzzy measures. Last but not least fuzzy integrals generalize most of fusion operators considered in the context of fuzzy sets (see Fig. 2).

4 Soft Fusion through the Fuzzy Integral

The concept of fuzzy integral is due to Sugeno, who presented in his work [33] a mathematical approach for the simulation of multicriteria evaluation taking into consideration some cognitive aspects. Sugeno's hypothesis is that the process of multicriteria integration undertaken by human beings transcend the linear combination of the different criteria with numerically expressed priorities, i.e. weighting sum strategy. Thus fuzzy measures, a generalization of classical measures through the consideration of subjectiveness, were established as the mean of expressing the *a priori* importance of the integrands. A new type of measures led to a new integration: the fuzzy integral.

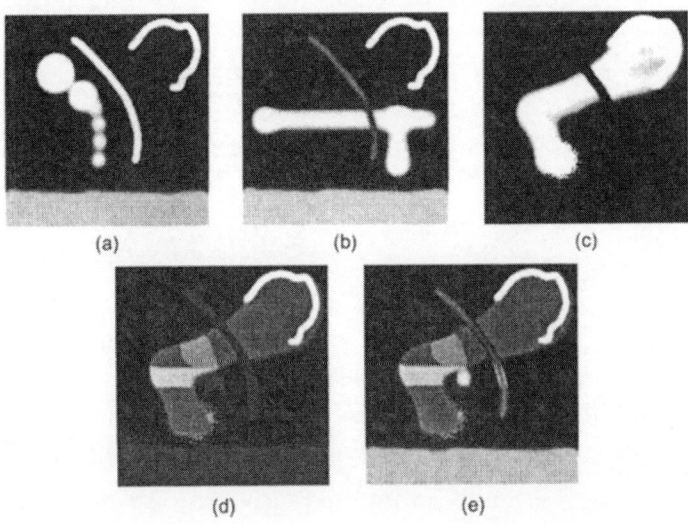

Fig. 3. Exemplary effect of the fuzzy measure modification in the integration of color channels through a fuzzy integral. (a-c) Input channels: (a) red, (b) green and (c) blue. (d-e) Results of the integration through the Sugeno Fuzzy Integral. (d) Result for a fuzzy measure: $\mu_R = 75, \mu_G = 40, \mu_B = 150, \mu_{RG} = \mathbf{100}, \mu_{RB} = 170, \mu_{GB} = 200, \mu_{RGB} = 255$. (e) Result for a fuzzy measure: $\mu_R = 75, \mu_G = 40, \mu_B = 150, \mu_{RG} = \mathbf{220}, \mu_{RB} = 170, \mu_{GB} = 200, \mu_{RGB} = 255$. The change of the fuzzy measure coefficient μ_{RG}, (d) vs. (e), modifies the result of the yellow areas, where red and green channels agree, i.e. present high values.

4.3 Some Properties of Fuzzy Integrals

Different properties of the fuzzy integral, which will be recalled in the application descriptions, can be found in this section. First the response of the operator in front of incoming data, what is called in filter theory the transfer function and in fuzzy sets related operator research known as level surfaces or curves, will be described (see figures 4 and 5).

Both types of integral present monotone non decreasing functions in the feature space $[0,1]^N$. They divide the feature space in dominance areas, due to the sort operation previous to the aggregation itself (2). These dominance areas, which receive the name of canonical regions of the hypercube [10], are defined by the hyperplane that joints the origin and the point where all the features present their maxima (see figure 4b).

The Sugeno Fuzzy Integral (see figure 4) presents so-called constant regions in its level curve [10], whose hypervolume depends on the values of the fuzzy measure coefficients. Here the output value of the fuzzy integral is constant and equal to the value of the corresponding fuzzy measure coefficient (see figure 4b).

430

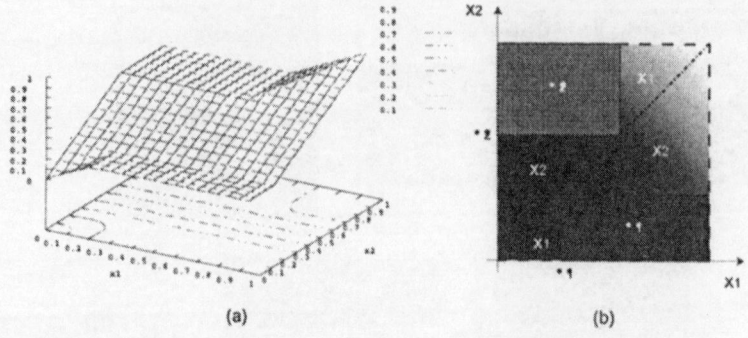

(a)   (b)

Fig. 4. Level curve of the Sugeno Fuzzy integral. (a) Response for $\mu_1 = 0.2, \mu_2 = 0.6$. (b) Projection on the feature plane.

In the Choquet Fuzzy Integral the action of the fuzzy measure coefficients is to modify the gradient of the channel being evaluated by that coefficient (see figure 5). The larger the coefficient is, the larger is the gradient for that variable. Thus the variations of the features with larger fuzzy measure coefficients will be reflected stronger in the fuzzy integral result than these variations of the features with smaller ones.

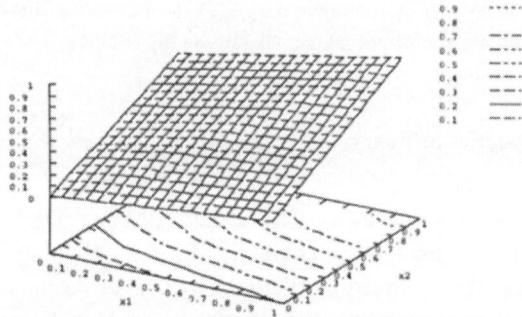

Fig. 5. Level curve of the Choquet Fuzzy integral for $\mu_1 = 0.2, \mu_2 = 0.6$.

The result of the fuzzy integral ranges from that of the minimum operator to that of the maximum [10]. Thus it covers all the range of reasonable results for a fusion operator. As formerly mentioned this result depends on the value of the fuzzy measure coefficients. Furthermore the fuzzy measure coefficients are implicitly responsible for the generalization relationships between fuzzy fusion operators (see Fig.2). For instance a minimum operator can be expressed as a Sugeno Fuzzy Integral where all coefficients except that for the set of all information sources are 0.

where M and N are arbitrary aggregation operators for fuzzy AND and OR respectively. For theorem, demonstration and deeper explanations the mentioned work is referred [5].

Being the fuzzy integrals monotonic operators limited in the range $[0, 1]$ for regular fuzzy measures, it seems possible to use the fuzzy integral for the computation of similarity relations to the Ideal and Anti-Ideal points of the hypercube. One open question in this definition is the differentiation between M and N. The work on applications [31] has shown that a practicable strategy is to use the fuzzy integral respect to a fuzzy measure μ for the computation of the distance to the Ideal, while the so-called dual fuzzy measure μ^* is used for the distance to the Anti-Ideal. The duality relationship between fuzzy measures is usually defined between possibility and necessity measures or between belief and plausibility measures [10]. A pair of dual fuzzy measures (μ, μ^*) satisfies:

$$\mu^*(\cdot) = 1 - \mu(\cdot^c), \tag{8}$$

where $(\cdot^c)$ states for the complement set.

4.4 Intelligent Localized Fusion Operators

Intelligent Localized Fusion (ILF) is a new paradigm for fuzzy fusion in image processing presented by the author [30]. Fuzzy Integrals become so-called Intelligent Localized Fusion Operators (ILFO) through the local definition of fuzzy measures.

Usually only one fuzzy measure is defined when being used in a fusion operation through the fuzzy integral. The fuzzy measure, as already mentioned, are used to quantify the *a priori* importance of the information channels and of its coalitions. Traditionally such quantification has been made globally. On the other hand different fuzzy measures are defined in the ILF paradigm. This fuzzy measures are locally distributed through a mask that points out the fuzzy measure to be used in each *loci* (see Fig. 7).

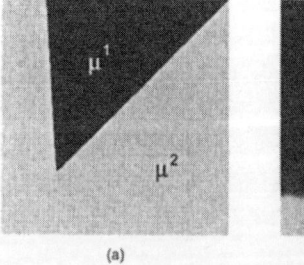
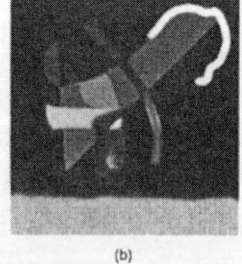

(a)   (b)

Fig. 7. Exemplary usage of an ILFO with the image in Fig.3. In the ILF paradigm different fuzzy measures are used in a fuzzy integral. (a) Example of a mask whereby the different fuzzy measures, e.g. μ^1 and μ^2, are localized. (b) Fusion result where the μ^1 adopted the values used for obtaining Fig.3d and μ^2 those used for obtaining Fig.3e

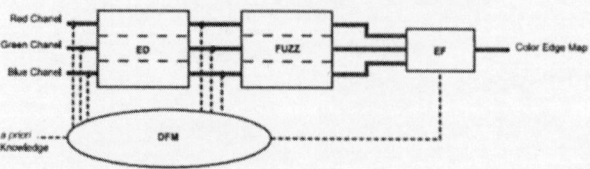

Fig. 8. Scheme for color edge detection with Intelligent Fusion Operators, ILFOs. ED: Edge Detection. FUZZ: Fuzzifier. EF: Edge Fusion. DFM: Determination of Fuzzy Measures

Fig. 9. Results with the framework for Color Edge Detection employing different values of the fuzzy measures coefficients and Sugeno Fuzzy Integral (SFI). (a) Input image (original in color). (b) Color edge map for values $\mu_R = 0.6$, $\mu_G = 0.25$, $\mu_B = 0.8$ (fuzzy-λ measures). (c) Fusion with SFI as minimum. (d) Fusion with SFI as maximum.

channels is undertaken in order to suppress background pixels, acting as a noise filter. The obtained results are shown in Fig.9.

A comparison of the different results shows the quality of using the fuzzy integral and thus of the parameterization of the fusion through the fuzzy measures. The operator used for Fig.9a shows no loss of edges with a more natural aspect than the result of a maximum in Fig.9b. The determination of the fuzzy measures was done empirically.

The employment of an ILFO opens new perspectives (see Fig.10). Through the usage of a mask, where the shadow regions present a different fuzzy measure than other areas , the detection of shadow false edges can be avoided.

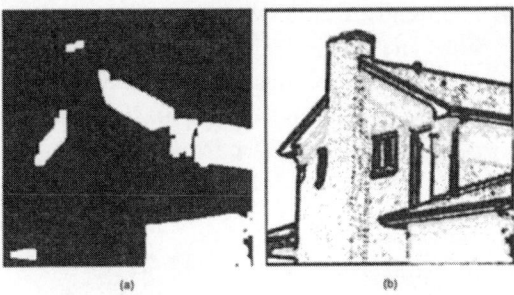

(a)   (b)

Fig. 10. Result of the fusion through a SFI-based ILFO, where the avoidance of shadow false edges is achieved. (a) Mask for the fuzzy measures. (b) Color edge maps after fusion.

5.2 Bio-inspired Multisensory Fusion for Image Segmentation

In this work a bio-inspired system for the fusion of color and infrared images is implemented. The goal of the system is the segmentation of the images in the input channels taking into consideration texture and color information for visual inspection. The processing of the color image is inspired in the processing undertaken in the visual primary system [28]. Moreover a parallelism between multisensory fusion in human brain and the final segment fusion is used (see Fig.11) [28]. The fuzzy integral is used for the fusion of color edge maps, for the fusion of color and texture information, and finally for the fusion of the segmented color and infra-red images. The fusion of color edge map, which result from the application of Gabor filters, follows the framework presented in the former section.

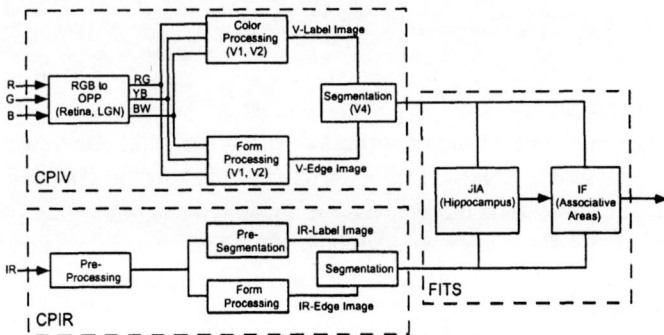

Fig. 11. Bio-inspired framework for the fusion of visual and infrared images. CPIV: Visual Image Processing Channel. CPIR: Infra-red Image Processing Channel. FTIS: Segmented Textural Information Fusion. RGBtoOPP: RGB to Color Opponencies. JIA: Joint Image Analysis. IF: Information Fusion. Inspiring biological modules from the primary visual system and high level areas are enclosed.

436

Since this work is currently in progress, only preliminary results of the fusion of segmented images is shown (see Fig.12). The color and infrared images were segmented through a Watersheed Transformation [6] after bio-inspired pre-processing. The segmented images are fused with the fuzzy integral, which achieves the successful fusion of reinforcing area labels (see Fig.12e).

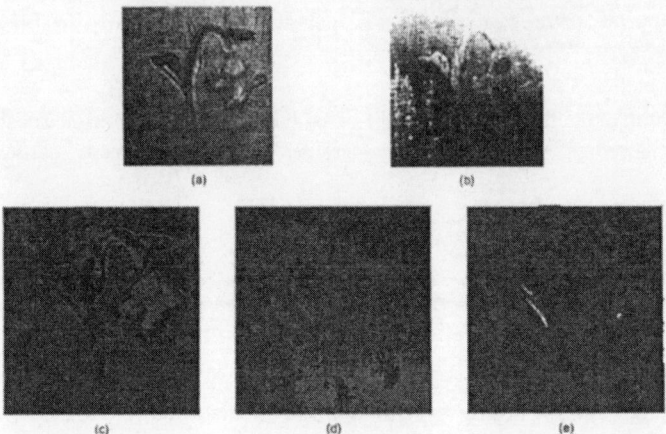

Fig. 12. Preliminary results with bio-inspired framework for the fusion of visual and infrared images. The results are obtained from a system for the automated visual inspection of textiles. Here a failure in the printing process has to be detected. The result of such faulty process can be observed in the color image (a) as a white thin band on the flower leave. In the IR image (b), in black. (c) Segmented color image. (d) Segmented infrared image. (e) Fusion of both segmented images.

5.3 Automated Visual Inspection of High Reflective Materials

The framework presented in this section is conceived to be used as pre-processing system in the automated visual inspection of materials with a high reflectivity. Different images of the object under inspection taken under different conditions (of illumination or shuttle time) are fused through a fuzzy fusion operator. The goal of the fusion in this case is the avoidance of the highlights produced by the reflective surface. The system was applied in both identification and detection tasks (see Fig.13). A detailed explanation on the application of the here presented framework can be found in [32].

In the first application the final goal of the system was to detect the absence of some item in a chocolate pack (see Fig.13a). A pre-processing was necessary in order to avoid the reflectance of the plastic bundle. It can be observed in Fig.14 that the softer the used fuzzy fusion operator is, the better is the performance of the system in terms of image contrast. The utilization of genetic algorithms for the determination of the fuzzy measure improved the results (see Figs.14d and 14e).

Fig. 13. Two of the input images in the pre-processing system of high reflective material images. (a) Image used in an identification system of chocolate items in a pack. (b) Image used for the detection of failures in lamp body.

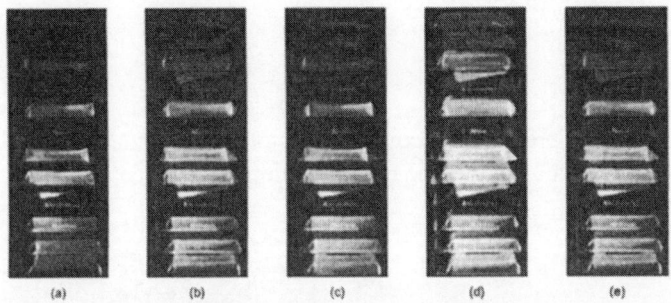

Fig. 14. Results of the pre-processing system for the avoidance of highlights in the identification problem. Different fuzzy fusion operators were used. (a) Result with minimum operator. (b) With OWA. (c) With Choquet Fuzzy Integral (CFI). (d) With automated parameterized OWA. (e) With automated parameterized CFI. The automated parameterization was realized through genetic algorithms.

The framework was also used in a system, whose final goal was the detection of structural faults in lamp bodies. The starting idea in this case was not to suppress all the reflections but to use the reflections on the faults as help for the detection. The usage of the ILFO paradigm allows this by pre-processing different image regions selectively. Thus a mask was used where the highlighted areas of smaller size are considered to be candidate faults. The results are encouraging, although further work have to be invested for the practical usage of the framework (see Fig.15).

5.4 Document Analysis through Data Fusion

As already mentioned the fuzzy integral was first used in image processing as image segmentation tool []. This work is the starting point for the developments here presented.

The existence of constant regions (see section 4.3) in the level curve of the SFI was exploited in a system for the classification of archive cards. The system can

438

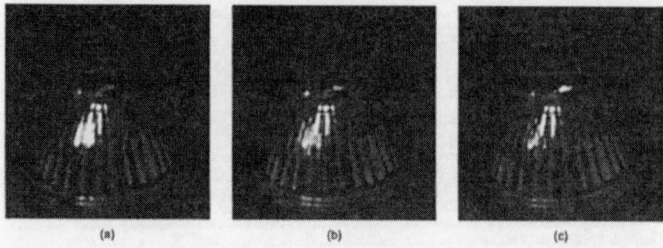

Fig. 15. Results of the pre-processing system for the avoidance of highlights in the detection problem. The fault to be detected can be visualized as a triangular lighter area in the upper part of the lamp. Different fuzzy fusion operators were used. (a) Result with weighted minimum operator. (b) With OWA. (c) With ILFO (Choquet Fuzzy Integral with localized fuzzy measures). The usage of an ILFO does not suppress the highlight on the fault in order to facilitate its detection.

be used for the suppression of textured backgrounds in the archive cards, which facilitates the extraction of the alphanumeric information (see Fig.16).

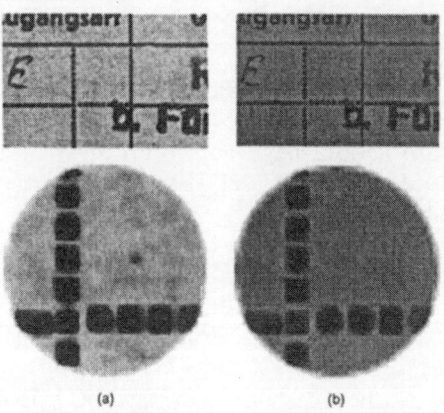

Fig. 16. Usage of the Sugeno Fuzzy Integral for background-foreground separation on color images of archive cards. The textured background is substituted by a unique grayvalue. (a) Input image and detail. (b) Output image and detail.

The same principle can be used for the segmentation of cards with different colors. The obtained results are shown in Fig.17. After a morphological dilation of the input images the SFI operates over the input channels with coefficients of the fuzzy measure adapted for each color. Thence the obtained label images are fused. This kind of segmentation presents some errors due to the incomplete suppression of text in the pre-processing stage (see Fig.17c).

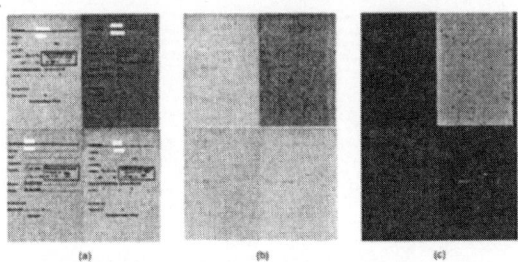

Fig. 17. Usage of the Sugeno Fuzzy Integral for segmentation of archive card color images. (a) Input image with four cards. (b) Image after morphological dilation for the suppression of text. (c) Label image resulting from the segmentation.

A bit more complex system was used for the segmentation of stamps on documents. The used strategy was to compute the fuzzy integral with two different fuzzy measures. These were selected in order for the stamp color color cluster to be maximally affected by the change. The difference image of these results was fused with the input image through a logical AND. The CFI, which presents a smoother response than the SFI, was proven to present a better performance. The obtained results are depicted in Fig. 18.

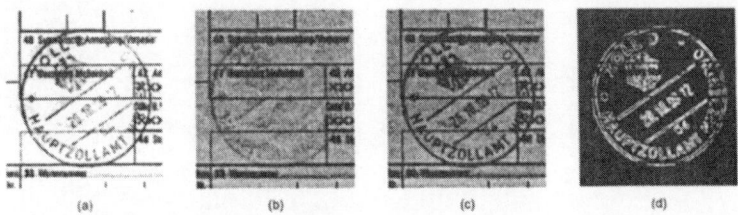

Fig. 18. Usage of the Sugeno Fuzzy Integral for segmentation of stamps. A fuzzy integral respect to two different fuzzy measures. The change of fuzzy measure affects the stamp color cluster, leaving the other components of the image unmodified. (a) Input image. (b) Fuzzy Integral result with the first fuzzy measure. (c) Fuzzy Integral result with the second fuzzy measure. (d) Final result, where the difference image between (b) and (c) was used as mask for the input one (a).

5.5 Color Morphology

Mathematical Morphology constitutes a well-known discipline for the analysis of spatial structures which has found an extended application in image processing. The basic element for the definition of a morphological operation is the existence of a ranking scheme, which is applied over the pixels underlying an structuring element known as mask. In grayvalue morphology the concepts of maximum and minimum

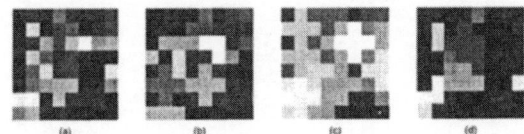

Fig. 19. Results of directed morphological operations with a cross mask on a random color image. (a) Input image, where pixel color is randomly defined. (b) Red-dilation with $\mu_R = 128$, $\mu_G = 1$, $\mu_B = 1$, $\mu_{RG} = 200$, $\mu_{RB} = 200$, $\mu_{GB} = 1$, $\mu_{RGB} = 255$. (c) Green-dilation with $\mu_R = 1$, $\mu_G = 128$, $\mu_B = 1$, $\mu_{RG} = 200$, $\mu_{RB} = 1$, $\mu_{GB} = 200$, $\mu_{RGB} = 255$. (d) Blue-erosion with $\mu_R = 1$, $\mu_G = 1$, $\mu_B = 128$, $\mu_{RG} = 1$, $\mu_{RB} = 200$, $\mu_{GB} = 200$, $\mu_{RGB} = 255$.

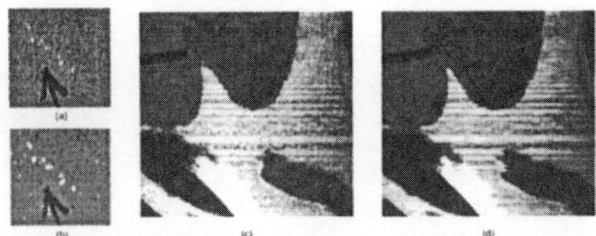

Fig. 20. Results of the here presented color morphology for the automated visual inspection of textiles. (a) Input image. (b) Result on (a) of a non-directed dilation with a cross mask. (c) Input image. (d) Result on (c) of an erosion with a linear mask directed to the fault color.

operators. Beside them there is a great number of softer operators mostly developed in the context of operators research related to fuzzy sets.

The transition from hard- to soft-fusion is progressive. Fuzzy fusion operators generalize most known aggregation operators. The complexity and computational cost of fuzzy fusion operators is larger than that of traditionally used ones. Nevertheless Soft Data Fusion presents a flexible theoretical framework, suitable for the implementation of robust image processing industrial applications.

The here presented tutorial elucidated the mathematical background of fuzzy fusion operators in order for them to become broader applied. It was shown that the fuzzy integral can be employed for image processing with very different purposes. The new introduced ILFO's constitute a good example of this flexibility.

Some of the still open research questions on Soft Data Fusion are: the research on computationally efficient algorithms for the implementation of the different operators, the automation of their parameterization, and extending its employment in a larger number of real applications.

References

1. M.A. Abidi, R.C. Gonzalez, eds. (1992). *Data Fusion in Robotics and Machine Intelligence*. San Diego: Academic Press.

created and the so-called *projection matrix*, the one that achieve the dimensional reduction, is obtained from all the database face images. In the off-line phase are also obtained the so-called *mean face* and the reduced representations of each database image. These representations are the ones to be used in the recognition process.

2.1   General Approach

Figure 1 shows a block diagram of a generic eigenspace-based face recognition system. A preprocessing module performs a normalization of the input face images and then a subtraction of the *mean face* ($\bar{\mathbf{x}}$). Normalization is necessary to initialize each face vector $\mathbf{x}$ with the same energy.

After that, $\mathbf{x} - \bar{\mathbf{x}}$ is projected using the projection matrix $\mathbf{W} \in R^{N \times m}$ that depends on the eigenspace method been used (see section 2.2). This projection corresponds to a dimensional reduction of the input space, starting with vectors $\mathbf{x}$ in R^N (with N the vector image dimension), and obtaining projected vectors $\mathbf{q}$ in R^m with $m < N$ (usually $m << N$). Depending on the eigenspace approach been used, the topology of the original face space would be preserved or not, and the reconstruction of the face vector $\mathbf{x}$ will be possible. For face recognition tasks it is not critical the reconstruction ability of the projection.

The *Similarity Matching* module compares the similarity of the reduced representation of the query face vector $\mathbf{q}$ with the reduced vectors $\mathbf{p}^k \in R^m$ that represent the faces in the database (see figure 1 again). By using a given criterion of similarity (see section 2.3), this module determines the most similar vector $\mathbf{p}^k$ in the database. The class C^k of this vector is the result of the recognition process, i.e. the identity of the face. In addition, a *Rejection System* for unknown faces is used if the similarity matching measure is not good enough. The rejection parameter of this system could be determined using the *Bayesian Optimal Criterion* proposed in [4].

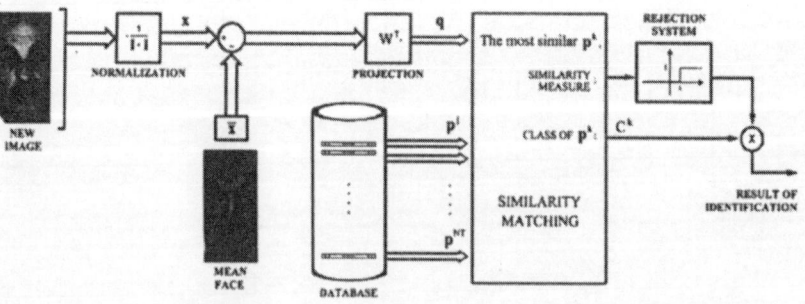

Fig. 1. Block diagram of a generic eigenspace-based face recognition system

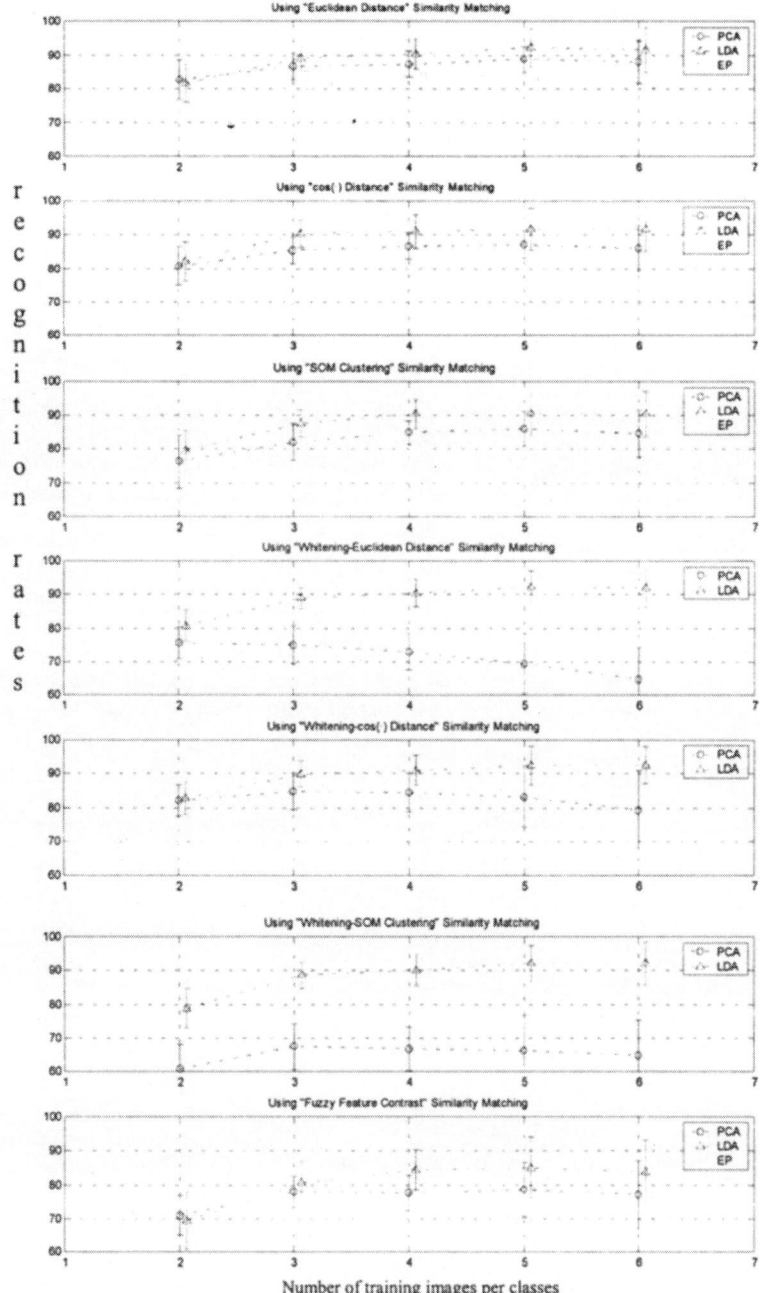

Fig. 8. Graphics of the mean recognition rates using different numbers of target images per class, and taking the average of 20 different target sets.

For each simulation we used a fixed number of target images, choosing the same types of images per class (in the Yale Face Database terminology the 10 images per class are: center-light, w/glasses, happy, left-light, w/no glasses, right-

complement of existing computational methods. Of course, an appropriate visual representation is required, which can be achieved by means of dimensionality reduction or multivariate projection methods combined with interactive visualization of the data [8]. Typical database representation, e.g., as an Excel sheet is not easily amenable to human perception and understanding. This illustrated in Fig. 1 together with the alternative, human-adapted visual representation of the same database. Thus, dimensionality reduction

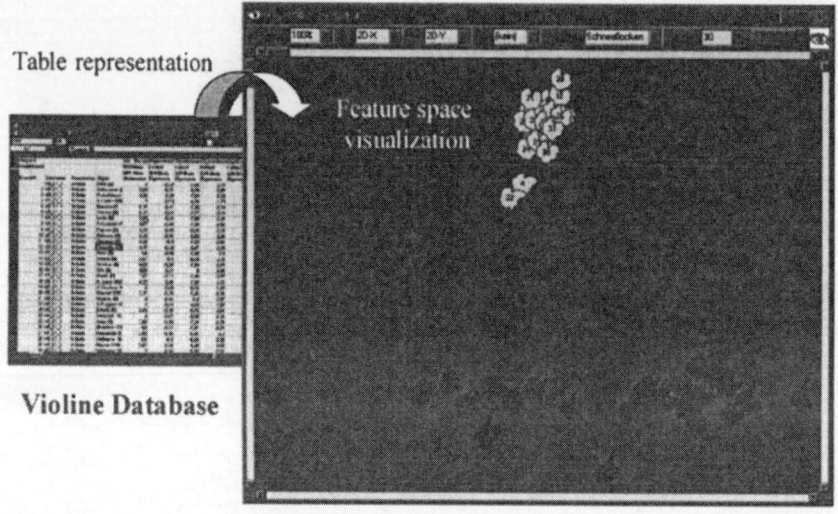

Fig. 1. Exploitation of human perceptive capabilities by appropriate presentation of multivariate data employing dimensionality reduction and interactive visualization.

is an ubiquituous problem and together with multivariate data visualization a topic of interest and interdisciplinary research for more than three decades. The availability of increasingly powerful computing hardware and storage media with unprecendented storage capacity results in application databases also constantly growing in size and numbers. Especially high economic interest applications, e.g., data mining and knowledge discovery applications, give renewed strong incentive to the field.

The aim of this tutorial is to introduce to the subject and its interdisciplinary application potential. Thus, a taxonomy of relevant methods from past to present was elaborated as baseline for a focused survey. Methods are briefly reviewed and qualitatively and quantitatively compared for classification and visualization purposes in an unifying approach. As a baseline for the quantitative comparison, appropriate cost functions for feature space assessment will be introduced, also ordered according to an elaborated taxonomy. Enhancements and benefits of interactive visualization and underlying visualization concepts, e.g., interactive navigation, for exploratory data analysis and knowledge aquisition will be introduced. Concept and architecture of two dedicated tools based on the introduced method spectrum, e.g., the Weight-Watcher and the Acoustic Navigator, will be presented and key applications will serve for further elucidation of the approach and its potential. The most practical methods and tools from the tutorial are available on the web as free demosoftware in the QuickCog system.

The limited page count does not allow a self-contained presentation of the complete tutorial in this paper. Thus, if required, further background information to the accompanying comprehensive set of slides can be taken from existing publications, e.g., mainly from the survey given earlier in [1]. In this paper, following the structure of the earlier survey and tutorial, the focus will be on the presentation of some extensions, new ideas, and new applications.

2 Feature Space Assessment

Both for assessment of dimensionality reduction methods with regard to achieved distance, signal, or topology preservation as well as achieved discriminance gain in the mapping, and for the mapping computation itself appropriate cost functions must be available. These can be grouped in unsupervised and supervised methods, as given in the taxonomy in Fig. 2. Unsupervised

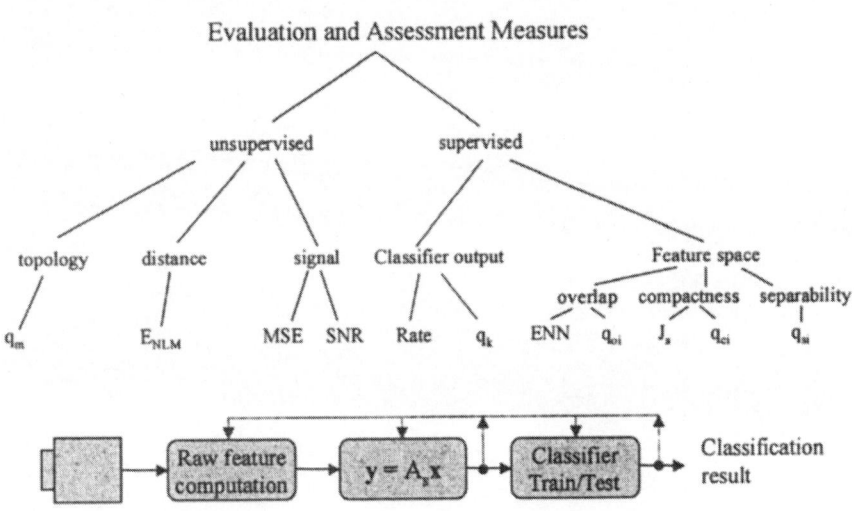

Fig. 2. Taxonomy of evaluation and assessment methods.

measures are particular relevant for visualization purposes, as they assess the preservation of data structure after dimensionality reduction in terms of

interactive human analysis. Interactive visualization techniques, e.g., interactive navigation, diverse component, grid, and attribute plots support human perception and analysis [1]. Figure 4 (upper left) shows the underlying multivariate data visualization architecture. Especially the features for accessing

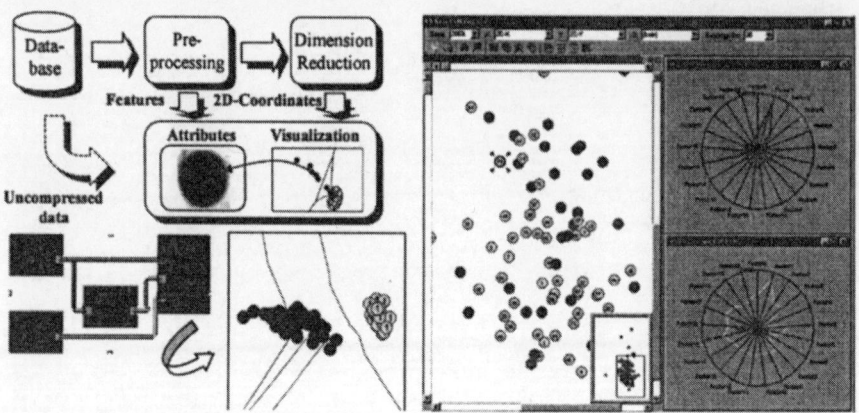

Fig. 4. Feature space reduction and interactive visualization: Architecture and dedicated tools WeightWatcher and Acoustic Navigator.

database contents from the top-level map should be pointed out here as unique characteristics of the approach. Two implementation have been conceived so far, the general purpose tool WeightWatcher (WW) in QuickCog and the dedicated Acoustic Navigator [12] with enhanced interactive features (Fig. 4). Further interactive enhancements are on the way, e.g., interactive selection, labeling, and extraction of arbitrary data from the map.

5 Applications

The methods and tools described and assessed in the tutorial have been employed in numerous scientific and industrial applications. Examples of applicability are given in

- Rapid Prototyping in the design of recognition systems
- Analysis of medical database [7]
- Analysis of psychoacoustic sound databases with the extension towards synthesis in sound engineering [12]
- Analysis in manufacturing automation, e.g., microelectronic manufacturing processes
- Analysis and design of integrated circuits with regard to design centering and yield optimization

For the case of rapid and transparent recognition system design a brief example will be given. A vision system was designed for a medical robot in an object recognition task. Dimensionality reduction and interactive visualization approach helped to assess the current system's capability in terms of feature space discriminance and occurrence of outliers. This is illustrated in Fig. 5. Additionally, the backtracking capability from the resulting interactive

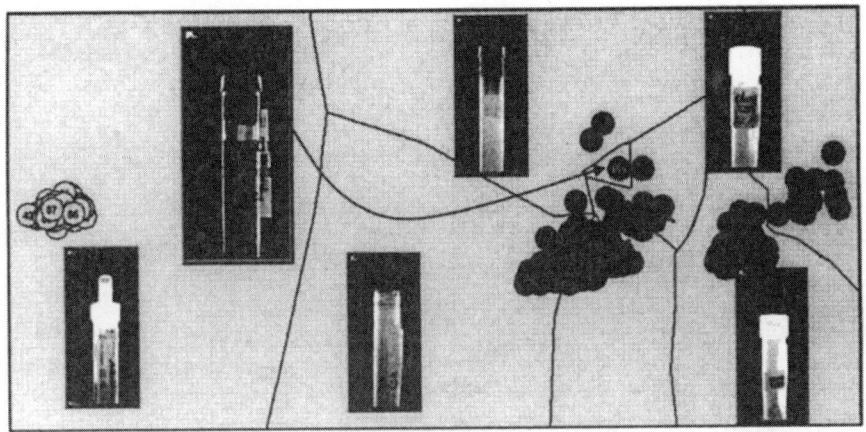

Fig. 5. Feature space for vision system of medical laboratory robot.

map is illustrated by invoking the original image of a selected object for each class from the underlying database. Thus, occuring problems, e.g., misclassifications, and underlying causes can be easily made overt. This alleviates trouble-shooting in system design and increases design speed, reliability, and overall productivity.

The second, very simple example is representative for the group of problems related to the analysis and optimization of microelectronic manufacturing processes. The multivariate, nonparametric approaches discussed in this tutorial can complement the prevailing spectrum of classical methods in this domain. This will be briefly elucidated by a database obtained from a Multi-Project-Wafer run comprising 10 wafers. Device and process parameters are monitored on four sites per wafer, 59 parameters each. All available chips of a specific design were dies from wafer 5 and the chips showed unsatisfactory behavior during testing. It was not evident, whether a processing problem or a design problem was responsible. Thus, the database was subject to the dimensionality reduction and interactive visualization approach. Fig. 6 shows the first projection of the MPC database on the left and on the right as multicomponent plot, where wafer 5 sites are colored in a dark and the remaining wafer sites in a light shade. Obviously, no anomaly or unusual structure can be observed in the data. The observed lack of obvious structure can be due to absence of abnormity or to occlusion by a majority of normal parameters and high intrinsic dimension. Thus, AFS was employed in the next step based on the wafer number as class affiliation to find parameters support-

470

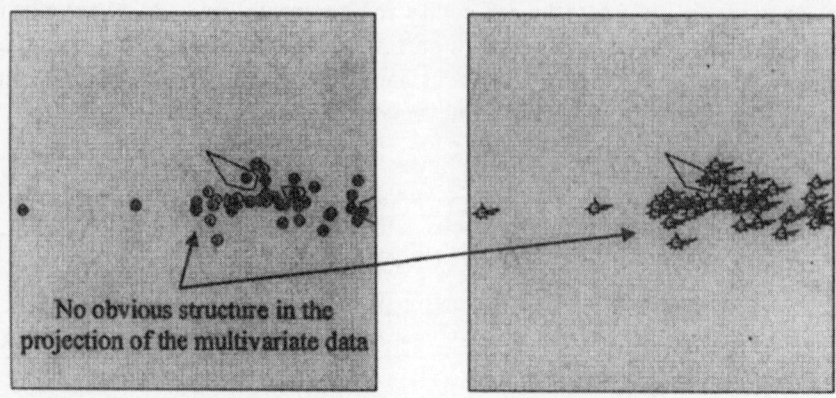

Fig. 6. First-cut projection of the MPC database.

ing the abnormity hypothesis. The two measures q_{oi} and q_{si} were employed, and a reduction from 59 to 7 and 3 parameters was achieved. The result is illustrated in Fig. 7, where now a subcluster can be identified in the visualization based on the selected parameters. Obviously, there is some support

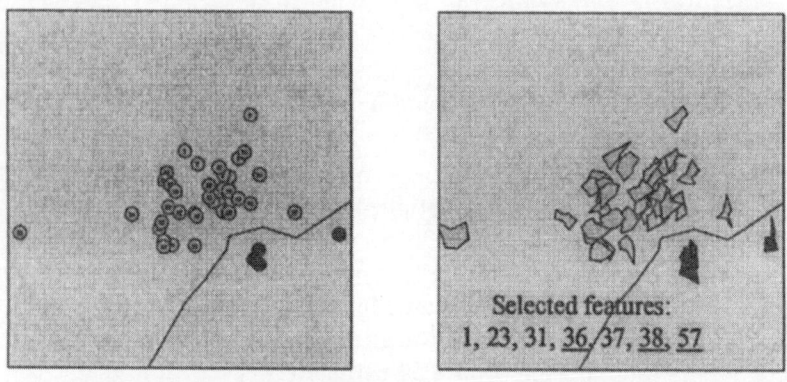

Fig. 7. Projection of the MPC database after selection.

for the hypothesis that wafer 5 is different from the other wafers. However, it must be validated, whether the identified parameters can be responsible for the observed chip behavior. In this case, which was instead of its simplicity not a toy problem, the parameters distinguishing wafer 5 from the remaining lot have no physical relevance with regard to the observed problem. Thus, a processing problem can be ruled out. This answer was found very fast by the methodology described in [1] and in this tutorial. In general, the outlined ap-

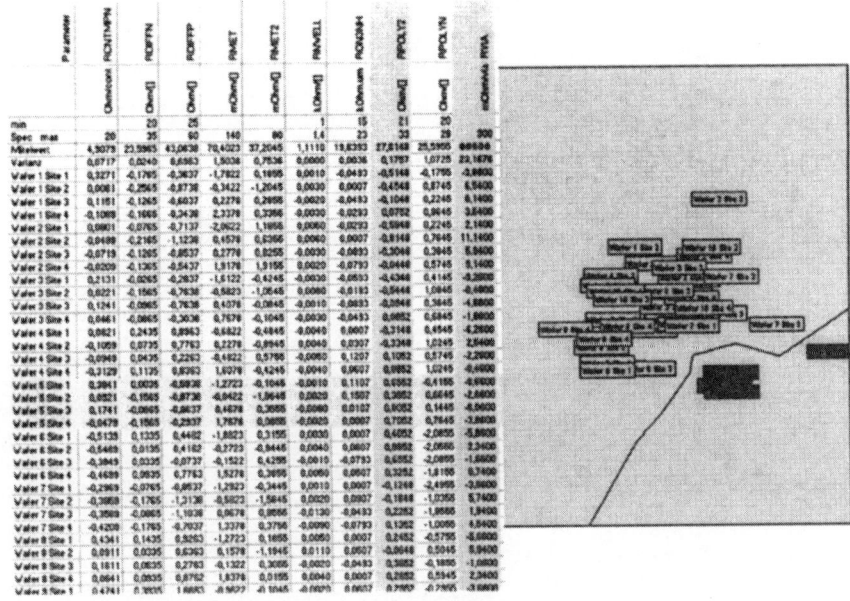

·**Fig. 8.** Checking selected variables for wafer 5 in MPC database.

proach offers fast, efficient, and transparent access to multivariate, complex data met in today's manufacturing processes. This example just scratches the surface, other nonobvious information can be found and employed for process optimization and centering.

6   Conclusion

In this tutorial, an introduction to and a focused survey of the principles and techniques for dimensionality reduction and interactive visualization has been given. Relevant techniques were qualitatively and quantitatively assessed and compared in an unifying approach. Basic problems of dimensionality reduction and some advanced concepts to cope with those were presented. An improved quality measure for compactness and an accelerated hierarchical projection method were briefly introduced in this paper. Two dedicated tools, the WeightWatcher in QuickCog and the Acoustic Navigator were presented. Methods and tools were demonstrated in several applications, e.g., for the high potential application analysing microelectronic process data.

In future work, improvements of hierarchical projections and mapping techniques in terms of speed, efficiency, mapping reliability and user convenience will be pursued together with the extension of the interactive visualization features. The survey given here must be extended due to numerous recently published approaches, e.g., in the TNN SI on data mining [3] or [6].

Part V

Agents, Multimedia and Internet

Developing Agent-Based Personalized Recommender Systems: An Experimental Study

Wei-Po Lee and Chih-Hung Liu

Department of Management Information Systems,
National Pingtung University of Science and Technology,
Nei-Pu, Pingtung, Taiwan

Abstract. The prosperity of electronic commerce has changed the traditional trading behaviors and more and more people are willing to conduct electronic shopping. However, the exponentially increasing information provided by the Internet enterprises causes the problem of overloaded information, and this inevitably reduces the customer's satisfaction and loyalty. One way to overcome such a problem is to build intelligent recommender systems to provide personalized information services. By analyzing the information provided by a customer, his browsing history, and the products he purchased through the Internet in the past, a personalized recommender system is able to reason about a customer's personal preferences and then provides the most appropriate information services to meet the needs of the customers. In this work, we develop an agent-based recommender system in which an evolutionary approach is proposed to learn a customer's preferences. In order to assess our approach, a prototype is built for DVD film recommendations. Experimental results and analysis show the promise of our approach.

1 Introduction

In recent years, the prosperity of electronic commerce has changed the traditional trading behaviors. Companies have been digitalized to promote the corresponding performance, and due to the on-going advance of Internet technology, more and more people are tempted to conduct electronic shopping. Therefore, Internet enterprises must develop further their new business portals to expand their markets and create more business opportunities. Though there have been some successful cases in electronic commerce, such as Amazon, Dell, CISCO, and eBay, etc., unfortunately more others are proven failures. It shows that advanced hardware facilities for Internet infrastructure are not the only decisive factor to guarantee a successful business in electronic commerce. In fact, the Internet enterprises must provide more other extra or value-added services for the consumer to promote the transactions. One popular way used by Internet enterprises is to provide more information related to their products. For example, an Internet book store may provide introductory notes of a book, biographical information about the author, and comments or reviews of the book, so that a consumer is able to know more about the books he is likely to buy, and then easily spot the book he would like to purchase. Also, there are some Internet companies playing the roles of information agents: they gather and classify information for various Internet products, and then evaluate products based on the comments made by the experts or on the feedback from the consumers. Companies of this kind attempt to act as the portals of products and provide the services of querying or browsing for the consumer. However, the exponentially increasing information along with the rapid expansion of the business Web sites causes the problem of overloaded information. Owing to the prodigious amount of information provided by different sites, the consumer has to spend more and more time on the net to find the information needed. Sometimes the contents of the Web pages are irrelevant to the consumer's expectation, but the consumer can not help but read them in order to filter them out to get what he or she really wants.

One way to overcome the above problem is to develop recommender systems to provide personalized information services. By analyzing the information (interests) provided by a consumer, his browsing history, and the products he purchased through the Internet in the past, a personalized recommender system is able to reason about a consumer's personal preferences and then provide personalized services to meet his requirements. On the one hand consumers can immediately access the information they are interested in, to save their time spent on reading the electronic documents; on the other hand, enterprises can get to know the customer's personal preferences and then develop most appropriate marketing strategies to attract customers of specific types and efficiently deliver the information interesting to them. The customer's satisfaction and loyalty can thus be increased, and the increase of visiting frequency by the consumers can further create more transaction opportunities and benefit the Internet enterprises.

In this paper, we describe how we adopt an agent-based methodology to develop a personalized DVD recommender system. Our system includes five agents: a *monitoring agent* to observe a consumer's browsing behavior and records what he is reading; an *information agent* to gather a customer preferences in an explicit way; an *information management agent* to manage the customers' profiles; a *learning agent* to analyze and learns a customer's interests; and a *performance agent* to give recommendations and to evaluate the performance of the system. This paper describes the functions of different agents and gives more details about the learning agent that employs an evolutionary approach to model a user's personal preferences. Experimental results show that our learning agent can learn a customer's interests in a short period of time and give sensible recommendations. In addition, the important parameters that influence the performance of recommendations are explored, and the k-nearest neighbor method is implemented for performance comparison. Furthermore, we also show that when a customer's interests change and that causes the performance of the recommender system declines, the learning agent can be resumed again and continues to learn the most up-to-date preferences for the customer.

2 Background

2.1 Personalized Recommender Systems

A personalized recommender system can provide information services in different ways. For the type of systems that keeps recording and analyzing a customer's previous preferences, the customer's personal information is first collected, then the system reasons about the customer's preferences by analyzing and modeling the personal information available. After that, the recommender system construct a computational model for the customer to predicate his preferences for other items of the same application domain. Therefore, developing a high performance learning mechanism to precisely model the consumer's preferences is one of the main issues in constructing intelligent personalized recommender systems [7][13]. In fact, the work of recommendation can be regarded as classification; it is, in a sense, using the information already known to build a model to predict the events unknown [2]. The model can be considered as a function mapping its input (the consumer's profile and a specific product item) into the output (a specific value within its own range). Thus the learning methods used to solve the problem of classification can also be employed to construct the model of a consumer's preferences.

2.2 Related Work

Some learning approaches have been applied to construct customers' profiles to deal with the problem of information overload. In [1], a statistic-based approach *tfidf* (term frequency and inverse document frequency [12]) was used to build a user's profile to recommend Web pages, and in [15] a similar approach is used to organize e-mail for users. In [6] the reinforcement learning method was employed for Web page recommendations and in [10], for book recommendations. In addition, a *k-nearest neighbor* method was used in [3][5] to find other customers who share the similar interests for a specific consumer. Consequently the system can recommend products to this consumer according to the evaluations from those who with similar tastes. In this work, we use an evolutionary approach to model a consumer's interests for DVD film recommendations.

There are also some recommender systems do not recognize a customer; instead it uses the ephemeral information provided by the customer with the built-in product information to look for the optimal products. Recommender systems of this type aim to assist the customer to find out what he really wants, when he is only able to identify the type of product he needs and describe the features or specific functions of the product. The key issue in developing recommender systems of this kind lies in the estimation of product similarity, and the case-based reasoning approach is usually taken to measure the similarity by calculating the weighted sum of different product features. [14] and [16] present some applications that are developed based on this approach.

3 Developing Agent-Based Personalized Recommender Systems

3.1 System Overview

The major tasks of a personalized recommender system include collecting a customer's personal interests, building a model to describe the information collected, and managing a customer's personal information. In our work, an agent-based methodology is thus adopted to develop a personalized DVD film recommender system, as it can solve problems in a distributed way. In this way each agent performs a specific work and different agents can work simultaneously to achieve the overall task. Fig. 1. illustrates the aspect of our recommender system.

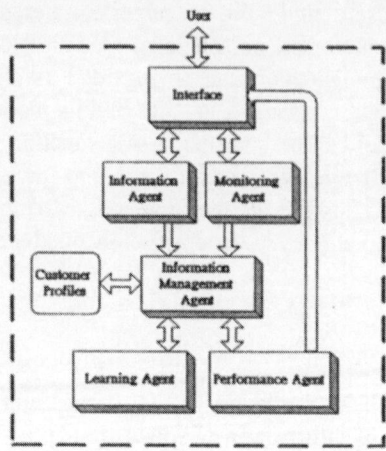

Fig. 1. The aspect of our agent-based personalized recommender system

In Fig. 1., the *monitoring agent* provides a browser-like environment in the front end for user's browsing, and in the back end records the contents the customer has read and categorizes them into different classes. Fig. 2. presents the environment in which the monitoring agent can record a customer's behavior when he is browsing the on-line movie site *allmovie*. To recognize how a customer is interested in the product presented in a page, the monitoring agent measures the time he stays to view that page. In addition to collecting information in the above implicit way, our recommender system also includes an *information agent* that provides two explicit ways to obtain a customer's personal preferences on items in the application domain. The first is to present the customer a form with especially designed table and to allow him directly point out what and how he is interested in. For example, in our later experiments of recommending DVD films, a customer can directly give the films he likes and dislikes to the information agent as samples for further analysis. Fig. 3. shows the form. Alternatively, the information agent could also employ an interactive method to explicitly collect a customer's interests. In this case, the agent requests the customer to rank the items listed in a questionnaire iteratively, and records the results to reason about a customer's personal preferences. As this work evaluates products in a binary way, the method of ranking items is not provided. The monitoring agent and the information agent both play the role of information collectors, and a customer can activate them through the interface to gather his personal preferences about the products.

Fig. 2. Information being gathered by the monitoring agent

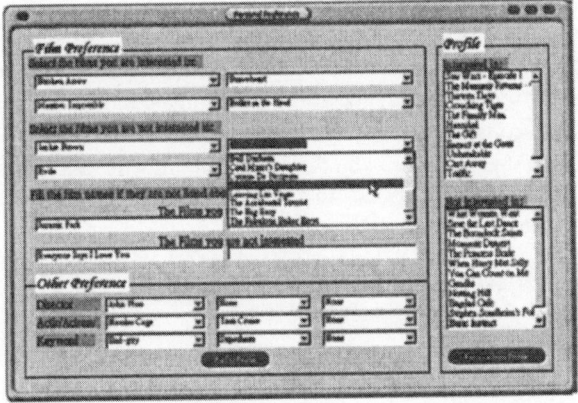

Fig. 3. Information being gathered by the information agent

The *information management agent* mainly takes the responsibility of analyzing and transferring a collected product item into a set of features (or attributes) which are pre-defined by the system to best represent that type of products. For example, we can define some features, such as category, author, publisher and price, to represent a book. After a book item is collected by the above agents (could be in the form of a Web page), it can then be represented as a vector in terms of the corresponding features. This agent also establishes and maintains a personal profile for each customer in which a profile mainly contains two parts: a personalized prediction model constructed by the learning agent and a set of product items with numerical values quantitatively indicating how much a customer likes them. The next section of experiment will include an example to describe the customer profile in more detail. Also a customer's personal profile will be updated when his preferences change. Since the customers' profiles are maintained by the information management agent, the only way for other agents to access the information recorded in the consumers' profiles is thus through the intermediary of this agent.

The *learning agent* is the kernel of our recommender system; it builds the model of prediction for a customer, based on the information recorded in his profile. As is mentioned, this model can in fact be considered as a classifier that recognizes an unfamiliar item as one of the categories representing different degrees of preferences on the products. Hence different machine learning approaches can be applied on the products, with their corresponding customer preferences, recorded in the profile as training examples to construct the classifiers. Once a classifier is constructed, it is then stored in the profile and used to predict how a customer is interested in other products in the same application domain. In our current implementation, the learning agent acts in a passive way. It is activated by the information management agent whenever necessary—at the time when a new customer registers to the system or when the performance of the recommender system declines due to the changes of a customer's preferences. An evolutionary approach is employed in our work to learn a customer's preferences and the details are described in a later section.

Contrast to the learning agent, the *performance agent* operates in a more active way: it keeps acting ever since a customer enters the system. With the learnt customer model, this agent can decide whether to recommend a specific product item to the corresponding customer or not. As a customer's interests may change occasionally, the system must continuously track his behavior and feedback in order to adapt to his most recent preferences. The performance agent is thus responsible for tracking the recommending performance and reporting the result to the information management agent. Based on the result, the information management agent can then determine if it is necessary to evoke the learning agent to establish a new model for this customer again.

3.2 Learning Customer's Preferences by GAs

When using evolutionary computation techniques to solve a problem, the primary step is to choose the proper representation for an individual. On the one hand, a genetic representation must be able to express explicitly the features of the solution of the problem to be solved; on the other hand, it must be suitable to be manipulated by the genetic operators to obtain the solution efficiently. In our system, an individual (chromosome) is defined to include two strings of numerical values. The first string indicates which product features should be taken into account in modeling a customer's preferences; it is represented as $<a_1, a_2, ..., a_n>$ in which each gene a_i is an integer of 0 or 1. The second string means the relative importance of each feature listed; it is represented as $<b_1, b_2, ..., b_n>$ in which each gene b_i is a floating point number between -1 and 1 as the associated weight for the corresponding feature term. As the learning agent will build a customer's model from the products ranked, the string length n is thus the number of distinguishable feature appearing in the training examples. Each individual has a same length, hence the standard genetic operators, crossover and mutation, can thus been directly employed to create new individuals. In our GA mechanism, the operations dealing with binary and floating point numbers (as described in [9]) are applied independently to two different strings.

To predict how a customer likes a certain product P, we firstly transfer the product into a n-dimension vector to match the feature description defined above, and then calculate the weighted sum of the product and the customer's model as:

$$\sum_{i=1}^{n} p_i \times a_i \times b_i$$

where p_i is the i-th component in the binary vector form of product P and it indicates whether the i-th product feature appears in P; a_i and b_i are the i-th components of the string classifier defined above; and n is the length of the classifier. The possible range of the above weighted sum can be divided into multiple sub-intervals in which each sub-interval represents a certain degree of preference, and then the degree a customer likes a specific product can be measured by examining which sub-interval a weighted sum lies in. The reason we use multiple discrete intervals here rather than a continuous one is to provide a simple way for a consumer to indicate his preference on a product. For instance, we can divide the overall range into five sub-intervals as increasing (or decreasing) scales of preference for a product (i.e., 1 to 5), and then request a customer to express how he feels about that product by giving a scale value.

In the learning phase, the learning agent uses the produces recorded in the customer's profile as training examples, and uses the evolutionary mechanism to select the appropriate features from those associated with the training examples and determine their corresponding weights as a consumer's model of preferences. During the evolutionary process, each individual (model) is used to predict how a customer likes the training examples by the weighted sum calculation already described above. The difference between the actual interval produced and the desired interval pre-indicated is defined as the error of prediction for each training example, and its complement as the accuracy. Therefore the average accuracy over all training cases can be defined as the fitness of an individual. The individual evolved is then recorded in the consumer's profile and used as the personal model (classifier) of his preferences for further product recommendations.

4 Experiments and Results

In order to assess the proposed approach, in this section we describe how it is used to develop a recommender system. In the experiments, we concentrate on evaluating our learning and performance agents. For the learning agent, we first investigate how a prediction model can be evolved by our GA mechanism, and then explore the effects of some relevant parameters used. After that, a series of experiments were conducted to compare our GA approach with the well-known k-nearest neighbor method. For the performance agent, we analyze how it monitors the system's performance that varies in response to the changes of a user's interests, and how it can adapt to a customer's most recent preferences.

4.1 The Customer's Profile

A DVD film can be described by the features such as *genres*, directors, actors, and so on, but in this work we choose to use the plot of a DVD film as the feature because it often is the most decisive factor a general audience used to determine whether he likes a specific film or not. In the preliminary experiments, we attempted to use the introductory note of a film as its corresponding plot, and employed the traditional text analysis approach (i.e., *tfidf*) to extract feature words to constitute a feature vector for a DVD film, as our previous work in document categorization [8]. We have found that, however, the introductory note is generally too short to extract representative terms. Therefore we decide to use the keywords associated with a film (available in the on-line film database) to represent the plot and to be product features. For example, the film *Jurassic Park* is represented as <attack, clone, dinosaur, experiment, genetic, ...> in our system. The keywords appearing in the training examples are defined as product features and used to constitute a chromosome (i.e., a_i in section 3.2).

For simplicity, in our current implementation the system only recognizes whether a customer is interested in a specific DVD item or not, rather than the degree in which he favors the item. Both implicit and explicit ways for information collection are provided and the product examples collected are maintained in a customer's profile for later training. Fig. 4. illustrates the aspect of the customer's profile and its functional relationships with different agents.

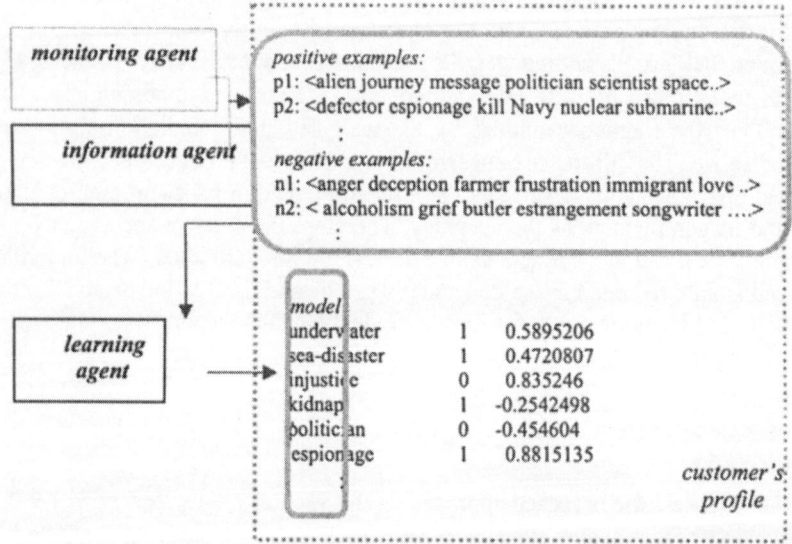

Fig. 4. The aspect of a typical customer's profile and its functional relationships with different agents. In this figure, the *positive* and *negative* include the training examples, and the *model* is the evolved result.

4.2 Evaluation of the Learning Agent

To evaluate the performance of our learning agent, we implement a GA mechanism and use it to evolve models of preferences for consumers. The training examples used in the GA runs are those films recorded in the consumers' profiles. In the first set of experiments, 60 training examples were used: a half of them were positive examples and the other half, negative. Here, we used a threshold value 0 to distinguish products: if the weighted sum calculated by the criteria described in section 3.2 is higher than the threshold, then the corresponding product is predicted as customer-like; it is customer-dislike otherwise. The proportion of the training examples predicted correctly is defined as the fitness of the specific model evaluated. Fig. 5. is a typical experiment illustrating how the evolutionary system converged. The upper curve in Fig. 5. shows the accuracy of the best individual evolved in each generation; and the lower curve, the average accuracy of all population members. It indicates that an accurate solution can be evolved successfully.

As is mentioned, the models learnt from the GA mechanism represent the consumers' preferences on DVD films; they then can be used to recommend different DVD films to the appropriate consumers. To further evaluate the results evolved above, we collected 40 DVD films not contained in the training set and used the evolved customer models to determine whether these test cases should be recommended to the corresponding consumers. Ten sets of evolutionary runs with tests were conducted. The test results show that in average 73% accuracy was obtained. This may be satisfactory as there were only 60 training examples used in the learning phase and they distributed in different genres.

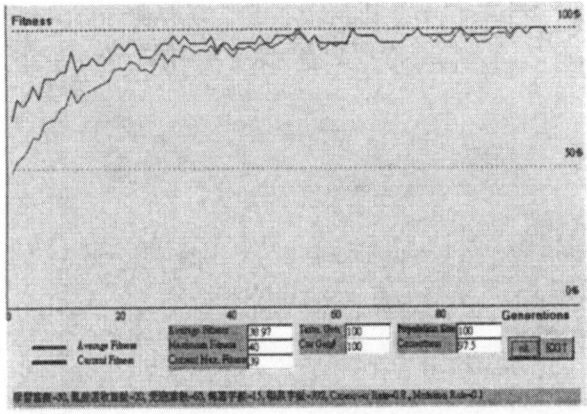

Fig. 5. A typical experiment run shows the converging behavior of the learning agent.

4.3 Effects of Some Relevant Parameters

To investigate the effect of the number of training examples used, we conducted a set of experimental runs with numbers of examples 20, 40, 60, and 80. The curve GA-0 in Fig. 6. shows the results in which each data was averaged from ten runs. It indicates that the prediction performance can be improved if more training examples were used. Though using sparse samples to reason about a customer's preferences in a large space could result in a relatively loose user model, it is nevertheless a more realistic situation.

In addition, we explored how different threshold values T used for separating the customer-like and customer-dislike items can influence the performance. Two sets of experiments with different threshold values (1.5 and 3.0) were performed; in each set four different numbers of training examples were used as above. Here, an individual in the training phase regarded a product as customer-like if the corresponding weighted sum was higher than T; and as customer-dislike if it was lower than –*T*. But in the testing phase the original threshold value 0 was used. The curves GA-1.5 and GA-3 in Fig. 6. show the results for threshold values 1.5 and 3 respectively. As can be observed, with the distance constraints in the learning phase, better results can be obtained in all cases of different numbers of training examples. This is because keeping two classes of items away with a certain distance can reduce the effect of overfitting, and more robust models can thus be evolved. Without the constraint, sensitive classifiers could be produced through learning—in those cases some weighted sums derived from positive and negative examples could have similar values close to the threshold used, and therefore the new items become relatively difficult to distinguish.

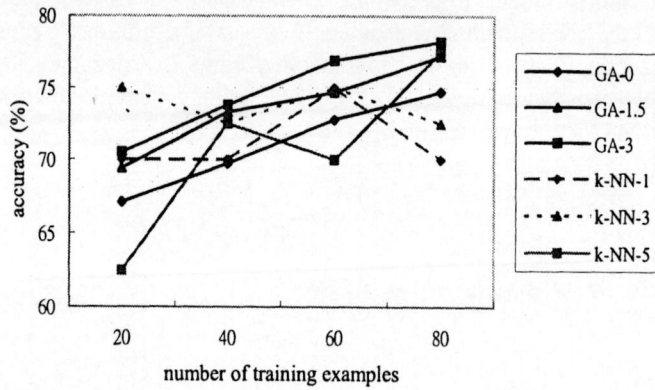

Fig. 6. The comparison of GA and *k*-nearest neighbor method; values are averaged over ten experiment runs.

4.4 Comparison of GA and *k*-NN

We have shown that our learning agent is capable of evolving customers' preference models and have also investigated the effects of some important parameters. In order to examine the performance of the evolved models further, we implemented the well-known *k*-nearest neighbor method [4], which was shown to give the best performance in Pazzani's work on Web page recommendations [11], for comparison. Three different values of *k* (1, 3, and 5) were used, and for each *k* different number of reference examples were provided as in the previous GA experiments. The curves *k*-NN-1, *k*-NN-3, and *k*-NN-5 in Fig. 6. illustrate the results. As can be seen, in our recommendation cases, GA with separation distance gives better results than the *k*-NN method.

From this figure, we can also observe that the prediction performance obtained from the tests is not as good as the performance in training in which 100% accuracy can be achieved. As is explained in the above section, using sparse samples to model customer preferences could lead to loose results. In addition, there may be another reason for that: in our current implementation, the products are categorized into two extremely different classes of customer-like and customer-dislike; and this binary decision making strategy makes products with similar characteristics to be assigned to completely different classes by the customer. Therefore, the performance can be improved if the products are evaluated in terms of different levels of preferences.

4.5 Analysis of the Performance Agent

In this recommender system, a customer's personal model will not be updated until the corresponding recommendation performance declines. In this situation, when 5 wrong cases occur during the most recent 10 recommendations, or 3 consecutive cases appear in our current implementation, the customer's preferences will be re-learnt: the most recent 10 cases collected will substitute the first 10 samples for training. Fig. 7. shows an example of how the system performance changes during the 60 consecutive recommendations for a certain customer. In this figure each data point indicates the number of correct cases accumulated from the last 10 recommendations. In this experimental example, the customer changed his preferences into completely different ones at the twentieth recommendation step and it caused the decline of system performance. Then the new preferences were learnt again and the performance was improved. The customer changed his interests again toward slightly different ones at the fortieth recommendation step. As it did not worsen the system performance so that the customer's model was not modified.

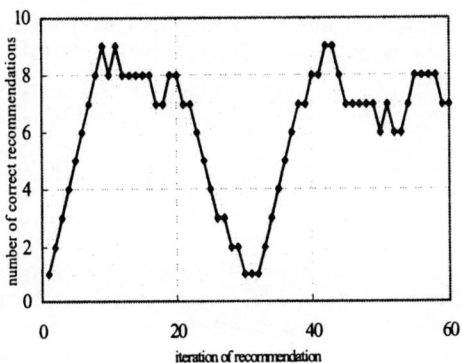

Fig. 7. An example illustrates how the system performance changes during the 60 consecutive recommendations.

5   Conclusions and Future Work

In this paper, we explain the need for Internet enterprises to provide personalized information services in making a successful Internet business, in addition to developing or improving the software and hardware equipment directly related to the Internet infrastructure. We also suggest that developing personalized recommender systems is a promising way to achieve the task. Therefore in this work, we present an agent-based methodology to develop recommender systems in which each agent is responsible for a certain sub-task, such as information gathering, customer modeling, information managing, and performance evaluating, related to the overall recommending work. To assess our approach, we use it to develop a personalized DVD recommender system. In this system, an evolution-based method is employed to learn a customer's preferences on DVD films and the evolved customer model is used to recommend the DVD items new or unfamiliar to the customer. A series of experiments are also conducted to examine the effects of different parameters, and the k-nearest neighbor method is implemented for performance comparison. In addition, the performance agent is evaluated to show how the system can re-build a customer's model when he changes the preferences. Experimental results have shown the promise of our proposed approach.

The recommender system presented in this work employs a binary decision making strategy that asks the customer to indicate whether he likes or dislikes a certain film. As is described, this strategy separates the products into two completely different categories. Under such circumstance, products that are slightly different could be dispatched to different groups and such a categorization makes no difference with that of two complete different products. A possible way to remedy the above situation is to evaluate products in terms of different levels of preferences; for example to ask the customer to rank products in a five-scale method. This will reduce the evaluation bias and more reliable and robust models can thus be obtained. We have been now conducting more experiments to address this issue.

Our work presented here points to some prospects for future research. One is to investigate other attributes of DVD films that could also be helpful in modeling a customer's preferences, in addition to the keywords about the plots. Also it is worthwhile to examine more other learning methods for performance comparison and apply appropriate methods to different situations. Another important issue is to explore the problem of scalability--using sparse samples to reason about a customer's preferences in a large space of product items.

References

1. M. Balabanovic and Y. Shoham. Fab: Content-Based Collaborative Recommendation. *Communication of the ACM*, 40(3): 66-72, 1997.
2. C. Basu, H. Hirsh and W. Cohen. Recommendation as Classification: Using Social Content-Based Information in Recommendation. *Proceedings of National Conference on Artificial Intelligence (AAAI-98)*, pp.714-720, 1998.
3. J. Breese, O. Heckerman and C. Kadie. Empirical Analysis of Predictive Algorithms for Collaborative Filtering. *Proceedings of International Conference on Uncertainty in Artificial Intelligence*, pp.43-52, 1998.
4. B. V. Dasarathy. *Nearest Neighbor (NN) Norms: NN Pattern Classification Techniques*. IEEE Computer Society Press, CA, 1990.
5. N. Good, J. B. Schafer, J. A. Konstan, A. Borchers, B. Sarwar, J. Herlocker and A. Riedl. Combining Collaborative Filtering with Personal Agents for Better Recommendations. *Proceedings of National Conference on Artificial Intelligence (AAAI-99)*, pp.439-446, 1999.
6. T. Joachims, D. Freitag and T. Mitchell. WebWatcher: A Tour Guide for the World Wide Web. *Proceedings of International Joint Conference on Artificial Intelligence*, 1997.

7. T. Kindo. H. Yoshida, T. Morimoto and T. Watanabe. Adaptive Personal Information Filtering System that Organizing Personal Profiles Automatically. *Proceedings of International Joint Conference on Artificial Intelligence*, pp.716-721, 1997.

8. J.-H. Liu, J.-J. Lu, and W.-P. Lee. Document Categorization by Genetic Algorithms. *Proceedings of IEEE International Conference on Systems, Man, and Cybernetics*, pp.3869-3873, 2000.

9. Z. Michalewicz. *Genetic Algorithms + Data Structures = Evolution Programs*. Second Edition, Springer-Verlag, 1994.

10. R. J. Mooney and L. Roy. Content-Based Book Recommending Using Learning for Text Categorization. *Proceedings of the ACM International Conference on Digital Libraries*, pp.195-204, 2000.

11. M. Pazzani, J. Muramatsu, and Billsus. Syskill & Webert: Identifying Interesting Websites. *Proceedings of National Conference on Artificial Intelligence (AAAI-96)*, pp.54-61, 1996.

12. G. Salton. *Automatic Text Processing*. Addison-Wesley, Reading, MA, 1989.

13. J. B. Schafer, J. Konstan and J. Riedl. E-Commerce Recommendation Applications. *Journal of Data Mining and Knowledge Discovery*, Jan., 2001.

14. S. Sen and K. Hernandez. A Buyer's Agent. *Proceedings of International Conference on Autonomous Agents*, pp.156-162, 2000.

15. R. Segal and J. Kephart. MailCat: An Intelligent Assistant for Organizing E-Mail. *Proceedings of International Conference on Autonomous Agents*, pp.276-282, 1999.

16. S. Shearin and H. Lieberman. Intelligent Profiling by Examples. *Proceedings of ACM International Conference on Intelligent User Interface*, 2001.

An Agent That Learns to Support Users of a Website

Fabio Abbattista[1], Aldo Paradiso[2], Giovanni Semeraro[1], and Fabio Zambetta[1]

[1] Dipartimento di Informatica, Università di Bari, Italy
  e-mail: fabio@di.uniba.it, semeraro@di.uniba.it, fzambetta@programmers.net
[2] GMD, Darmstadt, Germany
  e-mail: paradiso@darmstadt.gmd.de

Abstract. The terms agent, intelligent agent and 3D agent are becoming more and more frequently used in literature. A key issue in the Web community is that a web site *must* be equipped with a virtual agent able to support user. In this paper we propose the SAMIR system, a tool for animating 3D intelligent agents, mainly founded on a genetic algorithms based learning system, namely an XCS, and on a FFD based technique for the facial deformation of the character. SAMIR is based on an object-oriented architecture, and it adopts the MPEG-4' Facial Animation Parameters as a description format for facial expressions and animations.

1. Introduction

Men's life is rapidly shifting towards a "new virtual dimension", compared to the more traditional physical one, where users interact and communicate in ways made possible by e-mail, web, and chat technologies.

But all this power requires a price: common users are usually forced to use simple 2D interfaces unable to mimic immersive scenarios, such as the ones depicted in science fictions books like "Snow crash" [10] and "Neuromancer" [4].

The idea these books introduced was that the World Wide Web becomes a totally interconnected 3D virtual environment where people can retrieve information in a natural fashion, simply by navigating into its "paths".

Aside to this fascinating idea many research efforts are involved in developing Networked Virtual Environments (NVEs) where people, embodied by 3D avatars, may interact with other people or autonomous computer-guided 3D characters.

A similar idea opens huge interdisciplinary research fields dealing for example also with artificial intelligence topics. Efforts are employed to develop convincing intelligent autonomous personalities, with the help of psychologists, in order to provide characters with typical human features and behavior, and to investigate possible social interactions in these virtual environments via speech or body communication [3].

Besides, computer graphics enables avatars and character facial and body animations, lip-syncing and abilities to grasp, run, climb and walk into NVEs.

In the next section we introduce the SAMIR system. Section 3 details the module managing the behavior of the system, where section 4 reports some preliminary results. Finally, section 5 addresses the future work.

2. The SAMIR System

The SAMIR (Scenographic Agents Mimic Intelligent Reasoning) system describes a digital assistant where an artificial intelligence based Web agent is integrated with a purely 3D humanoid, robotic, or cartoon-like layout [1].

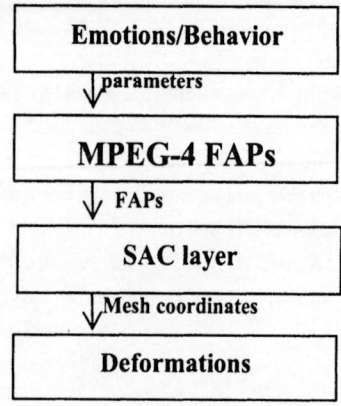

Fig. 1. The SAMIR model

Such a system is based on a multi-layered model (see Fig. 1). This model describes the generation process of the character's behavior and animation. The behavior/emotion, however generated, produces a set of parameters which need to be represented in MPEG-4 FAPs, which in turn will be applied onto the 3D facial model. The face is subdivided as a set of so-called Standard Anatomical components (SAC), which will be introduced later. Finally, different deformation algorithms are applied to SACs in order to generate facial animations. The layers are structured in such a way that they are functionally independent, and therefore the coupling rate between them is minimized, allowing us to design an architecture such that its components may be implemented and tested in a parallel fashion.

The architecture of SAMIR is shown in Fig. 2.

The **Event Interpreter** is the module devoted to the conversion of events occurring in the course of the user interaction. These events have been classified in User Triggered Events

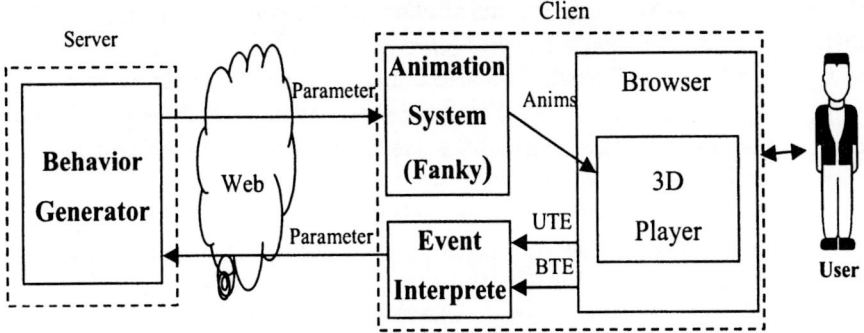

Fig. 2 - SAMIR system architecture

(UTE), corresponding to the dialogue log between the user and SAMIR, and Browser Triggered Events (BTE), representing user's choices and/or commands issued to the Web browser, without SAMIR's intermediation. The Event Interpreter module extracts, from UTE and BTE, relevant chunks of information (e.g. keywords, questions, user's errors, user's interests and preferences, etc.) and converts them in parameters suitable to be processed from the Behavior Generator module.

The **Behavior Generator** module, located on the server, generates the character behavior as well as its emotional response, inducing correspondent facial expressions, expressed in an abstract manner by MPEG-4' Facial Animation Parameters (FAP).

The aim of the Behavior Generator module is to discover and to apply the *animation rules*, on the basis of inputs coming form the Event Interpreter.

The output of the Behavior Generator is sent to the client, typically embedded into a Web browser used by the user. An **Animation System** interprets the parameters received and produces animation streams, which are shown and rendered by a 3D player, embedded into a Web browser. Details of the animation system are reported in [7].

3. The Behavior Generator

The Behavior Generator aims to improve the naturalness of the dialoguing agent. Two main goals of the Behavior Generator are: i) to profile users accessing the system according to their technical skill and ii) to display appropriately facial expressions according to the context of the current dialogue.

The Behavior Generator is composed by two main modules: the User Classifier Module and the User Interaction module.

The aim of the **User Classifier** module is the automatic assignment of one out of some predefined classes to each user, based on information drawn from previous sessions. This is known as *interaction modeling* [2]. In particular, the classification can be exploited to associate each class of users with an interface that is adequate to the degree of familiarity with the system, in order to speed up the process of understanding the organization and the content of Web site and to assist the user in retrieving the desired information.

In the SAMIR system, actually, we select three user classes, based on user skill: *Novice, Expert* and *Teacher*. These classes correspond to three different classes of users accessing a Digital library Web site developed in our University. The Novice and Expert classes represent, respectively, inexperienced and experienced students, while the Teacher class corresponds to the university teaching staff.

Furthermore, it is likely that, the user acquires familiarity with this service along the time, hence the system must be able to track out potential changes of the class the user belongs to. This problem requires the ability to identify the user and log his interaction. Thus, each new user receives an identity code - *User ID* – to be employed on any further access. Correspondingly, the system creates and associates a log file to each User ID, in which historical data on the interactions are stored.

By examining the data stored in each log file, it is possible to extract some features (number and frequency of errors, number and frequency of queries, etc.) that are useful to recognize the user model and to train a learning system in order to induce (a decision tree and/or) a set of rules used by the system to autonomously perform a classification of the users interacting with the Web site (see [9] for details of experimentations). User classification results are stored in a repository of users profiles and serve to choose the right interface each time the corresponding user accesses the site.

The **User Interaction** module is mainly based on a Classifier System. Learning Classifier Systems (LCS) are a machine learning paradigm introduced by Holland [5] in 1976. They are rule-based systems in which learning is viewed as a process of ongoing adaptation to a partially unknown environment through genetic algorithms and temporal difference learning.

An LCS receives in input the state of the environment via a set of detectors, treated as messages. These messages, processed by the LCS, determine the LCS's actions, by directing its effectors.

The learning module of SAMIR has been implemented through an XCS [11], a new kind of LCS. The main differences from the original LCS implementation are in the definition of classifier fitness and in the relation with reinforcement learning.

In our system, rules are expressed in the classical format *if* <*condition*> **then** <*action*>; each <*condition*> represents a combination of several possible events, such as:

- User classification (don't care, novice, teacher, expert) coming from the Users Classes repository;
- User request to the agent;
- Operation performed by the user in the Web site;
- Amount of user's errors;

The last three events come from the Event Interpreter module (see Figure 2).

The <*action*> of each fired rule represents the next expression that Fanky (the animation system) should display to the user in the course of the interaction. In particular, each expression is represented by a linear combination of the six fundamental expressions (surprise, sadness, joy, fear, disgust and anger) [8], in which each one of the fundamental expressions is included with a percentage ranging from 0 to 100.

So the *<action>* part of our rules provides to the Animation System the percentage of each of the six fundamental expressions, to be used to compose the desired expression of our character. The binary string representing the *<action>* part of a rule is composed by six groups of 4 bits each one. Each group of bit represents the percentage of one of the six fundamental expressions to be combined to produce the desired expression.

For example an expression composed by 40% of joy and 60% of surprise is coded into the following binary string:

0100	**0000**	**0110**	**0000**	**0000**	**0000**
% of Surprise	% of Sadness	% of Joy	% of Fear	% of Disgust	% of anger

4. Experimental Results

The experiments performed had a double aim: i) to verify the effectiveness of the XCS in learning a set of predefined interaction rules, ii) to verify the ability of the XCS in discovering new interaction rules.

In a preliminary phase we defined a set of 30 interaction rules in which different combinations of kinds of users (novices, experts, etc.) and different situations in the course of an interaction (actions performed, users requests and users errors) have been considered. This set of pre-defined rules represented the training set, that is the minimal know-how that SAMIR should possess to start its *work* in the Web site.

To evaluate the performance of the system, in the course of the training phase, the following procedure has been adopted for the credit assignment:

```
If    system's   action   match   expected   action   then
credit=100
Else credit=k/d(system's action, expected action).
```

where k is a constant and the distance measure (d) has been defined as the mean squared distance among the 6 components of the expression (surprise %, sadness %, joy %, fear %, disgust %, anger%).

In other words, a perfect match between the expected action and the action performed from the system corresponds to the best result, otherwise, the credit assigned to the rule is inversely proportional to the distance between the expected action and the action performed from the system.

For the sake of brevity, the next table shows only some rules learned by the system. The second and third columns represent, respectively, the input to the system (the *<condition>* part of the rule) and the expected output (the *<action>* part of the rule). The fourth and fifth columns show, respectively, the system's output and the corresponding *<condition>* part.

The first row represents a perfect match between the expected action and the action performed from the system. In the second row the rule does not represent a perfect match but the two actions (the expected one and the system's action) are quite *similar*. Last but not least, the third rule represents a new rule, discovered from the system.

Figure 5 shows the expressions corresponding to the action part of the last two rules in Table II, as displayed by the Fanky module.

Table 1. Results of the training phase

Rule	Input	Expected Output	System's Output	System's Condition
1	0001001001100	0 0 0 0 60 40	0 0 0 0 60 40	0#0#001001100
	00011110110 00	00 60 00 0	00 30 00 0	00011110110 00
3	01011000010 01 (A novice user makes few errors)	-	70 0 30 0 0 0	-

Due to the inherent features of XCS, SAMIR has been able to learn quite effectively the pre-defined rules of behavior and to generalize some new behavioral pattern that could update the initial set of rules. In such a way, SAMIR is comparable with a human assistant that, after a preliminary phase of training, will continue to learn new rules of behavior on the basis of personal experiences and interaction with human customers. The next step will involve the ability of different agents to share their learned rules in order to suggest to their *colleagues* some new behavioral patterns. Our idea is to exploit the concept of memetic algorithms as a *tool* for spreading useful (or fitted) knowledge and concepts [6] among the staff composed of virtual agents.

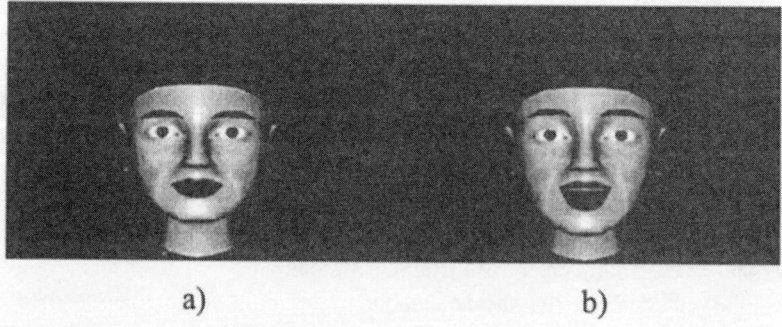

a)   b)

Fig. 3. Expressions corresponding to the last two rules in Table II, as displayed by the Fanky module. a) Expression for rule #2; b) Expression for rule #3.

5. Future work

The system will be improved by adding several features, like:

Speech capabilities: the integration of speech will complete the expressiveness of the character. We plan to integrate a TTS system, synchronize the speech with labial movements of the character, and merge the visemes (i.e. the visual counterpart of phonemes) with concurrent facial displays.

Other improvements consist in refining our model by considering other information derived from user profiles, like gender, background, education, and possibly mood. We also plan to perform a set of experiments to test the reaction of final users with different facial models (male, female, cartoonish, human-like, etc.).

Moreover, we will add an Editor of characters in order to provide users the ability to design their own agent. The editor will take into account the user profile and it will run in different modality: *novice*, the user will be provided with a set of predefined facial component (Pug nose, Turned-up nose, Aquiline-nose, Broad brow, Low brow, etc.); *expert*, the user will be provided with the set of facial points to be arranged according to his/her preferences.

Finally, as already stated, we would like to extend the system in order to involve more agents, able to cooperate and share their knowledge.

References

1. Abbattista, F., Semeraro, G., Zambetta, F.: S.A.M.I.R.: Scenographic Agents Mimic Intelligent Reasoning, in: Proceedings of the 7° Congress of Italian Association for the Artificial Intelligence, Milan, September 2000.
2. Banyon, D., and Murray, D.: Applying User Modeling to Human-Computer Interaction Design. Artificial Intelligence Review, 7 (1993) 199-225.
3. Cassell, J.; Sullivan, J.; Prevost, S.; Churchill E.: Embodied Conversational Agents. The MIT Press, Cambridge, Massachusetts, 2000. ISBN 0-262-03278-3.
4. Gibson, W.: Neuromancer, HarperCollins, 1995.
5. Holland, J.H.: Adaptation. In R. Rosen and F.M. Snell (eds.), Progress in Theoretical Biology, New York: Plenum, 1976.
6. Lynch, A., Thought Contagion, Basic Books, 1996.
7. Paradiso A., Zambetta F., and Abbattista F.: Fanky: A tool for animating 3D intelligent agents, in: Proceedings of Intelligent Virtual Agents 2001 (IVA'01), Madrid, September 2001, (in press).
8. Schlosberg, H.: The Description of Facial Expressions in Terms of Two Dimensions, in: Journal of Experimental Psychology, Vol. 44, No. 4, October 1952.

9. Semeraro, G., Ferilli, S., Fanizzi, N. and Abbattista, F.: Learning Interaction Models in a Digital Library Service, in: Proceedings of the 8th International Conference on User Modeling, Sonthofen, July 2001.
10. Stephenson, N.: Snow crash. Bantam Books, 2000.
11. Wilson, S.W.: Classifier Fitness based on Accuracy. In: Evolutionary Computation 3/2, (1995), 149-175.

ANTS: Automatic Navigation of Terrain Systems

Tim Batchelor Hnd,

Alumnus of Bolton Institute . F4, 112 Park Rd, Bolton, Lancs. England
++44(0)1204 406228. loanwolffe.farzone@virgin.net

Abstract. This paper presents the preliminary stages of the development of a Generalized Modular System Model for the implementation of autonomous agent navigation through terrain systems and the development of cooperative behaviors of 3D agents in a VRML scene. The model is based on a generalized concept of the cooperative behaviors of the social, ant and builds on previous successful models by re-integrating features that simulate individual aspects of the behaviors of the various species.

The paper takes a practical approach to the development of intelligent behavior in the 3D scene. It addresses the needs of the VRML programmer intending to make agents exhibit terrain following and collision detection as well as more high level behaviors in an intelligent way. It proposes a platform for the model that will allow the resulting work to be immediately accessible to the majority of internet users and sets forth the constraints and limitations that this choice of platform imposes on the development of the model.

1   Introduction

The development of simulated methods of intelligent navigation is becoming increasingly relevant in a multitude of real world applications, many of which are mission critical, but the most entertaining and dynamic arena in which this topic is significant is in the development of 3D game play. Here it is used to provide a rich source of user interaction and competitiveness. In the real world such simulations support the development of low and high level behaviors in robotics and have implications in the development of advanced control systems. In this paper we examine the problem of one or many robot agents navigating through a highly variable environment, the problem is complex and requires the application of techniques from a variety of allied fields.

There is a significant body of previous work in the field of Robot Motion Planning, Kuffner et. al. [1] use a randomized path planner as part of their simulated environment for providing high level software control of humanoid robots. In [2] a coordinated approach using a probabilistic path planner is used to compute collision free paths for multiple robots in a static environment. In [3] Ratering and Gini used a Hybrid Artificial Potential Field to resolve the two component problems of getting the robot to its goal and avoiding any obstacles in the process. In [4,5 and 6] Goldenstien, Large and Metaxas develop a Dynamic Systems Approach to modeling low level behaviors for autonomous agents. With regard to movements in the simulated environment in [7 and 8] Lin et. al. develop and discuss algorithms for simple and advanced collision detection in a simulated environment, and in [9] the realization of arbitrary navigation modes is developed together with a Generic User Interface Framework.

In this paper we take a practical approach to the development of intelligent behavior in the 3D scene. We design system components necessary for the implementation of 'Intelligent Navigation' where the robot is present as an agent in the scene and begin to develop higher level behaviors of multiple robots in the simulated environment. We have an existing model that exhibits intelligent navigation in the portrayal of the remote control of a client avatar in the other user's representations of a multi-user environment. In this situation the client within the constraints of the navigation mode provides the 'intelligence'. However the methods that the navigation mode consists in are not immediately available in a form that can be built into the behaviors of agents within the scene. Therefore we construct algorithms that provide the necessary navigational constraints. We incorporate all of these components into a large-scale scenario, which models the requirements for higher level behaviors.

This paper is organized as follows; In section 2 we discuss the reasons for the initial choice of platform and detail the constraints that this imposes on the development of any large scale model. The design of the overall model is developed in section 3, its features are related to the elements of the general problem and their resolution. In this section we also aim to show that in constructing a system that satisfies the requirements of our model individual elements of the problem are resolved and that satisfying the requirements for the completed model completely resolves the general problem. In section 4 the current Implementation State of the model is presented as of the time of writing. We give details of what system components have been constructed and the results of initial tests of these components. We examine any results for the implication of any design changes that may be required to the component or to the overall model. In the final sections we present those conclusions we have been able to draw from our model at the current stage of development and discuss the future development of this project.

2   Initial Choice of Platform

The initial choice of software platform for this project is VRML [10]. Virtual Reality Modeling Language is an immediately net-ready format for the description of 3 dimensional geometry and its animation. It is universally available to users of most if not all hardware platforms. Thus a model constructed in full compliance with the VRML specification is not dependent on any specific hardware or software platform. VRML contains methods for specifying the geometry of the model, for scripting the interactions between components of the model and by the use of a suitable plug-in to a web browser, to display this interaction in 3 dimensions.

There is the issue of which VRML browser should be used, unfortunately full hardware and software platform independence is an ideal that has yet to be achieved in VRML. It is expected that users will want to interact with the model rather than simply observing it, further that users may also want to interact with each other whilst experiencing the model. The Blaxxun Contact [11] VRML browser plug-in provides facilities for both of these activities, having all the features of a standard VRML plug-in with the addition of a multi-user chat feature and facilities for shared actions and events. The only other Browser system that offers this level of interactivity is the Cortona plug-in [12], which has multi-user features in development as of the time of writing. However there are currently open source applications in development which promise to make multi-user connections to the VRML scene using the VRML EAI (the External Authoring Interface).

3 Design of the General Model

Models based on the social ant have been used successfully on numerous occasions to develop corresponding robot behaviors such as Cooperative Transport [14], returning to a base location by using light direction as a navigational aid [15], Trail Formation, foraging patterns and self-organizing systems [16]. In each of these the observed behavior of various ant species has been the inspiration for development of similar behaviors in robots. In [14] the cooperative behavior of worker ants during the retrieval of large prey items was the foundation for a robotic system capable of using a small 'swarm' of ant like robots to retrieve various sizes of box and deliver these to a goal location. This was achieved through a decentralized multi-agent system with control generated via locally sensed information. In [15] the ability of ants and bees to return to their nest led researchers to design a system that enabled a robot to return to its initial location after (presumably) achieving its goal. This used the direction of ambient light to provide the reorientation towards the initial location and environmental cues such as localized landmarks in the final stages. In [16] the ability of ants to self-organize without the need for explicit long-range communications is shown to be the product of a local chemical communications system and active directed random walking search patterns. These result in the formation of a collective trail system.

None of the approaches mentioned above completely models the behavior of all species of ants, but each approach successfully models one or more aspects of ant behavior and applies it to robotic systems in a meaningful way. Thus the design of our model in this paper is intended to generalize the behavior of ant species and from this to synthesize a consistent design for cooperative insect behaviors in multi-agent systems.

3.1 Ant behaviors in General

The following is a dramatized account of a day in the life of a generic species that we shall call 'Pseudo Eciton Formica'. This is not intended to present the actual behaviors of any specific species but is intended to provide a scenario that we can use to begin constructing our general model.

'A day in the life of Ant'

As the first rays of dawn warm the sides of the nest Ant prepares to leave on her first foraging run of the day. The slight but definitely reassuring scent of the soldier on patrol lures her out into the sunlight. Ant moves away from the nest towards the light, she has no particular trail to follow here, the scents of yesterday have long since faded. In stead she moves steadily away from the nest ranging this way and that, her antennae tasting the cool air for any scent of food or prey. As she moves she secretes the foraging scent from a gland on her abdomen which serves to mark the trail as she passes.

Sometime later she stumbles across the carcass of a small mammal, it has been almost completely consumed by whatever animal preyed upon it but some pieces of tattered flesh remain. She grasps the nearest piece in her mandibles and lifts it above her head. It weighs two or three times what she does but effortlessly she turns away from the brightening sun and makes a beeline for the nest. Behind her she trails the sweet scent of success.

As Ant approaches the vicinity of the nest she crosses an outgoing trail, much stronger than the trail she has left on her own outgoing journey, it indicates that several of her sisters must have passed this way. Orienting along the trail she proceeds rapidly in the direction of the nest. In short order she becomes aware of a carrier approaching, she passes it hardly noticing the faint recruitment scent from it, her instinct driving her to deliver her burden to the nest. The carrier on the other hand is affected immediately by the success scent of Ant's back trail, it changes course towards a potentially rich food source where its particular skills may be needed. By the time Ant has reached the nest with her burden many more workers have located the same food source using Ants back trail.

Ant deposits her burden with the feeders, these never leave the nest but spend their lives masticating food into pulp and feeding this to the grubs in their cells. Ant is soon back on the thickening trail to her prize. At the corpse several workers and the carrier are attempting to move the carcass without success, they have changed their positions around the corpse many times to no avail. As the numbers of workers around the food increases there comes a point when en-mass their attention changes from trying to drag the corpse to dismantling it. The concentrated efforts of many workers manage to sever a large piece, with the carrier's long legs supporting the bulk of the mass the group sets off dragging the morsel along the homeward trail.

The corpse has attracted other insect scavengers, a white cockroach moving towards the prize attracts the attention of the patrolling soldiers, the cockroach is seized in the powerful mandibles and stung by poison glands. The poison scent recruits workers to the fray and in short order the much larger cockroach is subdued and workers begin dismantling this new prize.

By the end of the day Ant has moved or help to move hundreds of times her own weight in food into the nest, sometimes following trails formed by the living bodies of her sisters spanning some breach or obstacle. She has been tireless and determined, even ruthless in her daily occupation.

The above has been synthesized and extrapolated from video footage, from the writer's own observation of column raiding black army ants in West Africa in the '70's and the many excellent descriptions in papers mentioned above. 'Pseudo Eciton Formica' is an entirely imaginary species that has been presented so that our model can be used to experiment with various aspects of the behaviors of actual species and their interaction.

3.2 Low Level Behaviors of the Model

Table 1 lists simple behaviors or actions, which we will term 'low level' behaviors and detail any condition that is a prerequisite for this action. Where possible the type of algorithm that is to be used to simulate this action is also indicated.

Table 1. Low Level Behaviors

Behavior		Algorithm	Prerequisite
L1	Forward move	Linear Interp	!6,7,8 or 9
L2	Backward move	Linear Interp	6

L3	Turn left	SLERP(to)	7,8 or 9
L4	Turn right	SLERP(to)	7,8 or 9
L5	Follow terrain	TERRAIN	1 or 2
L6	Detect Collision	COLLIDE	1 or 2,3,4
L7	Detect light dir.	Dlight	3 or 4
L8	Detect target/food	Dtarget/Dprey	3 or 4
L9	Detect Trail type	Dtrail	3 or 4
L10	Lay Trail type	Ltrail	1
L11	Grasp object	Ograsp	6 and 8
L12	Release object	Orelease	11 or 13 and !1 or 2
L13	Lift object	Olift	11

These 'low level' behaviors are the basic actions that our simulated robot ants will be able to perform, used in combination they should provide the means to achieve a variety of more complex activities. We will first look at actions that involve movement.

Low level movements have been generated in actual robots in a variety of ways but for the purposes of this simulation we are concerned primarily with methods for simulating movements of the agent in VRML. Any object in the VRML scene can be repositioned at any time by sending a new position vector to the appropriate 'eventIn' of its 'Transform node', like wise for the orientation of the object. This process produces an immediate change in the object's position or orientation. Animated movement in VRML is most commonly accomplished by a process called 'Linear Interpolation', see [10]. Essentially this is a method that successively causes the object to occupy locations (or orientations) between the specified values over a defined time period. This method is readily applied to L1,..4, however in 3 and 4 another method is also available. Known as 'Spherical Linear Interpolation' this method is able to generate orientation values from one orientation to another. The speed of the movement can be derived from or determined as a function of the rate of change of position or orientation over time.

Terrain Following

In any animated movement over a variable surface i.e. terrain the ability to remain in contact with the surface, or to maintain a given height over the surface can not be taken for granted. Previous work in the field of Robot Motion Planning has largely been confined to motion in an ideal (flat) surface. Researchers have assumed that their methods will work on a variable surface or that they can be readily extended to do so, but little experimentation appears to have been carried out. Using an ideal surface has the advantage that all algorithms need only to consider two dimensions while navigation on a variable surface demands that the third dimension be taken into consideration. Terrain following is a technique that allows orientation and position to be animated in two dimensions whilst maintaining the animated object in contact with the surface being traversed or at a set height over it.

The ability to follow terrain is a requirement for all VRML browsers in that it is needed to accomplish a basic 'walk' type of navigation. However it is not a requirement of the specification that this ability be available to animations, since it is expected to apply only to user navigation. If an agent is to simulate this type of navigation where the surface is a variable one then some kind of terrain following method will be required. Here is a terrain following algorithm that has been tested and works on a VRML 'elevation grid', the most efficient means of specifying variable terrain in VRML.

A1 Terrain Following Algorithm

```
For each input value of projected position
    here = input.x, input.y + robot height, input.z
    there = here.x, here.y - 1000, here.z
    inverted = false
    While (true) do
        thisNode = rayhit between here and there
        if (hit node returned in thisNode)
            Path = hitpath returned in thisNode
            If (we hit the terrain)
                new robot position =
            input.x,
                location.y + robot height,
                input.z ,
                return new positon
            else ( we did not hit terrain )
                continue ray
                if (inverted)
                    here = location +
                    a small Y offset
                else ( not inverted)
                    here = location -
                    a small Y offset
                end-if
                continue while loop
            end-if
        else ( hit node not returned)
            if (not inverted)
                here = its initial value
                there = its Y value + 1000
                inverted = true
                continue while loop
            end-if
        end-if
        robot position = input value
        break out of while loop
    end-while loop
next input value
```

The method works by projecting a ray from the robot position directly down the Y axis, when the ray intersects the surface the point of contact is used to adjust the robot position. If projecting a ray downwards does not give a contact with the surface a ray is sent upwards instead. The method also allows for the presence of the agent in the scene. If the robot geometry is contacted before the surface is detected the method continues the ray trace beyond the robot's geometry.

Collision Detection

Collision detection is another technique that is required in user navigation, however as with terrain following it is not a requirement of the specification that collision detection methods between agents and objects within the scene be implemented in the browser. In [7 and 8] Lin et. al. describe a collision detection algorithm which can and (presumably) has been built into VRML browsers, but until such time as browser manufacturers make this technique available to the

VRML programmer for object/object collisions or object/agent collisions we must rely on those methods that we can easily program into VRML.

From [8] it is clear that the simplest collision detection method for object/object collisions in VRML would look something like the following;

A2 Simple object/object collision detection algorithm

```
For each object/object pair
     draw the smallest non oriented box
     to surround each object in the pair
     For each member of the pair
          If any vertex of this object's box
             is within The bounds of the other
          object's box
                Then collision is true
          End-if
     Next member
Next pair
```

This method is highly inaccurate but considering that it is easily implemented and very fast it is acceptable as a first approximation, given that it can subsequently be refined using the methods outlined in [7 and 8]. In order to obtain an algorithm for agent/object collision that we can use, we modify the above algorithm and any refinement of it as follows;

A3 Simple agent/object collision detection algorithm

```
For each agent in the system
     Draw the smallest non oriented
     box around the agent
     For objects within range of the agent
          Draw the smallest non oriented
          box around the object
          If (any vertex of this object's box is within
             the bounds of the agent's box) or
          (any vertex of the agent's box is within
             the bounds of this object's box)
                then collision is true
          End-if
     Next object
Next agent
```

Within this system it is permissible to consider all agents other than the agent being tested as 'objects' for the purposes of collision testing. Note that the range parameter is available to eliminate tests that have no possibility of succeeding. A suggested value for the range parameter is to double the width of the largest object within the system.

This algorithm should be used whenever an Agent changes position within the system, and may also come into play whenever an object changes position. This is one reason for keeping this process as simple as possible since the required overhead for processing this algorithm expands with each new agent or object.

There is a further method of collision detection that may prove to be adaptable in the development of the other detecting behaviors in our model. Instead of using boxes to surround the objects the agent can send out rays as in the Terrain Following algorithm. This time the rays are sent out laterally to detect other agents, objects or obstructions. This will prove particularly useful in detecting

504

when a food item is in a position to be grasped. Used for forward facing collision detection, such as in identifying a target or obstruction the ray trace collision detection algorithm would look something like the following;

A4 Detect target or obstruction algorithm

```
For each change in position or orientation,
  or continuously
    Compute ray trace forward to range
    If (we hit something)
        If (hit node type = 'target') or
            (hit node type = 'object') or
            (hit node type = 'agent')
              return target name
        End-if
    End-if
Next
```

The collision or detection response is application dependent and the method can be extended to develop a model 'sensor' with a definite beam width and height however we again have a problem with computational overhead. For example if we wish to scan forward with a beam width of θ and a beam height of ψ we can do so using a raster scan method firing one ray per iteration.

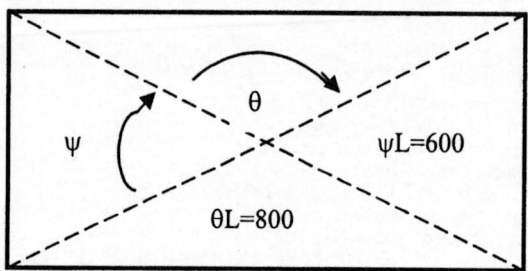

Fig 1. Vision 'sensor' field of view

In the figure θL is the number of vertical lines and ψL is the number of horizontal lines. Thus the total number of iterations is $\theta L \times \psi L = 800*600 = 480,000$. If this has to be done every time an agent moves it will significantly impact the speed of the model. We can reduce the number vertical and horizontal lines to 10 giving 100 iterations, this may be the beginnings of robot 'vision' for our model. We know that insect vision is strictly limited so this may well prove adequate within the confines of our model

Trail laying and Trail detection

Readers may be familiar with the 'hybrid potential field' method used in Ratering and Gini [3] to control a robot moving on an ideal surface. This method has problems when applied to a variable surface in 3 dimensions because it relies on a 3 dimensional grid of potentials to stimulate movements in 2 dimensions. Thus by extension it would require a 4 dimensional grid of potentials in order to stimulate movements in 3 dimensions. Their model could be extended to deal with the variable surface without extension to 4 dimensions but this would require that parts of the terrain that may present a local obstruction be treated as discontinuities. This would be fine for a robot with a flat footprint if we accept that it is limited in the degree of slope in terrain that it can deal with, however this is not a suitable method for modeling navigation in our ant agent.

We can use part of the method described in [3] to model the olfactory sense in our ants, we do this by using a 'continuous potential field' to model the strengths of each pheromone across our variable surface. When the trail is laid by our ant it generates a 'hill' in the continuous potential field at the ants location. This hill has a height and extent that models the intensity of the pheromone and is eroded over a time period corresponding to the time required for this pheromone to dissipate. A new field grid is required for each pheromone in the system. This provides for trail formation and deterioration as described in [16]. A crude directional method of trail detection can be constructed by polling the field strength of each pheromone at two points corresponding to the position of the robots ant's 'antennae' with respect to the agent position.

Object manipulation by the Agent

Object manipulation in VRML is generally accomplished by the use of a variety of sensor nodes, these provide an interface between the user movements of the mouse cursor and the object to which the sensor has been mapped. While this is less than completely satisfactory it does provide a means for the user to interact with objects in the scene. Once again these methods are not available to the agent. In the long term we have an opportunity in developing agent object manipulation to also resolve the problems with user object manipulation, but for now we simply need to formulate methods to achieve the low level object behaviors.

Ants use their mandibles (mouth parts) to lift, carry and dismantle prey or food items, therefore in this case what we need to model is the manipulation of an object by another object in the same scene. The second object in this case being the ant itself. This then is an appropriate point to discuss the physical design of our ant robot.

In the figure six two jointed legs are attached to the central thorax, the abdomen is attached to this in the rear. The head supports two opposed mandibles, two independently mobile antennae and two simple compound eyes. The ant is bi-symmetrical in its long axis. Variations on this design include versions with longer legs (carriers), with larger mandibles (soldiers) and with wings (males) plus the Queen with an extended abdomen for egg laying.

The physical model of the ant will be constructed as a VRML 'Avatar' this is a type of VRML prototype node that has features that enable the model to be animated as the representation of a user in multi-user environments. An avatar can appear to walk when moving and using the gesture interface can produce specific actions as required, such as closing its mandibles and raising its head. These can be timed to run simultaneously with animations corresponding to the movements of the object being manipulated.

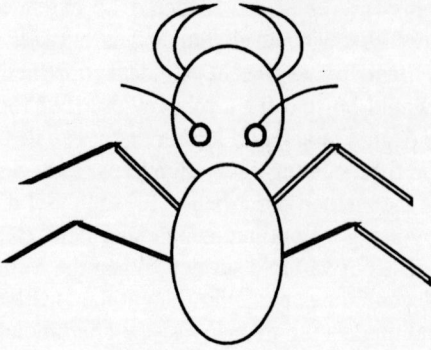

Fig 2. A simple Ant

The Avatar prototype instance is in turn wrapped in a 'robot' prototype which contains the agent code for low level behaviors, and provides the gesture, position and orientation input into the avatar prototype instance. If we trace a ray between the tips of the mandibles we can detect when an object is in position to be grasped, when it is held, and can thus be released or lifted. The ray trace method gives us a handle on the object that it hits so we can modify this object's position to correspond with the actions of the ant.

Detecting Light Direction

In VRML this can be done in one of two ways. If the lighting source is directional we can read the direction from the lighting source. If the light is not directional i.e. it is from a point source we can compute its direction by comparing our position to the position of the source and using vector math to determine our bearing with respect to the source, and vice versa. If necessary we can simulate the occlusion of the light source from objects in the scene by adding a visibility sensor to the light.

3.3 Higher Level Behaviors

Table 2 details high level actions that can be constructed using the low level actions detailed in table 1. The sequences assigned to each high level action are advisory at this stage and subject to change. They contain the basic actions that the higher level actions consist in and also actions that may be taking place at the same time as the given action. For example while a trail is being followed it may also be necessary to avoid static or moving objects. From observations it is difficult to determine that ants actively avoid either static or moving objects, they are able to climb over most obstacles and even over the living bodies of their sisters, however these actions are included here for the sake of completeness. The basic collision detection methods need to be used in our simulation if only to insure that our robot ants do not occupy the same space as each other in the scene. Higher level object avoidance behaviors may come into play when large numbers of ants are travelling both ways on the same trail.

Table 2. High level Actions

High Level Actions		Sequence
H1	Avoid static object	L8,(L3 or L4),L1
H2	Avoid moving object	L8,(L3 or L4),L1 or L2
H3	Seek static target	L8,(L3 or L4),H2,L11
H4	Seek moving target	L8,(L3 or L4),L1,L6,L11
H5	Seek goal position	L7,(L3 or L4),L1,L6,(H1 or H2)
H6	Follow trail	L9,(L3 or L4),L1,L6,(H1 or H2)
H7	Reposition	L12,L2,(L3 or L4),L1,L6,(H1 or H2)L11,L13

If all of the above actions are to be combined together in a general model of ant behavior some form of behavior arbitration may be required. For example when transporting a food item this behavior must take priority over the recruitment signals of a new find and if attempts to transport an item fail successively with larger numbers of ants behavior must switch to a dismantling operation [14].

The overall motivation of the ant colony must be to reproduce, but in ants as in bees this reproduction is done entirely by the queen. It is therefore reasonable to postulate some pervading pheromone that motivates the various casts of ant to perform the tasks for which they are designed. This would tend to be supported by swarming behavior where the entire colony follows the queen to a new nest site. With regard to the change in behavior from unsuccessful dragging attempts at prey to dismemberment, this could be due to a peak in a 'frustration' pheromone or perhaps it is the result of the 'poison' pheromone emitted when a live prey resists attack. With the result that the ants are driven into a frenzy of effort against the resisting item.

It has been shown that trail formation can occur without memory other than information represented in the environmental pheromone residue and without visual clues [16]. It has further been shown that autonomous agents can cooperate to move an item to a goal point [14]. However the fact that light direction was used to locate the goal point [15] is no direct proof that this is the case in all ant species, given the apparently simple visual apparatus ants in general posses.

The Final model will be implemented in a modular format consistent with the recommendations of [17]. This is intended to allow users of the model to determine by experimentation the relative importance of these high level behaviors in generating successful foraging techniques for various species.

4 Current state of the Model

The complete generalized model is currently unimplemented. The terrain following technique mentioned in section 3 has been constructed and tested [18]. Some modifications need to be made for inclusion into the ant colony model, specifically the terrain following algorithm assumes that the robot is humanoid and stands with its Y axis parallel to the global Y axis of the scene. An ant avatar on the other hand will have its long axis as the Z axis of its local coordinate system, its Y axis must be kept 'normal' to the surface over which it is travelling.

Thus the algorithm needs to be modified to take the normal to the surface given by the ray trace method and apply this to ant robot orientation for every change in position.

As regards robot navigation and its simulation in VRML a 'dumb' method of navigation has been implemented with humanoid avatars as robots. This drives the robot on a predetermined course but allows for interaction with the user [19]. Essentially when the client comes within range of the robot it attacks, its patrol is interrupted for the duration of the encounter. Various parameters control the range at which the robot can detect the client, if the robot responds to movement or simply to the presence of the client and if the attack is to be repeated while the client is within range. This will be incorporated into the 'soldier' ant in the final model.

5 Conclusions

Previous work has concentrated on modeling specific aspects of ant/robot navigation for reasons that are understandable. A model incorporating these aspects back into a generalization of ant behavior will be difficult to implement but not impossible. However the constraints of building this model under the VRML specification may prove overly restrictive, in the long term though the benefits of having a universally accessible model will prove useful to both the Robot Motion Planning community and to the Artificial Life community. In addition the techniques that are developed in VRML to generate Autonomous Agent Behaviors will prove very useful to the Multi-user VRML developers and in particular these techniques should be of direct use to the emerging VRML Games interest group [20].

Contributions are invited for the behavioral model of 'Pseudo Eciton Formica' its environment and ecosystem to the writers e-mail address above. An HTML version of this paper will be maintained during the development process, together with relevant VRML examples at the following web address:

http://members.netscapeonline.co.uk/tjbtc/site/home/Tutorials/

This work is not currently funded by any grant or organization.

References

[1] Kuffner, J. Jr.,Kagami, S., Inaba, M. and Inoue, H. Dynamically-stable Motion Planning for Humanoid Robots. Dept. of Mechano-informatics, The University of Tokyo. http://www.jsk.t.u-tokyo.ac.jp/~kuffner/humanoid

[2] Svestka, P. and Overmars, M.H. Coordinated Path Planning for Multiple Robots. Dept. of Computer Science, Utrect University, Netherlands. http://citeseer.nj.nec.com/

[3] Ratering, S. and Gini, M. Robot Navigation in a Known Environment with Unknown Moving Obstacles. Dept. of Computer Science, University of Minnesota. Autonomous Robots 1(2) 1995, 149-165 http://www-users.cs.umn.edu/~gini/navigation.html

[4] Goldenstein, S., Large, E. and Metaxas, D. Dynamic Autonomous Agents: Game Applications. Center for Human Modeling and Simulation, Computer Science Department, University of Pennsylvania. http://www.cis.upenn.edu/~siome/publ.html

[5] Goldenstein, S., Large, E. and Meṭaxas, D. Nonlinear Dynamic Systems for Autonomous Agents Navigation. Complex Systems and Brain Sciences, Florida Atlantic University (Large 1999). http://www.cis.upenn.edu/~siome/publ.html

[6] Goldenstein, S., Large, E. and Metaxas, D. Nonlinear dynamical system approach to behavior modeling. VAST lab., Center for Human Modeling and Simulation, Computer and Information Science Department, University of Pennsylvania. Published in 'The Visual Computer (1999) 15:349-364 (c) Springer-Verlag 1999'. http://www.cis.upenn.edu/~siome/publ.html

[7] Hudson, T.C., Lin, M.C., Cohen, J., Gottschalk, S. and Manocha, D. V-COLLIDE: Accelerated Collision Detection for VRML. Department of Computer Science, University of Carolina . http://www.cs.unc.edu/~geom/collide.html

[8] Lin, M.C., Cohen, J., Gottschalk, S. and Manocha, D. Collision Detection: Algorithms and Applications. Department of Computer Science, University of Carolina. http://www.cs.unc.edu/~geom/collide.html

[9] Volk, T., and Grahn, H. Blaxxun Interactive Eisenheimerstr 61-63 80687 Munich, Germany with Althoff, F and Lang, M. Insitute for Human-Machine Communication, Technical University of Munich. A Generic User Interface Framework. (Blaxxun Interactive 2000) http://www.blaxxun.com/

[10] The VRML '97 specification , VRML org, technical info, specifications, VRML '97; http://www.vrml.org/ technicalinfo/specifications/vrml97/

[11] The Blaxxun Contact Plugin, Blaxxun Interactive http://www.blaxxun.com/c/s?cat=7&sub=5&url=/products/contact/index1.html

[12] The 'Cortona plugin', Parallelgraphics, http://www.parallelgraphics.com/products/

[13] Microsoft DirectX (drives D3D hardware acceleration) http://www.microsoft.com/directx/default.asp

[14] Kube, R.C., and Bonabeau, E. Cooperative transport by ants and robots. Edmonton Research Centre, Syncurde Canada Ltd, and Santa Fe Institute, Santa Fe USA. (1998). http://citeseer.nj.nec.com/

[15] Nehmzow, U and McGonigle, B. Robot Navigation by Light. Robotics Laboratory, Laboratory for Cognitive Neuroscience, Psychology Department, Edinburgh University. Presented at the European Conference on Artificial Life, ECAL 1993. http://citeseer.nj.nec.com/

[16] Schweitzer, F., Lao, K. and Fereydoon Family. Active Random Walkers Simulate Trunk Trail Formation by Ants. BioSystems 41 (1997) 153-166. http://citeseer.nj.nec.com/

510

[17] Bryson, Joanna. Cross-Paradigm Analysis of Autonomous Agent Architecture. Division of Informatics, The University of Edinburgh. http://citeseer.nj.nec.com/

[18] Batchelor, T.J. ,Terrain following notes. http://members.netscapeonline.co.uk/tjbtc/site/home/ Tutorials/terrain.htm

[19] Batchelor, T.J. Multi-user robot enemies : http://freespace.virgin.net/loanwolffe.farzone/ Projects/Combat2/

[20] Batchelor, T.J. Homepage of the 'VRML Games' webring http://freespace.virgin.net/loanwolffe.farzone/

Greta: A Simple Facial Animation Engine

Stefano Pasquariello[1], Catherine Pelachaud[2]

[1]Kyneste S.p.A, Rome, Italy
[2]Department of Computer and Information Science University of Rome "La Sapienza", Rome, Italy

Abstract. In this paper, we present a 3D facial model compliant with MPEG-4 specifications; our aim was the realization of an animated model able to simulate in a rapid and believable manner the dynamics aspect of the human face.

We have realized a Simple Facial Animation Engine (SFAE) where the 3D proprietary facial model has the look of a young woman: "Greta". Greta is the core of an MPEG-4 decoder and is compliant with the "Simple Facial Animation Object Profile" of the standard. Greta is able to generate the structure of a proprietary 3D model, to animate it, and, finally, to render it in real-time.

Our model uses a pseudo-muscular approach to emulate the behaviour of face tissues and also includes particular features such as wrinkles and furrow to enhance its realism. In particular, the wrinkles have been implemented using bump mapping technique that allows to have a high quality 3D facial model with a relative small polygonal complexity.

1   Introduction

MPEG-4 is an ISO/IEC standard developed by MPEG (Moving Picture Experts Group), the committee that also developed the well known standards MPEG-1 and MPEG-2.

MPEG-4, whose formal ISO/IEC designation is ISO/IEC 14496, was finalized in October 1998 and became an International Standard in the first months of 1999. The fully backward compatible extensions under the title of MPEG-4 Version 2 were frozen at the end of 1999, to acquire the formal International Standard Status early 2000. Some work, on extensions in specific domains, is still in progress.

MPEG-4 provides the standardized technological elements enabling the integration of the production, distribution and content access paradigms of: digital television, interactive graphics applications (synthetic content) and interactive multimedia (World Wide Web, distribution of and access to content).

Among other items, MPEG-4 defines specifications for the animation of face and body models within a MPEG-4 terminal [4, 9]. We realized a Simple Facial Animation Engine (SFAE) that has the look of a young woman: "Greta", it is the core of a MPEG-4 decoder and it is compliant with the "Simple Facial Animation Object Profile" defined by the standard.

2 Facial Animation Coding in MPEG-4 Standard

According to the MPEG-4 standard, the face is a node in a scene graph that include face geometry ready for rendering. The shape, texture and expressions of the face are generally controlled by the bitstream containing instances of Facial Definition Parameter (FDP) sets and Facial Animation Parameter (FAP) sets.

Upon initial or baseline construction, the Face Object contains a generic face with a neutral expression: the "neutral face". This face is already capable of being rendered. It is also immediately capable of receiving the FAPs from the bitstream, which will produce animation of the face: expressions, speech etc. If FDPs are received, they are used to transform the generic face into a particular face determined by its shape and (optionally) texture. Optionally, a complete face model can be downloaded via the FDP set as a scene graph for insertion in the face node.

2.1 The Object Profiles

MPEG-4 defines three profiles for facial animation object that allow different levels of configuration of the decoder:

- **Simple Facial Animation Object Profile**: The decoder has its own proprietary model that is animated by a coded FAP stream.
- **Calibration Facial Animation Object Profile**: This profile includes the simple profile. The encoder transmits to the decoder calibration data for some or all of the predefined feature points. The decoder adapts its proprietary face model such that it aligns with the position of these feature points. This allows for customization of the model.
- **Predictable Facial Animation Object Profile:** This profile includes the calibration profile. MPEG-4 also provides a mechanism for downloading a model to the decoder according to Section 2.3 and animating this model.

Actually, Greta, presented in this paper, is fully compliant with MPEG-4 specifications of the Simple Facial Animation Object Profile.

2.2 The Neutral Face

At the beginning of an animated sequence, by convention, the face is assumed to be in a neutral position. All the animation parameters are expressed as displacements from the position defined in the "neutral face" (see Fig. 1).

2.3 The Features Points

In order to define face animation parameters for arbitrary face models, MPEG-4 specifies 84 "feature points" located in relevant somatic places of the human face (see Fig. 1). They are subdivided in groups and labeled with a number, depending on the particular region of the face to which they belong. Therefore, some of the feature points (black labeled in the figure) are affected by the FAPs.

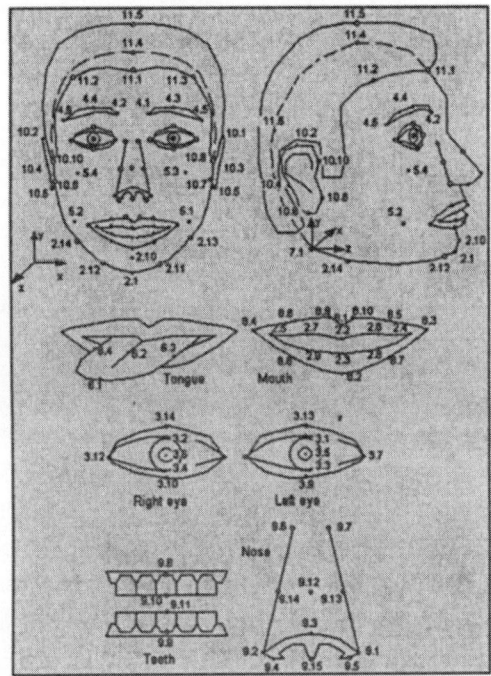

Fig. 1. The "feature points" in the "neutral face" (the black ones are affected by FAPs)

2.4 The Facial Animation Parameters (FAPs)

The FAPs are based on the study of minimal perceptible actions and are closely related to muscle action. The 68 parameters are categorized into 10 groups related to parts of the face.

FAPs represent a complete set of basic facial actions including head motion, tongue, eye, and mouth control. They allow the representation of natural facial expressions. They can also be used to define facial action units [5].

Technically, the FAPs define the displacements of the feature points in relation to their positions in the neutral face. In particular, except that some parameters encode the rotation of the whole head or of the eyeballs, a FAP encodes the magnitude of the feature point displacement along one of the three Cartesian axes.

The magnitude of the displacement is expressed by means of specific measure units, called FAPU (Facial Animation Parameter Unit). Except for FAPU used to measure rotations, each FAPU represents a fraction of a specific distance on the human face; so, is possible to express FAP in a general way by a normalized range of values that can be extracted or reproduced by any model.

2.5 The Facial Definition Parameters (FDPs)

The FDPs are responsible for defining the appearance of the 3D face model. These parameters can be used in two ways:

- To modify the shape and appearing of a face model already available at the decoder.
- To encode the information necessary to transmit a complete model and the rules that must be applied to animate it.

In both cases the animation of the model is described only by the FAPs. FDPs, on the contrary, are typically used only at the beginning of a new session. However, in our work is not necessary to use the FDPs because, actually, Greta is compliant with the Simple Facial Animation Object Profile.

3 The 3D Model of Greta

Greta is a Simple Facial Animation Engine (SFAE) compliant with the Simple Facial Animation Object Profile of the MPEG-4 standard [7]; so, our decoder has a proprietary model that has been directly animated by the FAPs.

3.1 The Polygonal Structure of the 3D Model

After a study of the best solution for the surface representation of the 3D facial model [10], we decided to use polygonal surfaces because modern graphic workstation are adept at displaying them and can update complex facial polygonal models in near real time. In particular we used OpenGL technologies to develop, display and animate our original 3D model (see Fig. 2-3).

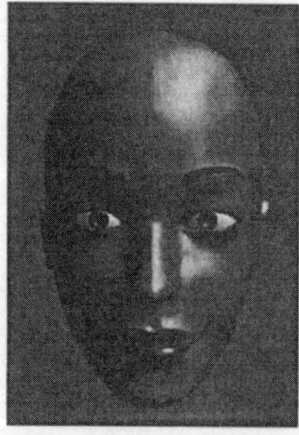

Fig. 2. Greta 3D model   **Fig. 3.** The internal anatomic components

In the human face there are specific regions that are dedicated to the communication of the information and to express the emotions, so they need to be well defined. We concentrated our efforts on giving a great level of detail in the most expressive regions of facial model:

The **mouth**: is the most mobile region of the human face; it takes part in the composition of all the face expressions and the lips generate visemes during the speaking action [1]. The polygonal structure (see Fig. 4) of this region should be conformable to the muscles fibers (in particular the *orbicularis oris*) around the mouth that make possible a great number of expressions and visemes [6].

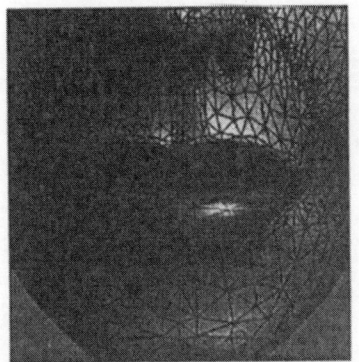
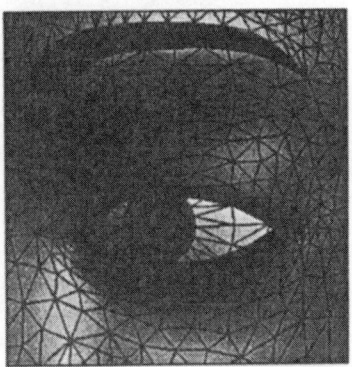

Fig. 4. The mouth of Greta   **Fig. 5.**The eye of Greta

The **eyes**: with brows are easily the most magnetic and compelling part of the human face. The polygonal structure (see Fig. 5) of this region should make easy coordinated linguistic and gaze communicative acts, blink, eyebrow and eyelid (upper and lower) moves.

3.2   Critical Regions of the 3D Model

In order to simulate the complex behaviour of the skin like the generation of wrinkles and furrows during the muscular action, we dedicated particular attention to the polygonal structure of two further regions of the facial model:

- The **forehead**: is a plane region of the face in the rest and neutral position, but, during communicative acts, the raising of the eyebrows generates on it regular horizontal wrinkles, that are perpendicular to the vertical pull of the forehead muscles. So we needed a regular horizontal structure for the polygonal lattice of the forehead (see Fig. 6).

516

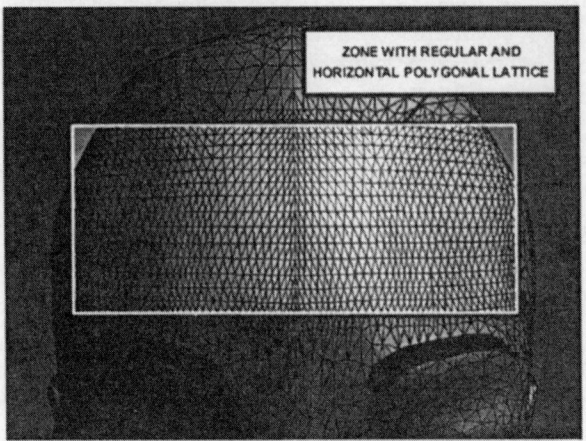

Fig. 6. The forehead of Greta with its polygonal structure

- The **nasolabial furrow**: is typically of the contraction of the *zygomatic mayor* muscle. For example during a smile, there is a generation of a clear furrow between the stretched skin near the mouth an the skin in the cheeks that, pressed up from below, bulges out like a tiny balloon being squeezed. We needed, in this part of the model, a well defined line of separation between the two skin regions, with a considerable polygonal density (see Fig. 7).

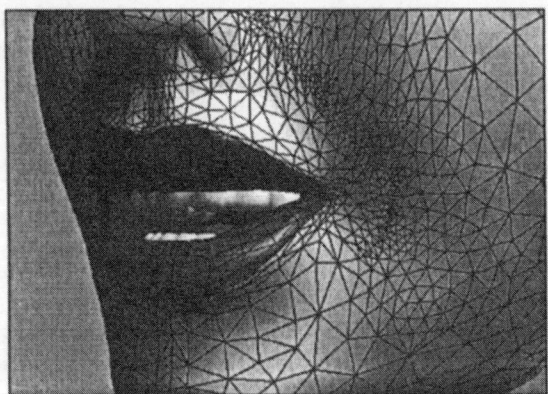

Fig. 7. The polygonal structure in the nasolabial furrow in Greta

3.3 Subdivision of the 3D Model in "Specific Areas"

In order to have more control on the polygonal lattice we subdivided the 3D model surface in "specific areas" that corresponds to the feature points affected by the FAPs. This subdivision was necessary to circumscribe and control the displacements of polygonal vertexes induced by the FAPs applied in the various feature points. For example, in the nasolabial furrow, the two skin zones on the opposing sides of the furrow belong to distinct specific areas. All the specific areas are classified and can be seen in Fig. 8-9.

Fig. 8. Subdivision of Greta in "specific areas" (front view)

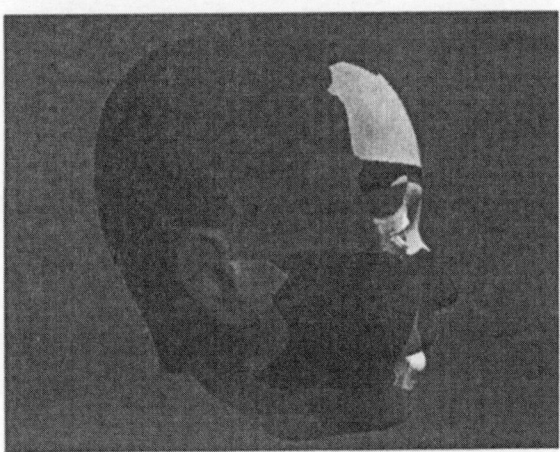

Fig. 9. Subdivision of Greta in "specific areas" (side view)

4 Simulation of the Skin Behaviour

Greta uses the FAPs to animate the skin of the 3D model in a realistic way, but it doesn't make a physical simulation of the muscles of the face and of the viscous-elastic behaviour of the skin, because the need of real-time calculation of the animation is a great restriction for this approach. So the FAPs activate functions that deform the 3D polygonal lattice during the time and make an "emulation" of the real behaviour of the face tissues [12].

518

4.1 Definition of "Area of Influence"

We have to define the "area of influence" of the feature point. The deformation, due to the FAP for a single feature point, is performed in a zone of influence that has an ellipsoid shape whose centroid is the feature point. The "area of influence" is the zone of the polygonal lattice that is within the ellipsoid; so we have a precise number of vertex that are affected by the displacement due to the FAP (see Fig. 10). It must be noticed how the area of influence of each feature points can overlap one another and, so, a vertex of the 3D model can be under the influence of various feature points and their FAPs.

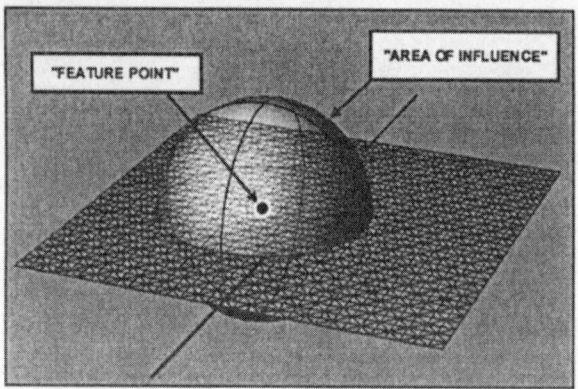

Fig. 10. The ellipsoid and the "area of influence"

4.2 Definition of "Displacement Function"

The FAP is like a muscular action applied in the key point [5, 8, 13]; so, the displacement of the vertexes of the polygonal lattice into the area of influence has to emulate the behaviour of skin under a traction applied just in the feature point.

The displacement Δx_j of a vertex P_j (x_j, y_j, z_j) due to a FAP operating in the x direction FAP_x is computed (without considering now the "specific area") as:

$$\Delta x_j = W_j * FAP_x \tag{1}$$

The weight W_j is based on the distance of the vertex from the feature point and it is computed by a function that can be seen in Fig. 11, where d''_j is a normalized distance. Therefore, in the sequence of Fig. 12-14 is possible to see effect of the displacement function centered in the white point on a plane polygonal lattice, with a growing intensity of displacement.

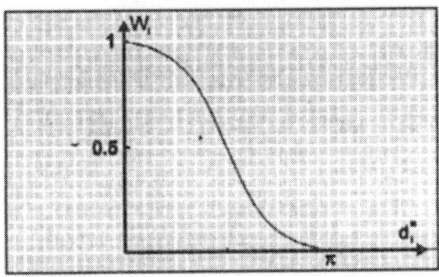

Fig. 11. The function of the weight W_j

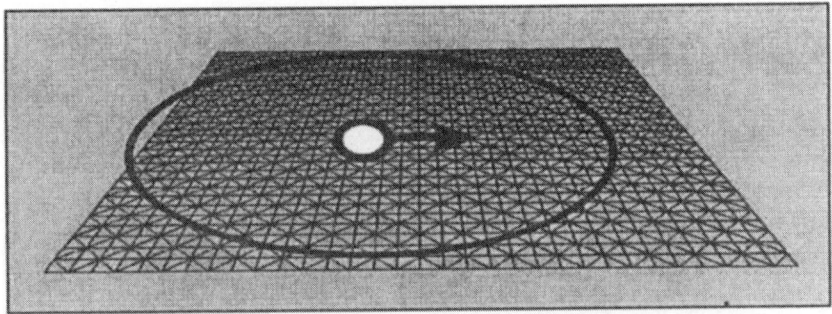

Fig. 12. Deformation within an area of influence (null intensity)

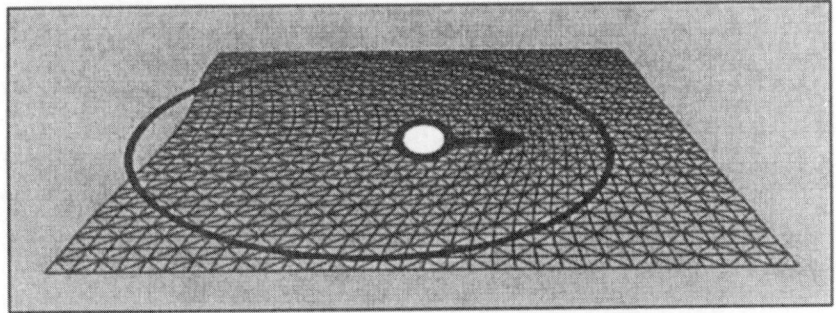

Fig. 13. Deformation within an area of influence (medium intensity)

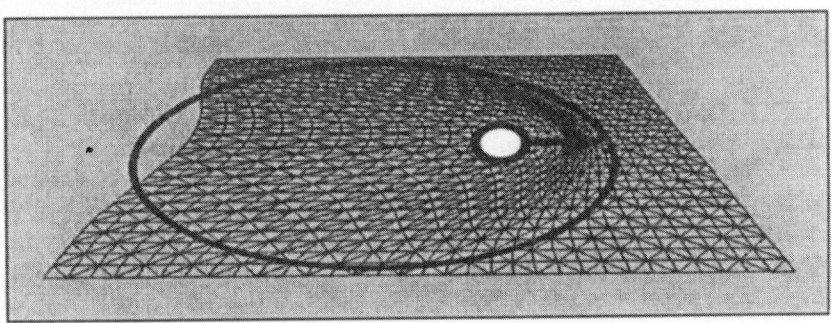

Fig. 14. Deformation within an area of influence (maximum intensity)

4.3 Simulation of Furrows and Bulges

Greta has the possibility to emulate the complex behaviour of the skin like furrow and bulges (i.e. the nasolabial furrow in Fig. 7 that occurs during the contraction of the *zygomatic mayor* muscle) [11, 12, 13]. The mechanism that generates the furrow in the skin is explained in Fig. 15; there is the traction of the mimic muscle that operates only onto a limited linear part of the skin and there are two distinct zones:

- The left zone, that is formed by the elastic skin stretched by the muscle.
- The right zone, that is formed by the accumulation of the skin due to the compression.

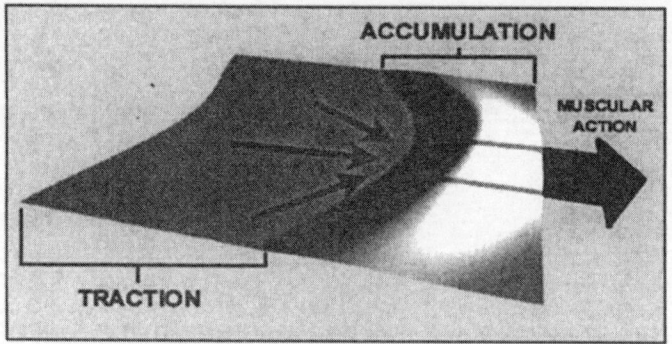

Fig. 15. The mechanism that generates the skin furrow

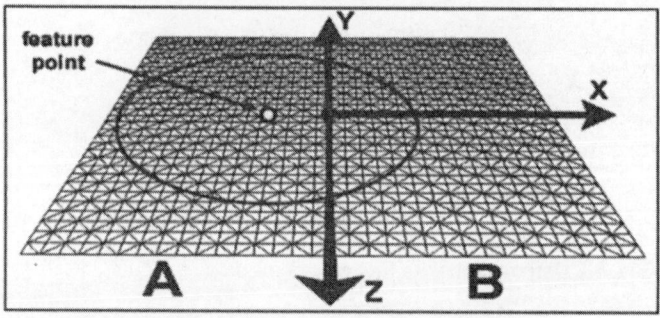

> **Fig. 16.** The two zones of skin and the feature point

If we have a simple situation like the Fig. 16 and we want a furrow along the z axis, the emulation of this behaviour of the skin is obtained by the sum of the horizontal displacement of the vertexes (Δx_i of a vertex P_i (x_i, y_i, z_i)) in the A and B zones (according to the laws previously explained and illustrated in Fig. 12-14) and the vertical displacement (Δy_i of a vertex P_i (x_i, y_i, z_i)) of the vertexes in the B zone, according to the following law:

$$\Delta y_i = \Delta x_i * K_1 * (0.5 * (1 + \cos(d_i))) * (1 - \exp(-d'_i / K_2)) \qquad (2)$$

where d_i is the distance of the vertex P_i from the feature point, d'_i is the distance between the vertex and the z axis, and K_1 and K_2 are constants. Therefore, in the sequence of Fig. 17-19 (if we don't consider the wrinkles) is possible to see the effect of the displacement functions (along x and y axis) on a plane polygonal lattice when a FAP is applied in the feature point (green labeled), with a growing intensity. Finally, in the Fig. 7 is possible to see the effect of the displacement functions on the Greta face during the action of the *zygomatic major*.

4.4 Simulation of Wrinkles

Greta can also emulate the behaviour of the skin in the zone of the forehead where, during the raising of the eyebrows, there is the generation of many horizontal wrinkles. The mechanism is similar to that seen in the previous section, in fact there is an accumulation of skin due to the compression of the raising eyebrows but there is a series of little horizontal creases because the substrate of the forehead is different from the substrate of the cheeks.

Greta uses a bump mapping technique based on per-vertex shading of the OpenGl visualization; it is an innovative technique that let to have an high quality of the animated facial model with a *small polygonal complexity*. In fact, in OpenGl based systems, the Gouraud algorithm, that makes possible the shading rendering mode, uses the normals applied in the vertexes to compute a smooth lightning on the surfaces [2]. So, using the regular disposition of polygons in the forehead region of the 3D model, is possible to perform a perturbation of the normals, *without moving the vertexes*, and have the bump mapping, that emulates the forehead wrinkles. This is possible, because the Gouraud algorithm performs an interpolation of colors between horizontal darkest and lightest zones, determined by the normal orientations that are controlled by our perturbation function. The function is rather complicated and, based on the regular disposition of vertexes in the forehead region, uses many periodic functions; but more explicatory are the Fig. 17-19 that describe a bulge (created by the techniques of the previous section) with realistic wrinkles and the Fig. 20 that shows the wrinkles on the forehead of Greta during the raising of the eyebrows.

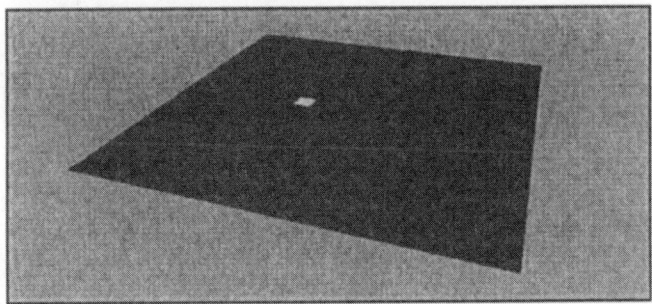

Fig. 17. Generation of a bulge with wrinkles (null intensity of the FAP)

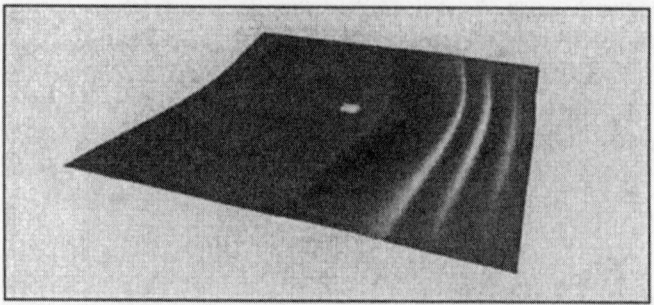

Fig. 18. Generation of a bulge with wrinkles (medium intensity of the FAP)

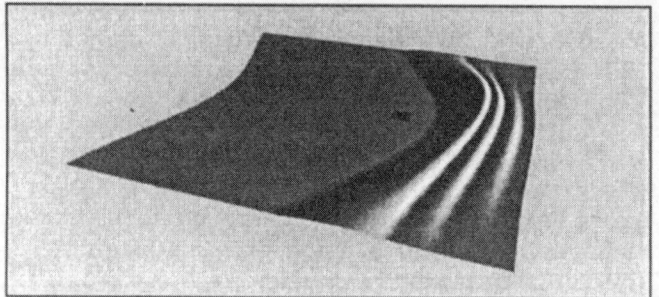

Fig. 19. Generation of a bulge with wrinkles (maximum intensity of the FAP)

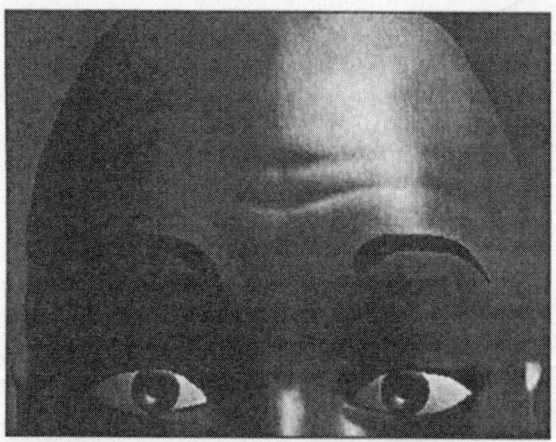

Fig. 20. Generation of the wrinkles on the Greta forehead

5   FAPs in Action

The action of the FAPs on the whole 3D model is obtained by the application of the previous explained techniques; the extension from the plane polygonal lattice to a complex 3D polygonal face is not difficult. After computing the displacement that all the FAPs produce for each vertex, is straightforward to calculate the new vertex coordinates and displaying the deformed Greta face using the OpenGl Gouraud shading. This procedure is performed for each frame of the animation.

5.1 Application of FAPs on the 3D Model

Generally the displacements Δx_j (we have a similar situation for Δy_j and Δz_j) of a vertex P_j (x_j, y_j, z_j) due to a single FAP operating in the x direction is computed using three kind of information:

- The intensity of the FAP in the related feature point.
- The position of the vertex in the "area of influence" of the feature point.
- The position within a "specific area".

The displacement Δx_j of a vertex P_j (x_j, y_j, z_j) in the k-th specific area due to a FAP operating in the x direction FAP_x is computed as:

$$\Delta x_j = W_j * W_k' * FAP_x \qquad (3)$$

The weight W_j is based on the distance of the vertex from the feature point and it is computed by the function seen in Fig. 11, the weight W'_k is defined by the designer after a calibration of the animation and describe how the intensity of the FAP influences the displacement of the vertexes of the k-th "specific area". For example, if we want that the vertexes of the eyelids don't be affected by the FAPs that control eyebrows is possible to set $W_k = 0$ during the computation of the Δx_j when the above mentioned FAPs operate; this is possible because the vertexes of the eyebrows and eyelids belong to distinct specific areas (see Fig. 8-9) and, so, the movements of the vertexes of the eyebrows don't affect the vertexes of the eyelids. This approach gives a great control on the displacement of the vertexes and allows a natural animation of the 3D face without visual artifacts during the rendering.

5.2 Application of auxiliary deformations

Greta uses auxiliary deformations to increase the realism of the animation in specific parts of the model. These auxiliary deformations perform a displacement of vertexes in the same manner of the FAPs, the only difference is that the intensity of the displacement is not controlled by the FAP stream but is a function of various weighted FAPs. These auxiliary deformations are used, during specific actions, to increase the realism of the behaviour of the face skin in two zones:

- The zone between the eyebrows during the act of frowning (see Fig. 21).
- The zone of the cheeks that swells out during the act of smiling.

Fig. 21. The frown of Greta

6 Conclusions

The facial model of Greta is made of 15,000 triangles. The high number of polygons allows us to maintain a great level of detail in the most expressive regions of human face: the eyes and the mouth. Indeed, these regions play an important role in the communication process of human-human conversation and in expressing emotions (see Fig 22-23).

The graphic engine of Greta has a conception similar to others MPEG based projects that were previously realized [7]. The novelty of Greta is the high quality of the 3D model, and the generation in real-time of wrinkles, bulges and furrows, enhancing the realism and the expressive look to the animated face. This work is part of an ongoing research on the creation of an embodied conversational agent, that is an agent able to communicate with other agent(s) or user(s) as well as to express its emotion [3].

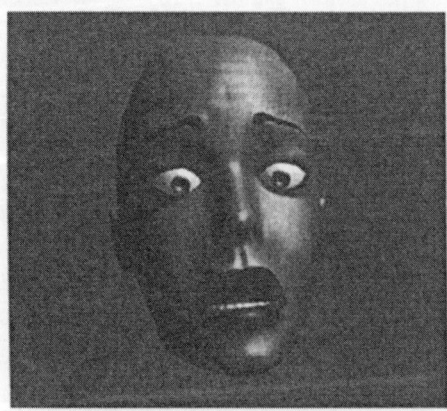

Fig. 22. The fear of Greta

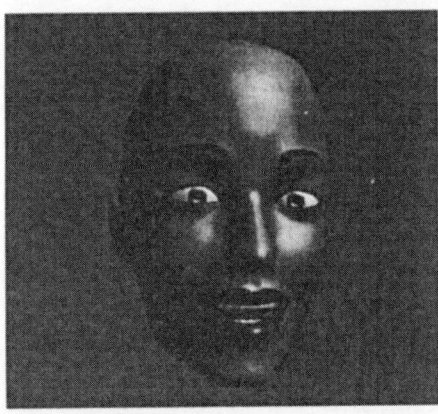

Fig. 23. The joy of Greta

References

1. Cohen MM, Massaro DW (1993) Modeling coarticulation in synthetic visual speech. In M. Magnenat-Thalmann and D. Thalmann, editors, Models and Techniques in Computer Animation, Tokyo, Springer-Verlag
2. Davis N, Davis J, Davis T, Woo M (1993) OpenGL Programming Guide: The Official Guide to Learning OpenGL, Release 1 ("Red Book"). Reading, MA: Addison-Wesley ISBN 0-201-63274-8
3. De Carolis N, Pelachaud C, Poggi I, De Rosis F (2001) Behavior planning for a reflexive agent. In IJCAI'01, Seattle, USA, August 2001
4. Doenges P, Lavagetto F, Ostermann J, Pandzic IS, Petajan E (1997) MPEG-4: Audio/video and synthetic graphics/audio for mixed media. Image Communications Journal, 5(4), May 1997
5. Ekman P, Friesen WV (1978) Manual for the Facial Action Coding System. Consulting Psychologists Press, Inc., Palo Alto
6. Faigin G (1990) The Artist's Complete Guide to Facial Expression. Watson-Guptill, New York
7. Lavagetto F, Pockaj R (1999) The facial animation engine: towards a high-level interface for the design of MPEG-4 compliant animated faces. IEEE Trans. on Circuits and Systems for Video Technology, 9(2):277-289, March 1999
8. Magnenat-Thalmann N, Primeau NE, Thalmann D (1988) Abstract muscle actions procedures for human face animation. Visual Computer, 3(5):290-297
9. Ostermann J (1998) Animation of synthetic faces in MPEG-4. In Computer Animation'98, pages 49-51, Philadelphia, USA, June 1998
10. Parke FI, Waters K (1996) Computer Facial Animation. A. K. Peters, Wellesley, MA
11. Reeves B (1990) Simple and complex facial animation: Case studies. In Vol 26: State of the Art in Facial Animation, pages 90-106. ACM Siggraph'90 Course Notes
12. Sederberg TW, Parry SR (1986) Free-form deformation of solid geometry models. Computer Graphics (SIGGRAPH '86), 20(4):151-160
13. Waters K (1987) A muscle model for animating three-dimensional facial expressions. Computer Graphics (SIGGRAPH '87), 21(4):17-24, July

Fast Text Compression Using Artificial Neural Networks

M. P. Sriram and A.Dinesh

E-mail: owen@rediffmail.com

CONTACT ADDRESS: (OLD NO.) 20, THIRD CROSS ROAD
R.A. PURAM, CHENNAI-28, TAMIL NADU, INDIA

Abstract. Neural networks have the potential to extend data compression algorithms beyond the character level n-gram models now in use, but have usually been avoided because they are too slow to be practical. We introduce a model that produces better compression than popular Limpel-Ziv compressors (zip, gzip, compress), and is competitive in time, space, and compression ratio with PPM and Burrows-Wheeler algorithms, currently the best known. The compressor, a bit-level predictive arithmetic encoder using a 2 layer, 4×10^6 by 1 network, is fast (about 10^4 characters/second) because only 4-5 connections are simultaneously active and because it uses a variable learning rate optimized for one-pass training.

1. Introduction

One of the motivations for using neural networks for data compression is that they excel in complex pattern recognition.

Standard compression algorithms, such as Limpel-Ziv or PPM or Burrows-Wheeler are based on simple n-gram models: they exploit the non-uniform distribution of text sequences found in most data. For example, the character trigram *the* is more common than *qzv* in English text, so the former would be assigned a shorter code. However, there are other types of learnable redundancies that cannot be modeled using n-gram frequencies.

For example, Rosenfeld combined word trigrams with semantic associations, such as *"fire...heat"*, where certain pairs of words are likely to occur near each other but the intervening text may vary, to achieve an unsurpassed word perplexity of 68, or about 1.23 bits per character (bpc) , on the 38 million word Wall Street Journal corpus. Connectionist neural models are well suited for modeling language constraints such as these, e.g. by using neurons to represent letters and words, and connections to model associations. We follow the approach of Schmidhuber and Heil of using neural network prediction followed by arithmetic encoding, a model that they derived from PPM (prediction by partial match), developed by Bell, Witten, and Cleary. In a predictive encoder, a predictor, observing a stream of input characters, assigns a probability distribution for the next character. An arithmetic encoder, starting with the real interval [0, 1], repeatedly subdivides this range $P(y = e|th)$, the probability that the next symbol is e in context th, i.e. $xth = 1$ and all other $xi = 0$. In our implementation, we predict one bit at a time, so y

is either 0 or 1. We can estimate $P(y = 1|xi) \simeq N(1)/N$, for each context xi, by counting the number of times, $N(1)$, that $y = 1$ occurs in the N occurrences of that context. In an n-gram character model (n about 4 to 6), there would be one active input ($xi = 1$) for each of the n contexts of length 0 to $n - 1$. (In practice, the length 0 context was not used). The set of known probabilities $P(y|xi)$ does not completely constrain the joint distribution $P(y|x)$ that we are interested for each character in proportion to the predicted distribution, with the largest subintervals for the most likely characters. Then after the character is observed, the corresponding subinterval becomes the new (smaller) range. The encoder output is the shortest number that can be expressed as a binary fraction within the resulting final range. Since arithmetic encoding is optimal within one bit of the Shannon limit of $\log 2\ 1/P(x)$ bits (where $P(x)$ is the probability of the entire input string, x),compression ratio depends almost entirely on the predictor. Schmidhuber and Heil replaced the PPM predictor (which matches the context of the last few characters to previous occurrences in the input) with a 3-layer neural network trained by back propagation to assign character probabilities when given the context as input. Unfortunately the algorithm was too slow to make it practical to test on standard benchmarks, such as the Calgary corpus. Training on a 10K to 20K text file required days of computation on an HP 700 workstation, and the prediction phase (which compressed English and German newspaper articles to 2.94 bpc) ran 1000 times slower than standard methods. In contrast, the 2-layer network we describe, which learns and predicts in a single pass, will compress a similar quantity of text in about 2 seconds on a 100 MHz 80486 (typically achieving better compression).

In section 2 we describe a neural network that predicts the input one bit at a time and show that it is equivalent to the maximum entropy approach to combining probabilistic constraints. In (3), we derive an optimal one-pass learning rate for independent, stationary inputs, and extend the model to the more general case. In (4), we find experimentally that neural network compression performs almost as well as PPM and Burrows-Wheeler (current the best known character models) in both speed and compression ratio, and better (but slower) than popular Limpel-Ziv models such as UNIX compress, zip, and gzip.

2. A Maximum Entropy Neural Network

In a predictive encoder, we are interested in predicting the next input symbol y, given a feature vector $x = x1, x2,..., xM$, where each xi is a 1 if a particular context is present in the input history, and 0 otherwise. For instance, we may wish to find
We consider first the case of independent inputs
$(P(x_i, xj) = P(x_i)P(xj))$, and a stationary source

in finding. According to the maximum entropy principle, the most likely distribution for $P(y|x)$ is the one with the highest entropy, and furthermore it must have the form (using the notation of Manning and Schütze)

$$P(x, y) = 1/Z \, \pi_i \alpha_i^{\,f_i(x,y)} \tag{1}$$

where $f_i(x, y)$ is an arbitrary "feature" function, equal to x_i in our case, a_i are parameters to be determined, and Z is a normalization constant to make the probabilities sum to 1. The a_i are found by *generalized iterative scaling* (GIS), which is guaranteed to converge to the unique solution.

Essentially, GIS adjusts the a_i until $P(x, y)$ is consistent with the known probabilities $P(x_i, y)$ found by counting n-grams.

Taking the log of (1), and using the fact that $P(y|x) = P(x, y)/P(x)$, we can rewrite (1) in the following form:

$$P(y|x) = 1/Z' \, \exp(\Sigma i \; w_i x_i) \tag{2}$$

where $w_i = \log a_i$. Setting Z' so that $P(y = 0|x) + P(y = 1|x) = 1$, we obtain

$$P(y|x) = g(\Sigma_i \; w_i x_i), \tag{3}$$

$$\text{where } g(x) = 1/(1 + e^{-x}) \tag{4}$$

which is a 2-layer neural network with M input units x_i and a single output $P(y = 1|x)$. Analogous to GIS, we train the network to satisfy the known probabilities $P(y|x_i)$ by iteratively adjusting the weights w_i to minimize the error, $E = y - P(y)$, the difference between the expected output y (the next bit to be ob-served), and the prediction $P(y)$.

$$w_i = w_i + Dw_i \tag{5}$$

$$Dw_i = \Box x_i E \tag{6}$$

where $\Box$ is the learning rate, usually set *ad hoc* to around 0.1 to 0.5. This is fine for batch mode training, but for on-line compression, where each training sample is presented only once, we need to choose $\Box$ more carefully.

3. Maximum Entropy On-line Learning

We now wish to find the learning rate, h, required to maintain the maximum-entropy solution after each up-date. We consider first the case of independent inputs ($P(x_i, x_j)$ $= P(x_i)P(x_j)$), and a stationary source ($P(y|x_i)$ does not vary over time). In this case, we could just compute the weights directly. If we let $x_i = 1$ and all other inputs be 0, then we have

$$p \; \Box P(y|x_i = 1) = g(w_i) \tag{7}$$

$$w_i = g^{-1}(p) = \ln p/(1 - p) \; \Box \ln N(1)/N(0) \tag{8}$$

where $N(1)$ and $N(0)$ are the number of times y has been observed to be a 1 or 0 respectively in context x_i. (The approximation holds when the $N(\cdot)$ are large. We will return to the problem of zero counts later). If we observe $y = 1$, then w_i is updated as follows:

$$\Delta w_i = \ln (N(1)+1)/N(0) - \ln N(1)/N(0) \; \Box 1/N(1) \tag{9}$$

and if $y = 0$,

$$\Delta w_i = \ln N(1)/(N(0)+1) - \ln N(1)/N(0) \ \Box -1/N(0) \tag{10}$$

We can now find $\Box$ such that the learning equation $\Delta w_i = \Box x_i E$ (which converges whether or not the inputs are independent) is optimal for independent inputs. For the case $x_i = 1$,

$$\Delta w_i = \Delta w_i E/E = \Box E \tag{11}$$

$$\Box = \Delta w_i/E \tag{12}$$

If $y = 1$, then

$$E = 1 - p = 1 - N(1)/N = N(0)/N \tag{13}$$

$$\Box = \Delta w_i/E = 1/EN(1) = N/N(0)N(1) = 1/\Box^2 \tag{14}$$

And if $y = 0$, then

$$E = -p = -N(1)/N \tag{15}$$

$$\Box = \Delta w_i/E = -1/EN(0) = N/N(0)N(1) = 1/\Box^2 \tag{16}$$

where $N = N(0) + N(1)$, and $\Box^2 = N(0)N(1)/N = Np(1 - p)$
is the variance of the training data.

If the inputs are correlated, then using the output error to adjust the weights still allows the network to converge, but not at the optimal rate. In the extreme case, if there are m perfectly correlated copies of xi, then the total effect will be equal to a single input with weight mwi. The optimal learning rate will be $\Box = 1/\Box\Box^2$ in this case. Since the cor-relation cannot be determined locally, we introduce a parameter $\Box L$, called the long term learning rate, where $1/m\Box\Box_L \Box 1$. The weight update equation then becomes

$$\Delta w_i = \Box LE/\Box^2 \tag{17}$$

In addition, we have assumed that the data is stationary, that $p = P(y|x_i)$ does not change over time. A better model is one where p periodically varies between 0 and 1 over time.

The last few bits of input history is a better predictor of the next bit than simply weighting all of the bits equally, as $\Box$ L does. impose a lower bound, $\Box_s$, on the learning rate, a parameter that we call the *short term* learning rate. This has the effect of weighting the last $1/\Box$ bits more heavily.

$$\Delta w_i = \Box S + \Box L/\Box^2)E \tag{18}$$

Bpc.

Finally, recall that $\Box = N(0)N(1)/N = Np(1 - p)$, which means that the learning rate is undefined when either count is 0, or equivalently, when $p = N(1)/N$ is 0 or 1. A common solution is to add a small offset, d, to each of the counts $N(0)$ and $N(1)$. For the case $d = 1$, we have Laplace's law of succession, which is optimal when all values of p are *a priori* equally likely 1) For text, their experiments suggest a value of d $\Box 0.1$, though we found that 0.5 works well. Thus, in (18), we use

$$\Box^2 = (N+2d) / (N(0)+d)(N(1)+d) \tag{19}$$

4. Conclusions

It was shown that it is practical to use neural networks for text compression, an application requiring high speed.

Among neural models, the best one found combines long and short term learning rates to achieve a balance between using the entire input history and favoring the most recent data to adapt to changing statistics. Although we optimized the neural network to predict text, it was found to give good performance on a wide variety of data types. It would probably be better to have the program automatically adjust $\square_S$ and $\square_L$, rather than set-ting them *ad hoc*, but this would just introduce another parameter to determine the adjustment rate. The problem of *ad hoc* parameters seems to be unavoidable. Our next aim is to experiment to find better compressors that exploit syntactic or semantic constraints, our aim of using neural networks in the first place.

Acknowledgements

We would like to sincerely thank Mr. Vallinayagam of St. Joseph's College of Engineering for guiding us throughout this project. We would also like to extend our heartfelt thanks the Prof. Murali Krishna of St. Joseph's College of Engineering without whose inputs this project would not have been a possibility.

References

1. Bell Timothy, Witten, Ian H. and Cleary John   G. - Modeling for text compression.
2. Burrows, M. and Wheeler, D. J. - A block-sorting loss less data compression algorithm.
3. Gilchrist Jeff, Archive Comparison Test. http://act.by.net/act.html
4. Rosenfield, Ronald - maximum entropy approach to adaptive statistical language processing.
5. Manning, Christopher D., and Schütze, Hinrich, - Foundations *of Statistical Natural Language Processing*.
6. Data Compression FAQ http://www.cs.ruu.nl/wais/-html/ na-dir/compression-faq.html .

(Flowchart follows on the next page.)

532

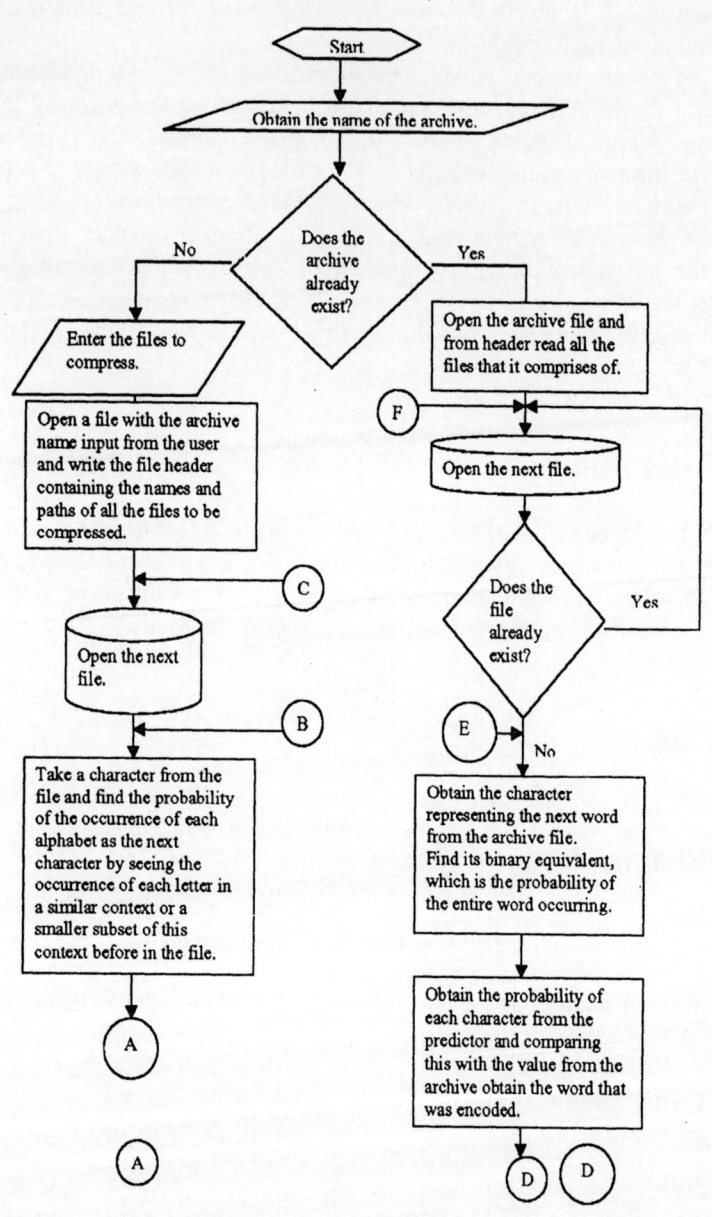

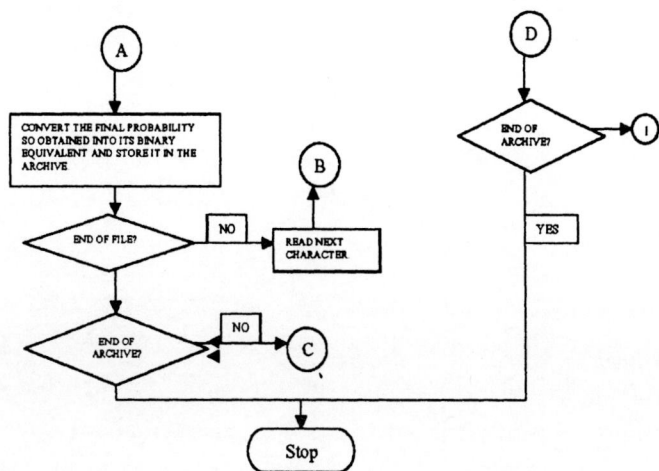

Visualization of Data Using Genetic Algorithm

M. Sarfraz[1], S. A. Raza[2]

[1]Department of Information and Computer Science, King Fahd University of Petroleum and Minerals, Dhahran 31261, Saudi Arabia
[2]Department of Management and Information Systems, King Fahd University of Petroleum and Minerals, Dhahran 31261, Saudi Arabia

Abstract. In order to obtain a good spline model from large measurement data, we frequently have to deal with knots as variables, which becomes a continuous, non-linear and multivariate optimization problem with many local optima. Hence, it is very difficult to obtain a global optima. In this paper, we present a method to convert the original problem into a discrete combinatorial optimization problem and solve it by a genetic algorithm. We also incorporate a corner detection algorithm to detect significant points (corner points), which are necessary to capture a pleasant looking spline fitting for 2D and 3D data. In case of too large data, a data reduction concept is also utilized to economize the computation in the algorithm design. As a curve and surface model, the parametric B-Spline model has been approximated to various 2D and 3D data. The chromosomes have been constructed by considering the candidates of the locations of knots as genes. The knots to the corresponding corner points have been kept fixed to minimize the computation cost. The best model among the candidates is searched by using Akaike's Information Criterion (AIC). The method determines the appropriate number and location of knots automatically and simultaneously.

Keywords: Visualization, data, genetic algorithm, B-spline, approximation.

1 Introduction

Data fitting and visualization with spline [4-7] is one of the important technologies in the area of computer graphics. If we have to make a good model from measurement data, having a complicated underlying function data, it is difficult to approximate it by a single polynomial. In this case, a spline is one of the most appropriate approximating functions.

The key to using a spline is the determination of good knots [4]. To obtain good approximation, one needs to place the knots as precisely as possible. In such cases, we have to deal with knots as variables. Then the problem becomes a continuous nonlinear and multivariate optimization problem with many local optima. Therefore, it is difficult to obtain a global optimum [5].

The underlying scheme, in this paper, is based upon the Genetic Algorithm (GA) approach in [5]. The GAs, introduced by Holland [3], are the search techniques based on the concept of evolution. Given a well-defined search space in which each solution is represented by a bit string, called a chromosome, a GA is applied with its three genetic search operators (selection, crossover and mutation) to transform a population of chromosomes with the objective of improving the quality of the chromosomes. The individual bits of a chromosome are called genes. Before the search starts, a set of chromosomes is randomly chosen from the search space to form the initial population. The three genetic search operations are then applied one after the other to obtain a new generation of chromosomes in which the expected quality over all the chromosomes is better than that of the previous generation. The process is repeated until stopping criterion is met. Finally, the best chromosome of the last generation is reported as a final solution.

Our work is the extension of the work in [5] where a spline model for open curves has been presented. We consider data arose from any scientific phenomena and a parametric form of B-Spline is used to achieve best approximation curve or surface. In order to aid our GA, a corner detection algorithm has also been used to determine the corner points. The achievement of corner points is of great importance. It helps to minimize the time for the visualization of the data as the number of iterations as reduced during the running of GA.

The organization of the paper is as follows. Section 2 gives a brief description of B-splines. The idea of corner detection is explained in Section 3. The curve and surface data fitting techniques are discussed in Sections 4 and 5 respectively. The practical results are demonstrated in Section 6 and Section 7 concludes the paper.

2 The B-Spline

The points on a parametric B-Spline curve are represented as ordered set of values: $Q(t) = [x(t), y(t)]$. The parameter t takes the values from a specified range; conventionally from 0 to 1. Parametric form of curve allows multiple values of y for single or more values of x. A parametric curve is approximated by *piecewise polynomial curves*. B-splines consist of curve segments whose polynomial coefficients depend on just a few control points. This behavior is called *local control*. Thus, moving a control affects only a small part of curve. This local behavior is due to the fact that each vertex is associated with a unique basis function. Let $T = \{t_0, t_1, \ldots t_m\}$ be a nondecreasing sequence of real numbers, i.e., $t_i \leq t_{i+1}, i = 0, \ldots m-1$. The t_i are called *knots*, and T is the *knot vector*. The ith B-spline basis function of degree p, denoted by $N_{i,p}(t)$ is defined as

$$N_{i,p}(t) = \frac{(t-t_i)N_{i,p-1}(t)}{t_{i+p-1}-t_i} + \frac{(t_{i+p}-t)N_{i+1,p-1}(t)}{t_{i+p}-t_{i+1}} \tag{1}$$

Given $n+1$ control points, the B-spline $S(t)$ is given by:

$$S(t) = \sum_{i=0}^{n} N_{i,p}(t)c_i, \tag{2}$$

where c_i is a B-spline coefficient.

3 Corner Detection

Corner detection is related to detection of high curvature points in planar curves. A number of approaches have been proposed by researchers [2]. We have used the technique suggested by Chetverikov and Szabo [2]. The details of this approach are as follows.

The proposed two-pass algorithm defines a corner in a simple and intuitively appealing way, as a location where a triangle of specified size and opening angle can be inscribed in a curve (see Figure 1).

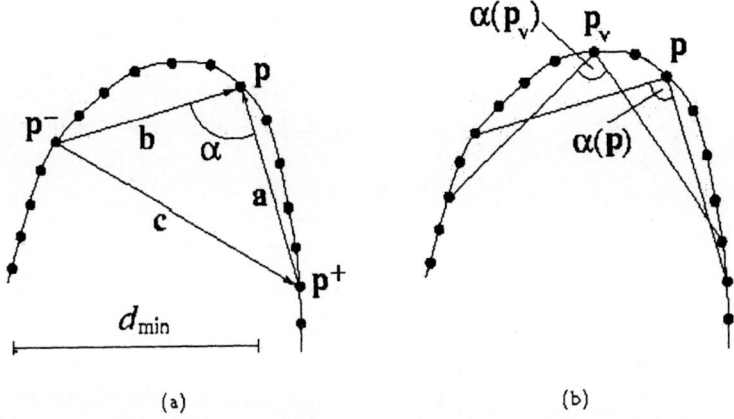

(a)   (b)

Fig. 1. Detecting high curvature points: (a) determining if p is a candidate point; (b) testing p for sharpness non-maxima suppression

A curve is represented by a sequence of points p_i in the image plane. The ordered points are densely sampled along the curve, but contrary to the other four algorithms, no regular spacing between them is assumed. A chain-coded curve can also be handled if converted to a sequence of grid points. In the first pass the algorithm scans the sequence and selects candidate corner points. The second pass is post-processing to remove superfluous candidates.

The First pass is as follows. In each curve point p the detector tries to inscribe in the curve a variable triangle (p^-, p, p^+) constrained by a set of simple rules:

$$d^2_{min} \le \left|p - p^+\right| \le d^2_{max},$$ (3)

where $\left|p - p^+\right| = |a| = a$ is the distance between p and p^+. The distance $\left|p - p^-\right| = |a| = a$ between p and p^- and α is the opening angle of the triangle. The latter is computed as follows:

$$\alpha = \arccos \frac{a^2 + b^2 - c^2}{2ab}$$ (4)

538

Variations of the triangle that satisfy the above mentioned conditions are called admissible. Search for the admissible variations starts from p outwards and stops if any of the conditions is violated. Among the admissible variations, the least opening angle $\alpha(p)$ is selected. $\pi - |\alpha(p)|$ is assigned to p as the *sharpness* of the candidate. If no admissible triangle can be inscribed, p is rejected and no sharpness is assigned. The Second pass of the algorithm is as follows. A corner detector can respond to the same corner in a few consecutive points. Similarly to edge detection, a post-processing step is needed to select the strongest response by discarding the non-maxima points.

A candidate point p is discarded if it has a sharper valid neighbor p_v: $\alpha(p) > \alpha(p_v)$. A candidate point p_v is a valid neighbor of p if $|p - p_v| \leq d^2_{max}$. The d_{min}, d_{max} and α_{max} are the parameters of the algorithm. The upper limit d_{max} is necessary to avoid false sharp triangles formed by distant points in highly varying curves. α_{max} is the angle limit that determines the minimum sharpness accepted as high curvature. In practice, we often set $d_{max} = d_{min} + 2$ and tune the remaining two parameters, d_{min} and α_{max}. The default values are $d_{min} = 7$ and $\alpha_{max} = 150°$.

4 Curve Fitting

The scheme used to convert the original continuous problem into a discrete optimization problem is an extension of the work in [5]. Each data point corresponds to a single gene in the bit string of a chromosome. In this formulation if a gene is equal to 1, we put a knot at the corresponding data point and if the gene is equal to 0 we do not (see Fig. 2).

If the given points lie in the interval $[a, b]$ then the appropriate number of knots are determined in the interval (a, b) called the *interior knots n*. The initial population, consists of K individuals of genelength L. The genes are randomly set to 0 and 1. However, the corner points are determined before the creation of initial population and the genes corresponding to those points are intentionally set to 1 in the initial population and in the population of the subsequent generations. The idea behind this scheme is not to lose those points as they are important in determining the outlines of the shapes.

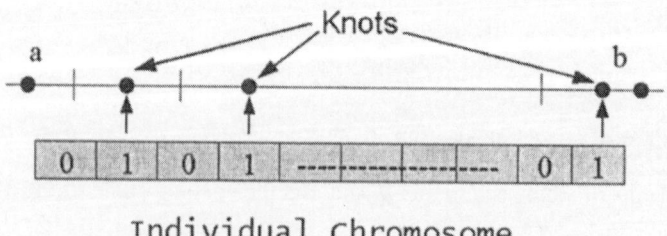

Fig. 2. Genetic formulation

In addition to the conventional genetic control parameters (crossover and mutation), another control parameter *knot ratio R* is also used as suggested by [1]. Akaike's Information Criterion (AIC) [1] is used as a fitness measure. By using

AIC we can choose the best model among the candidate models automatically. AIC is given by

$$AIC = N \log_e Q_1 + 2(2n + m),$$ (5)

where N is the number of data, Q_1 is given by the following equation:

$$Q_1 = \sum_{j=1}^{N} w_j \left\{ \{Sx_j(t) - x_j(t)\}^2 + \{Sy_j(t) - y_j(t)\}^2 \right\}.$$ (6)

The n is the number of interior knots, m is the order of the spline to be fitted on the given data. It should be noted that the smaller value of Equation 5 gives better fitness. The $Sx(t)$ and $Sy(t)$ are the x and y components respectively of the approximated spline $S(t)$ over the data F.and w_j is the weight of data, taken to be 1 for all data points in our case. The subscript of Q means the dimension of the data.

We also propose a parameter which we have named as *decimation*. This parameter enables the data to be selected interval wise without loosing the contour of the font as well as the corner points determined by the corner detection algorithm. This has been used in order to decrease the gene length of the chromosomes.

In the context of genetic algorithm, a Roulette wheel selection and a double point crossover has been used. The probability of crossover C is taken to be 0.7 and the probability of mutation M is taken to be 0.001, while $0 \le R < 0.5$ has been used. In case of the data in Fig. 3a, a decimation of 4 has been used while in case of the data in Fig. 5a, it has been kept as 2.

The summary of the algorithm is given below.

1. Input the data to be fitted.
2. Input the control parameters.
3. Find corner points using corner detection algorithm.
4. Create an initial population by using random numbers.
5. For each individual in the population make the bits corresponding to the significant points as 1.
6. For each individual compute data fitting and obtain the fitness value.
7. If total number of generations exhausted, stop the computation, otherwise go to Step 8.
8. Do selection by using the fitness values.
9. Do crossover and make the individuals of the next generation.
10. Do mutation and go back to Step 5.

540

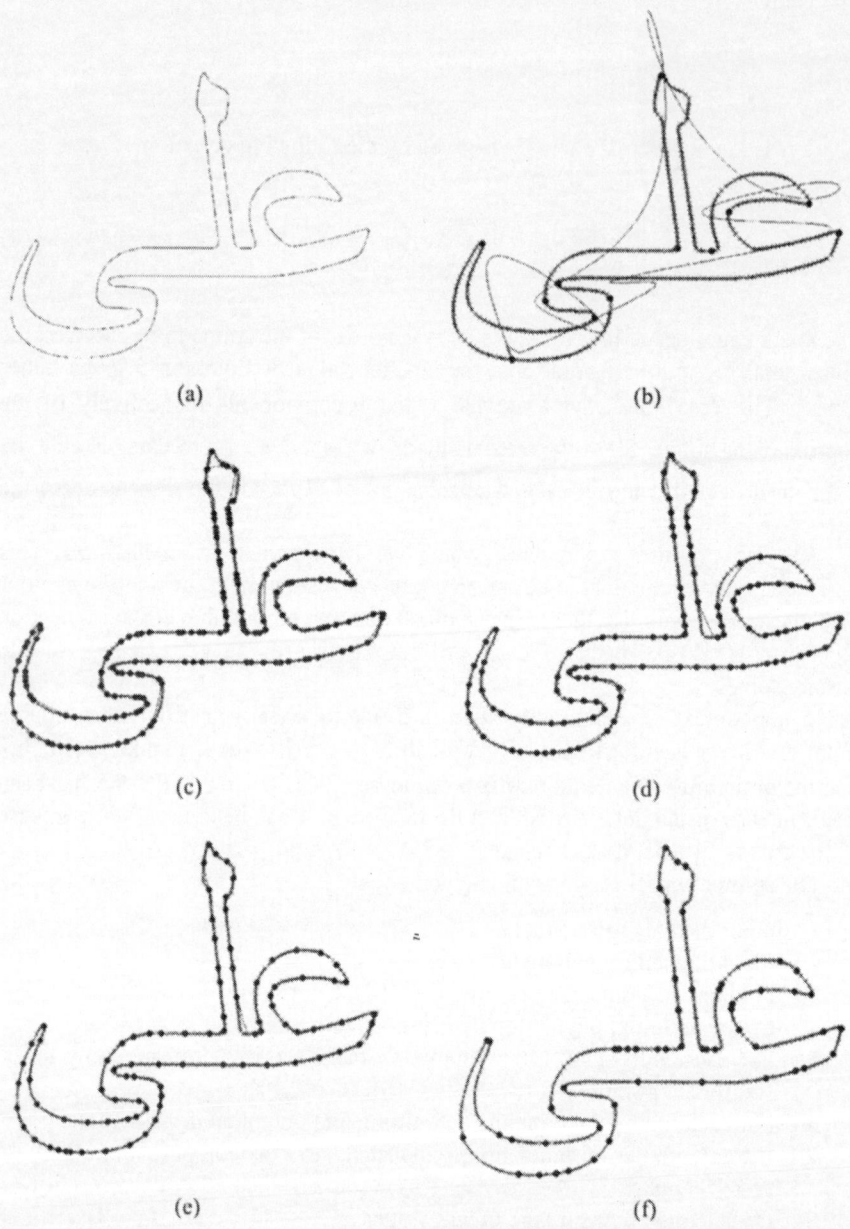

Fig. 3. (a) the outline data of an image; (b) spline approximated to the significant points before running GA; (c) at 25th generation; (d) at 50th generation; (e) at 75th generation; (f) after convergence

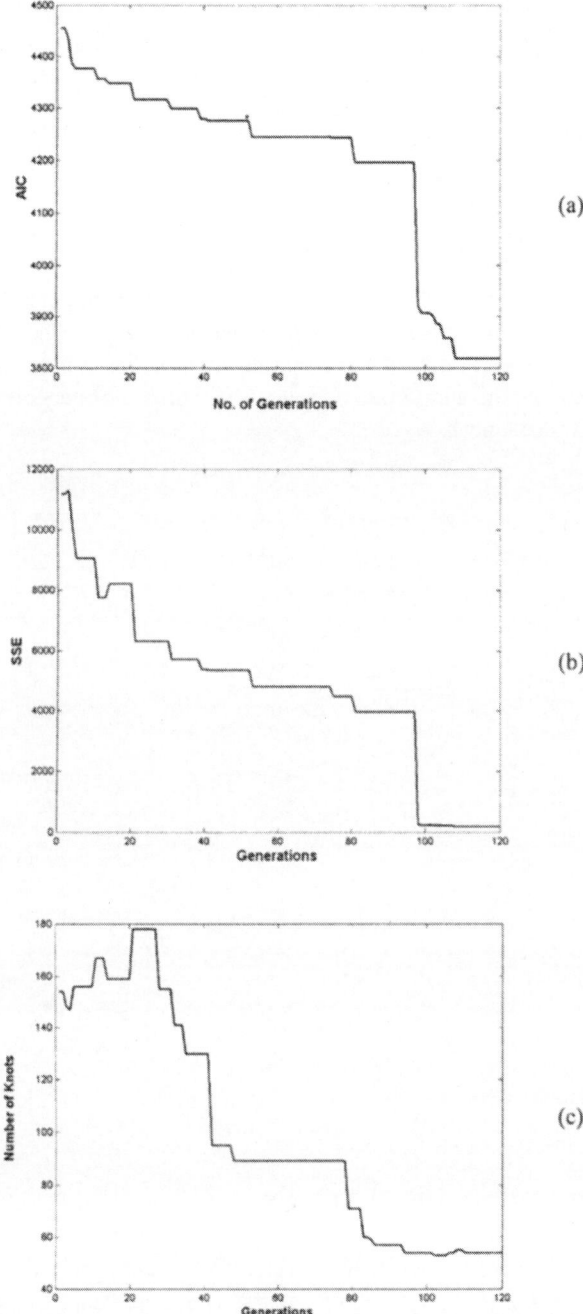

Fig. 4. (a) AIC vs number of generation; (b) sum square error vs number of generation; (c) knots vs number of generation

5 Surface Fitting

A step ahead of data fitting for curves is the surface fitting over a mesh of data. The curve scheme, in the previous sections, has been extended for the fitting of surfaces. The scheme we used in our algorithm for surface fitting is to calculate individual splines by considering the rows and columns of the mesh data one by one. This if the parametric surface is denoted by $S(t, u)$ with the parameters t and u in the two directions, then the calculations are done along parameter t and the parameter u one after the other. First of all, the calculations are done along the parameter t by taking the rows of the mesh data one by one. Then the columns are considered one by one for the calculations along the parameter u. This scheme makes the surface fitting problem similar to the curve fitting. The difference is that in this case we have to consider the optimization of the individual curves of the mesh data one by one and iterate over the total number of generations of our GA to get an optimized mesh.

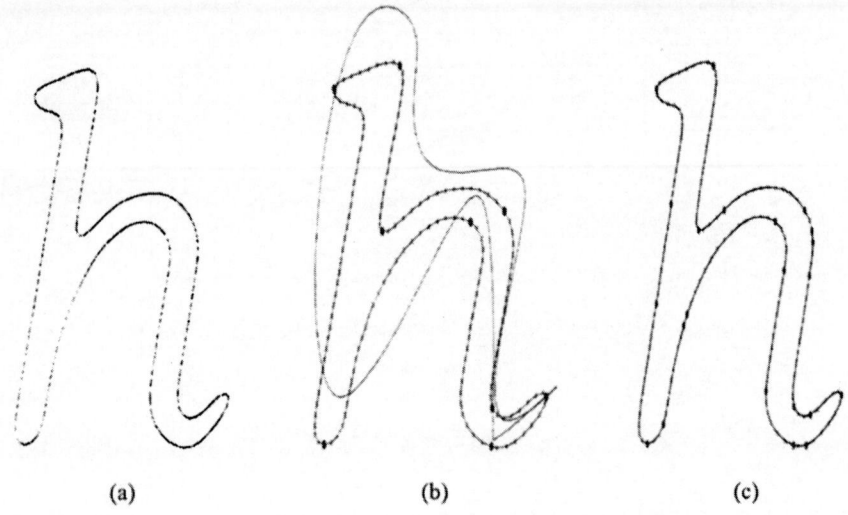

(a)　　　　　　　　　(b)　　　　　　　　　(c)

Fig. 5. (a) a data for an outline of a Font "h"; (b) detection of corner points and spline approximated to these points before running GA; (c) after GA Converges

6 Demonstration and Analysis

This section is meant to demonstrate the scheme tested on various 2D and 3D data. The algorithm uses a cubic parametric B-Spline. The method, however, is independent of the degree of the spline.

Figure 3(a) is composed of 1640 planar data points, which were reduced to 411 points after applying a decimation of 4. The corner detection algorithm detected 12 corner points (see Figure 3(b)). The genetic algorithm was run for 120 generations with a population size of 30. The subsequent figures (see Figure 3(c-f)) show the behavior of algorithm at various generations. The algorithm converged at 98th generation (see Figure 3(f)). The graphs of AIC, Sum Square

Error and Knots plotted against the number of generations have also been shown on which only the best values in a certain generation were plotted (see Figure 4).

Figure 5(a) represents another planar data detected from as an outline of a bitmap image of letter "h". This is composed of 320 points which were reduced to 160 after applying a decimation of 2. Similar to Figure 3, other results for the letter "h" have been shown in Figure 5(b-c).

A surface data (monotone data) composed of 30×30 points is considered in Fig. 6(a). The corner points together with convergence result are shown in Figure 6(b) whereas Fig. 6(c) is the shaded model of the resultant surface.

7 Conclusion and Future Work

This paper has introduced a data visualization technique using genetic algorithm and spline idea. Visualization of 2D as well as 3D data are the target of the scheme. The scheme presented is effective in the determination of the appropriate number of knots and their locations simultaneously for as large data as available. The genetic algorithm is partly aided by the corner detection for the determination of corner points. These corner points are important in capturing the shape of the data.

Some of the suggested future work directions may be as follows. Instead of NUBS, one can think of using the Non-uniform Rational B-spline (NURBS) to be used, incorporating the optimization of weights. This might help to reduce the number of iteration to converge the results. This work is in progress with the authors and expected to be sent for publication as a subsequent paper.

Use of Parallelism is another idea to be introduced for faster visualization of the data. In this case, the algorithm needed to be redesigned. It is expected that the load of computation will be concentrated in steps 5 and 6 of the algorithm in Section 4. We can apply parallel computing here to save computational time. The authors are currently looking for a practical implementation of such a parallel algorithm.

References

1. Akaike, H. (1974) A new look at the statistical model identification. IEEE Trans. Automatic Control, 19(6), 716-723.
2. Chetverikov, D., and Szabo, Z. (1999) A simple and efficient algorithm for detection of high curvature points in planar curves. Proc. 23rd Workshop of the Australian Pattern Recognition Group, 175-184.
3. Goldberg, D. E. (1989). Genetic Algorithms in Search, Optimization and Machine Learning. Addison Wesley.
4. Harada, T., Yoshimoto, F., and Aoyama, Y. (2000) Data Fitting using a genetic algorithm with real number genes. Proc. of the Iasted International Conference on Computer Graphics and Imaging.
5. Moriyama, M., Yoshimoto, F. and Harada, T. (1998) A method of plane data fitting with a genetic algorithm. Proc. of the Iasted International Conference on Computer Graphics and Imaging, pages 21-31.

544

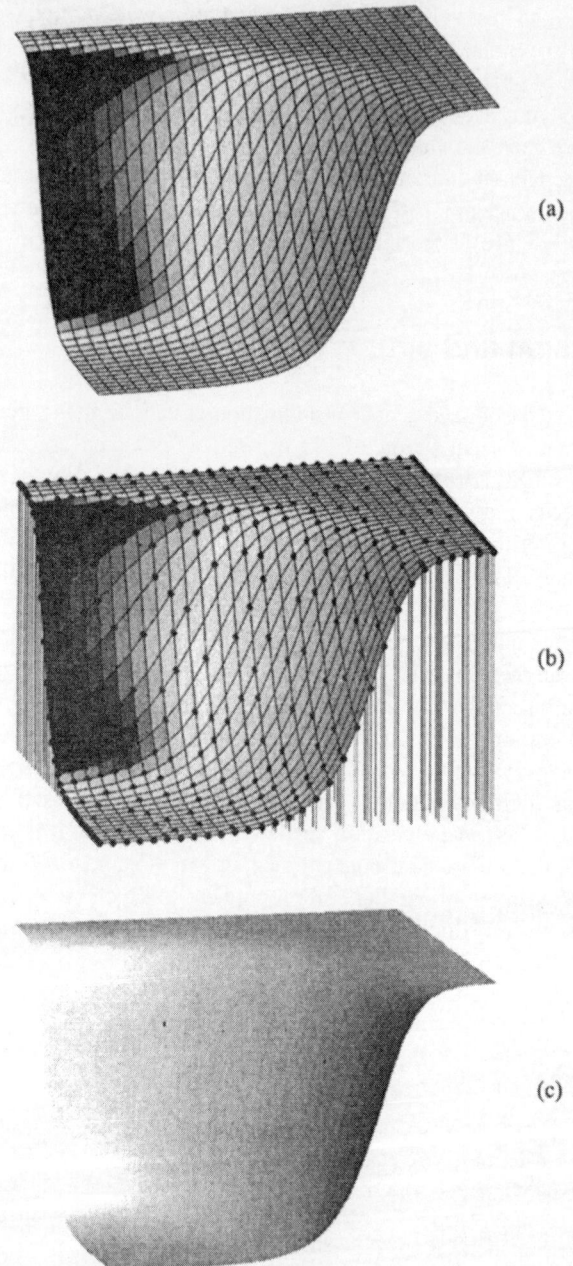

Fig. 6. (a) input surface after shading; (b) approximated surface after convergence; (c) output surface after shading

6. Sarfraz, M., (2000), A Rational Spline for Visualizing Positive Data, The Proceedings of IEEE International Conference on Information Visualization-IV'2000-UK, IEEE Computer Society Press, USA, 57-63.
7. Sarfraz, M., Butt, S., and Hussain, M. Z. (2001), Visualization of Shaped Data by a Rational Cubic Spline Interpolation, Computers & Graphics, 25(5).

Measuring Facial Emotional Expressions Using Genetic Programming

A. Loizides, M. Slater and W. B. Langdon

Computer Science, University College, Gower Street, London, WC1E 6BT, UK
{A.Loizides,M.Slater,W.Langdon}@cs.ucl.ac.uk
http://www.cs.ucl.ac.uk/staff/A.Loizides, /staff/M.Slater, /staff/W.Langdon
Tel: +44 (0) 20 7679 7209, Fax: +44 (0) 20 7387 1397

Abstract. Genetic Programming techniques can be used to produce regression equations that quantify emotional expressions on a facial model. The formulae give emotional scores based on the position of 25 automatically generated ílandmarksî on the face. The method shown here is an integrated part of a system that maps multidimensional data sets to naturalistic visual structures such as a face.

1 Introduction

[1] have previously proposed a new way of visualising multivariate data sets using an automatic mapping to 3D face models. The Empathic Visualisation Algorithm, EVA, provides an automatic mapping from semantically important features of the data to emotionally or perceptually signiÝcant features of the visual structure (such as the face). It is a prerequisite of this method to be able to quantify individual facial expressions, like happiness, sadness, anger and fear.

Measuring facial expressions is important in several other applications. In Virtual Reality exposure therapy for social phobia for example, facial expressions are signiÝcant components of the evaluative feedback used to generate social anxiety [2].

Here we propose a Genetic Programming [3] (GP) based method for measuring universally recognised emotional expressions based on movements of certain points on the surface of the face.

Section 2 gives the background work to facial expressions, followed by Section 3 that explains how we measure emotional expressions. The GP system used is described in Section 4 and results are shown in Section 5. The conclusion is in Section 6.

2 Background

[4] developed the Facial Action Coding System (FACS). This system provides a notation for recording and describing the expressions of the face by considering the combinations of muscles that are used to create them. Ekman uses action units or AUs such as the inner brow raiser and the upper lip raiser to describe muscles or

groupings of muscles that perform speciÝc actions in facial expressions. [5] identi-Ýed six key expressions of emotion which can be recognised across most cultures. These are neutral, anger, sad, happy, fear, surprise and disgust.

We can score the facial expressions by looking at the FACS system. We can use Ekmanís description of what deÝnes certain emotional expressions. For example, Ekman describes happiness as a contraction of the left and right zygomatic major muscles. From inspection we can set a contraction value for these two muscles which describes what could be considered a perfect smiling face. We can then compare this muscle set with those on a speciÝc generated face and produce a mean square error from the ideal.

This method could not be used in the EVA however, since it would have introduced a circularity into the method. Having generated faces using Ekmanís method (the different muscle contraction values of our facial model) we could not use the same method (muscle contraction values) to measure the emotional expressions in the face. The main reason for avoiding this, is because the emotional expressions used cannot be treated independently. Each emotional expression, is affected by a number of muscles and any single muscle usually affects a number of emotional expressions. There is no one-to-one relationship between muscles and emotional expressions.

Moreover, we wish to base the measurement of emotions on real people since it is real people who will draw conclusions from observing the facial models. Hence, the requirement for this new technique to measure emotional expression that is described below.

3   Measuring Emotional Expressions

An experimental set-up was created to quantify emotional expressions on the face e.g. scale of happiness-sadness, scale of anger-calmness, fear-relax. This is achieved using a number of points / ìlandmarksî on the face that are signiÝcantly inÐuenced by muscle contractions and express the various emotions measured. We use 25 such points: these are mainly points near the eyes and the mouth together in addition to a stable point that is used as à reference point. This reference point was selected to be the top of the nose. The number 25, of landmarks on the face, was selected arbitrarily using the trial-error method.

Two sets of data (randomly generated facial expressions) were created, the Ýrst consisting of 200 faces and the second of 150 faces, and positions of each of the ìlandmarksî, for each face, were recorded with respect to the reference point. In fact the distances of the 25 points from the reference point were recorded for each face. Each of the faces were subjectively assessed for emotional state by at least 3 different people for each face (from a pool of about 30 people) for both data sets. The answers of the subjects (the mean for each face) for the Ýrst data set only (training data set) were used to create symbolic regression equations. For the symbolic regressions we had, y_i being the response variable denoting the subjectsí assessments of the degree of, say, happiness on a $0 - 100$ scale, where 100 indicates maximum happiness and 0 maximum sadness with 50 being neutral. A similar response variable existed for each of the other emotions. On the other hand, we had 25 explanatory variables x_i, $i = 1, ..., 25$ representing distances of each of the landmarks on the face from the reference point. A separate symbolic regression was carried out for each type of emotion. Hence, a separate estimated regression equation was produced for each emotion.

The second data set (evaluation data set) was then used to verify the equations formed, by comparing users subjective measurements, to the results produced by the symbolic regression equations when applied to this second data set for each emotion. Highly signiŸcant positive correlations were found between the results from the equations produced and users subjective evaluations. In fact we use these estimated regression equations to measure, quantify, emotional expressions in the Empathic Visualisation Algorithm (EVA).

4 Training Data - GP Configuration

Table 1. Table for GP symbolic regression.

Objective:	Find a function of one independent variable y_i and 25 dependent variables, in symbolic form, that Ÿts a given sample of 200 data points (faces).
Terminal set:	$x_i, i = 1, ..., 25$ where x_i are the landmark points on the face.
Function set:	$+, -, \%, *$.
Fitness cases:	200 faces.
Raw Ÿtness:	The sum, taken over the 200 Ÿtness cases, of the RMS error between value of the dependent variables produced by the S-expression and the target value y_i.
Standardised Ÿtness:	Equals raw Ÿtness.
Hits:	Number of Ÿtness cases for which the value of the dependent variables produced by the S-expression comes within 0.01 of the target value y_i.
Wrapper:	None.
Parameters:	Population Size = 750, Generations = 70.
Success predicate:	An S-expression scores 20 hits.

In Figure 1 we can see the evolution of the GP for the different expressions: degree of happiness, degree of anger and degree of fear. We arbitrarily select the two faces shown on top of Figure 1 for presentation in more detail here. All the other cases show similar types of results. The measurements for the three different expressions are shown with degree of happiness identiŸed by the letter H, degree of anger identiŸed by the letter A and degree of fear by the letter F. Next to each graph there is a wider bar, rendered black, showing the average response for the emotional expression encountered, from a pool of at least 3 measurements. This is in fact the

548

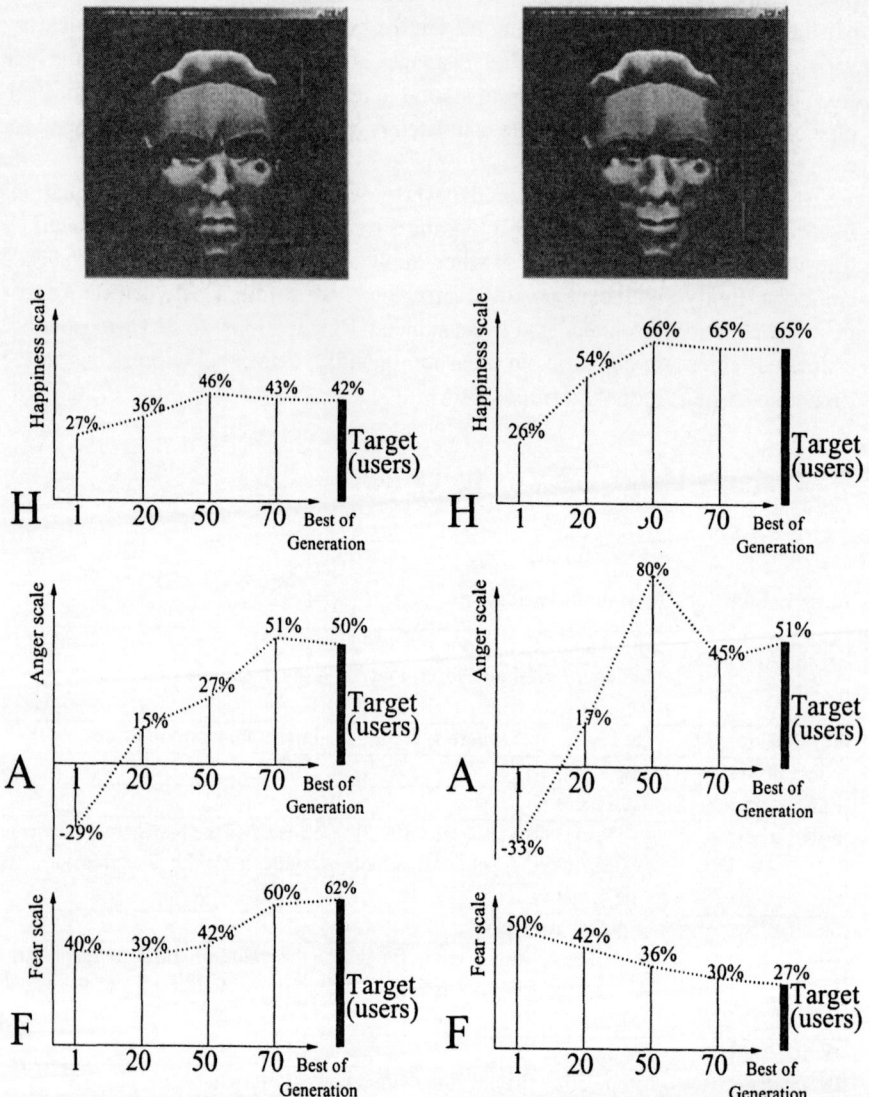

Fig. 1. Here we have for the 2 faces shown the corresponding measurements for degree of happiness (H), degree of anger (A) and degree of fear (F) for the best individual in generation 1, 20, 50 and 70 (last generation).

target result of the symbolic regression of the GP. The rest of the bars in each graph show measurements for the different expressions of the best of generation individual GP for generations 1, 20, 50 and 70. As can be seen, the difference between target measurements which are user evaluations and GP measurements decreases, as the number of generations increases. This decrease is also evident in Figures 2, 3 and 4 where for each expression we have a plot between the root mean square error (RMS) of best of generation individual against the actual number of generations.

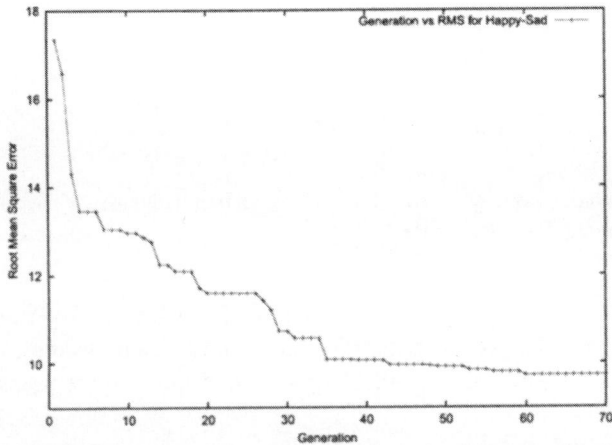

Fig. 2. The evolution of RMS error of best of generation individual against the number of generations for the Happiness-Sadness scale.

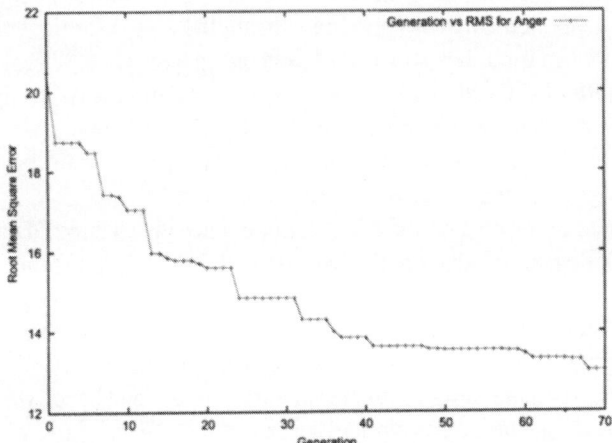

Fig. 3. The evolution of RMS error of best of generation individual against the number of generations for the Angry-Calm scale.

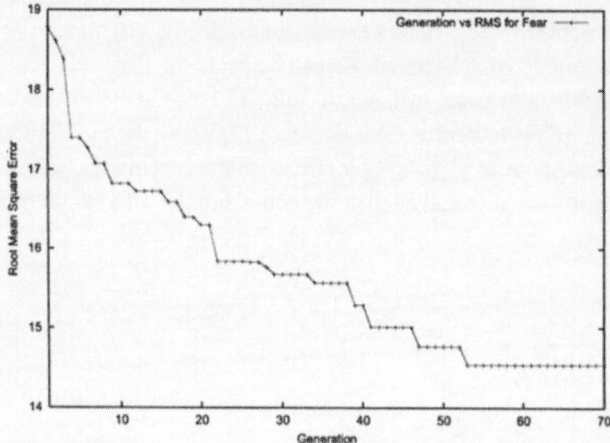

Fig. 4. The evolution of RMS error of best of generation individual against the number of generations for the Fear-Relax scale.

From these Ygures we can conclude that the GP is learning over time, and we can also see from the two examples given in Figure 1, the measurements for emotions of best of generations individual are almost identical to the userís subjective evaluations.

Therefore, the evidence suggests that the estimated regression equations produced by the GP are a good approximation to user subjective evaluations for the emotional expressions we encounter here.

5 Results

Is there evidence of positive correlation between X and Y, with X being users evaluations of the emotional expressions and Y being the scores for the emotional expressions produced from the estimated regression equations?

Here we are looking for evidence of positive correlation between the two variables (X and Y). For the Ýrst emotional expression, the happiness-sadness scale $r_1^2 = 0.85$ and the test statistic is $t_1 = 19.7$. For anger-calm scale $r_2^2 = 0.75$ and $t_1 = 13.8$ whereas for fear-relax we have $r_3^2 = 0.67$ and $t_3 = 10.5$.

The t-test is performed on the evaluation (and not the training) data set of 150 faces. On 148 degrees of freedom the critical $t = 2.6$ at 1% signiÝcance level.

Hence the tests are signiÝcant for all three cases and thus there is evidence of high positive correlation between variables X and Y. Therefore, there is evidence that the estimated regression equations will produce similar results to user evaluations of emotional expressions for the facial model we use and hence we can replace users with these equations. The similarity is evident in Ýgures 5,6,7 where the two measurements (user evaluations and estimated regression equations) are plotted together. As can be seen from the graph very clearly there is a positive correlation between the two variables.

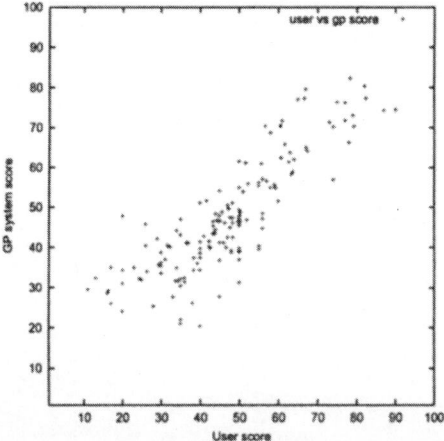

Fig. 5. Here we have the mean user measurement of happiness-sadness scale plotted against the value produced by the symbolic regression for the evaluation data set of 150 faces. As we can see from the graph the two sets of values are highly correlated and in fact $r^2 = 0.85$.

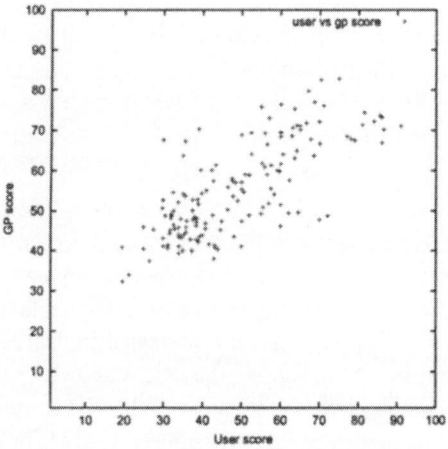

Fig. 6. Here we have the mean user measurement of angry-calm scale plotted against the value produced by the symbolic regression for the evaluation data set of 150 faces. As we can see from the graph the two sets of values are highly correlated and in fact $r^2 = 0.75$.

6 Conclusion

We have shown a technique to automatically measure emotional expressions of a particular face using GP techniques. Users are replaced by estimated regression equations for different emotional expressions. These estimated equations take into account the movement of certain points mainly around the mouth and eye with respect to a stationary point on a facial model.

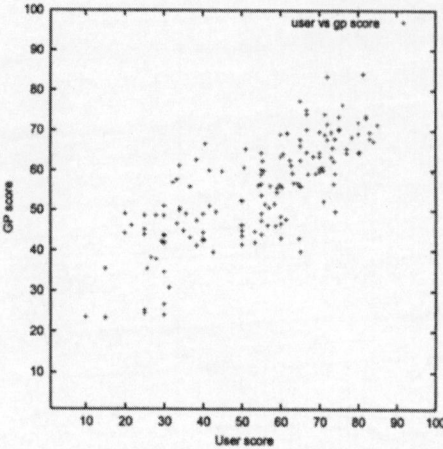

Fig. 7. Here we have the mean user measurement of fear-relax scale plotted against the value produced by the symbolic regression for the evaluation data set of 150 faces. As we can see from the graph the two sets of values are highly correlated and in fact $r^2 = 0.67$.

We use the ability to quantify emotional expressions in a computer generated manner shown in this paper, to allow us automatically map multidimensional data sets, such as accounting and Ýnancial reports, to faces and hence visualise the data in a more naturalistic way. We call this system Empathic Visualisation Algorithm (EVA).

In EVA, a GP is used to automatically derive a face which has emotional expressions representing aspects of interest in the data set to the user. In fact, these emotional expressions represent the emotional state users would have, if they were to analyse the data themselves. Hence the word empathic in the name of the system. For example, in Ýnancial data, if the user is very interested in degree of liquidity, then the more liquid the company is the happier the face will look. Furthermore, if the user is also interested in proÝtability, the more proÝtable the company the more calm (less angry) the face will look. The GP is organised so that for the faces derived, the emotional expressions take into account these user interests, or aspects of importance, which might also be conÐicting.

It is important that the method used to measure emotional expressions is not the same as the method used to produce the emotional expressions, otherwise it would result in a circularity in the system. An emotional expression is produced by manipulating muscle contractions. But a particular muscle might have implications for many emotional expressions. Hence, we can not use the same system (muscle contraction values) to also measure the facial expressions.

The method presented here, allow us to independently measure emotional expressions. We are still using Ekmanís FACS system in order to produce facial expressions but when it comes to measuring them, we use the estimated regression equations being derived by the GP.

References

1. Loizides, A., Slater, M.: The empathic visualisation algorithm, chernoff faces revisited. In: Technical Sketch, ACM Siggraph 2001 Conference Abstracts and Applications, ACM Siggraph (2001) 175 accepted.
2. M. Slater, D-P. Pertaub, A.S.: Public speaking in virtual reality: Facing an audience of avatars. In: IEEE Computer Graphics and Applications. Volume 19., IEEE (1999) 6ñ9
3. Koza, J.R.: Genetic Programming: On the Programming of Computers by Means of Natural Selection. MIT Press (1992)
4. Ekman, P., Friesen, W.V.: Facial Action Coding System (Investigatorís Guide). Consulting Psychologists Press, Inc., Palo Alto, California, USA (1978)
5. Ekman, P.: The argument and evidence about universals in facial expressions of emotion. Handbook of Social Psychopysiology (1979) 143ñ146

Fuzzy Based Web Server Workload Modeling and Prediction

Chin Wen Cheong

FOSEE, MultiMedia University, 75450 Malacca, Malaysia.

Abstract. In this paper, Web server workload is modelled by considering two most significant metrics, namely the burstiness factor and the bandwidth utilisation. The integration of the workload metrics are accomplished by using a fuzzy inference system(FIS) to lessen the complexity of the metrics relation. Due to the fluctuating of WWW requests, the utilisation states are viewed as a series of stochastic processes. The fuzzy Markovian prediction is study in two manners, firstly the utilisation states and secondly the Markovian property. The results of fuzzy approaches perform a more appropriate comparison

1 Introduction

The WWW to date is revolutionarily growing and some web services responsiveness are degrading due to unprecedented workload. The exponential growth of the clients' demand causes the imbalance workload distribution for homogenous and heterogeneous web servers. Traffic congestion becomes critical when the DNS scheduling [1] of a multi-server system fails to provide scalability and flexibility to handle the load traffic. Additionally, the existence of traffic burstiness phenomenon in the peak hour [3] at certain time scales has caused the web servers services degraded to a crawl state. Thus, in order to provide highly reliable service in the business oriented WWW as well as LAN, workload handling and forecasting are vital.

A great deal of literature works has been carried out in forecasting including statistical and artificial intelligence approaches. The statistical approach comprises of moving average, exponential smoothing, time series, regression and economic modeling[5]. By the way, the artificial intelligence (AI) concepts which include knowledge engineering, expert system, fuzzy logic, neural network (ANN) and genetic algorithm (GA) are introduced to the forecasting methods. However, the traditional as well as the AI forecasting methods are having several drawbacks where the statistical methods are ill defined to represent vague input data and human judgement. Likewise, despite genetic and ANN forecasting possesses powerful searching and learning of past data, GA as well as ANN are more suitable dealing with numerical data instead of linguistic values. Due to fluctuation of server workload distribution, the numerical data collected is vague. To increase the typicality and accuracy, the forecasting methods are expected to process numerical information incorporating with expert judgements for future workload estimation. Thus, fuzzy inference system seems to be a better candidate, since it is an effective approach to utilize linguistic rule derived from numerical data pairs. A lot of fuzzy forecasting models have been proposed such fuzzy self-regression by Feng and Guang[6], fuzzy neural approach to time series prediction by Nie [7] and forecasting method from Wang and Mendel[8].

This paper focuses upon two data predictions based on two fuzzy approaches, namely the fuzzy states and the fuzzy markovian memoryless property. Firstly, the prediction is based on fuzzy states where a fuzzy Markov model is formed to represent the transitions of the server resource utilisation. The future server workload states are based on predefined evaluation of state transition matrix. The judgement of fuzzy states are referred to the distribution of server utilisation index derived from a fuzzy inference system(FIS)[9], which is characterise by two Web servers' performance metrics namely the server's burstiness factor and bandwidth utilisation. Fuzzy rules are established by taking into account of different combinations of fuzzified workload metrics values with regard to pre-defined membership functions. A fuzzy algorithm is utilised to explore the steady states of the server workload after n transition periods.

Further, in second approach, the fuzzy memoryless property of Markov Chain is analysed to provide a better prediction. Sometimes the occurrence of a particular web server's utilisation state is related to the consequence states. The non-fuzzy markovian property is constrained to this situation where, any of the two consequence states must be independent. The fuzzy approach is able to capture the dependency of at least two or more consequence transition states. Further, the fuzzy markovian property will convert to its original property if non-existence of fuzzy circumstance.

2 Web server workload modeling

Great deals of studies have been look into the Web Server performance issues. Arun Iyengar[10] analyses the collected performance data and Log file from several real sites. The research emphasise on the dynamic file(CGI) analysis which invoke the accesses of database. The examined condition of the limited server process power shows that in order to optimise the Web server performance, the restriction should pinpoint the quantity of dynamic pages. Another Server HTTP Log analysis has been done by Martin F. Arlitt[11]. Six different Log data sets have been analysed for the duration from one week to one-year activities. The throughput studies have find out ten common characteristics such as successful requests, document types, distinct requests, file size distribution, etc.

2.1 Burstiness issues and analysis

The common Web server performance evaluation can be viewed in two metrics, the latency and throughput or bandwidth(capacity) of the requests. The measurements that describe the Web server performance are not restricted to[12]:
- Connections per second
- MegaBits per second

- Round trip time
- Errors per second

However, the overall Web service performance may be influenced by the hardware platform, operating system, network condition and server software. On the other hand, the burstiness of the WWW traffic is also one of the well-known factors that caused the traffic surge. Some studies have been carried out to model and identified the burstiness of the traffic. G. Banga, et.al[13] analysed the HTTP LOG file of web server and claimed that the HTTP request traffic is bursty with the burstiness being observable over different time-interval scales. The analysis also resulted that even a small amount of burstiness degrades the throughput of a Web server. In Mark E.C and Azer Bestavros[3] show that the subset of network traffic that is due to WWW transfers can show characteristics that are consistent with self-similarity. It is also shown that Web traffic contains bursts observable over four orders of magnitudes.

According to the mentioned literature survey, in order to model the web server utilization states, two metrics namely the throughput(Mbits) and the burstiness factor of the HTTP Log is analyzed. The traced data of CalgaryNet Server[14] is collected from time and date 00:00:00 August 28, 1995, through 11:23:23 August which is only a small portion of the original trace data exhibits busy traffic. This selected data shows the property of self-similarity with the Hurst parameter of estimated H equivalent to 0.54. The entry consists of information of one request per line with following format:

Hostname   [DAY MON DD HH:MM:SS:YYYY]   "Request"   Status code   Bytes

The historical data will be analyzed according to the throughput(Mbits per time) and the burstiness factor. A fuzzy system will be developed to integrate the two mentioned metrics into a so-called server utilization index. This index will indicate the Web server utilization resource for certain time window.

2.2   Burstiness factor

Consider an HTTP LOG composed of N requests to a Web server with a certain duration(T)[12]. Assume τ is the HTTP LOG requests inter arrival time of the Web server and λ is the average arrival rate of requests observed in the HTTP LOG, where $\lambda_{ave} = \tau / N$. Consider the epochs(where epochs is the division of requests inter arrival time and k equal subintervals of duration τ/k.

Assume A(n) is the number of HTTP requests that arrive in epoch n and λ_n is the arrival rate of requests during epoch n given by

$$\lambda_n = \frac{A(n)}{(\tau / N)} \qquad (1)$$

A^{above} and A^{below} are defined as the total number of HTTP requests that arrive in epochs in which the epoch arrival rate λ_n exceed and below the average arrival rate λ_{ave} observed in the HTTP LOG respectively. The burstiness factor, b is defined as the fraction of time during which the epoch arrival rate exceeds the average arrival rate of the HTTP LOG. The parameter b is shown as below:

$$b = \frac{number\ of\ epochs\ for\ which\ \lambda_n > \lambda_{ave}}{k}, \text{ where } 0 \le b \le 1 \qquad (2)$$

3 Fuzzy system in utilization index determination

The two metrics are the indication of the servers' busyness. To obtain a more proper representation of the servers' utilisation states, the two metrics are integrate to a so-called utilisation index by using a fuzzy inference system. Fuzzy inference system[15] is based on fuzzy set theory and fuzzy rules-based approach in decision analysis and variety of fields, especially dealing with uncertain and complex systems. The components of a fuzzy system include fuzzification(fuzzy input memberships), inference engine and defuzzification respectively. The computation of the FIS is manipulates using Matlab's fuzzy toolbox. The workload metrics are fuzzified into three fuzzy linguistic spaces as follows: Web server utilisation, $U = \{M_1, M_2\}$, where M_1 represents the burstiness factor(b) and M_2 represents the throughput(Mbits). The workload metrics are fuzzified into three fuzzy linguistic spaces as $M_1 = \{$low, moderate, high$\}$ and $M_2 = \{$weak, medium, strong$\}$.

3.1 Membership function

The M_1's M_2's and the output(U)membership functions are defined as the trapezoidal, bell-shape and triangular membership functions respectively. The shape determination of membership function has been an issue[16,17] among the researchers. Different types of MF determination have been proposed either by knowledge acquisition from an expert or group of experts, direct estimation, and even by automatic generation using artificial approaches such as artificial neural network, genetic algorithm, inductive reasoning, etc. However, in this paper, the chosen MF is just sufficient to capture the uncertainty of the attributes(b-factor, etc) and the fuzzy states are predefined. The sensitivity and the best fitted MF is not further analyse in this article. The membership function is illustrated in **Figure 1**.

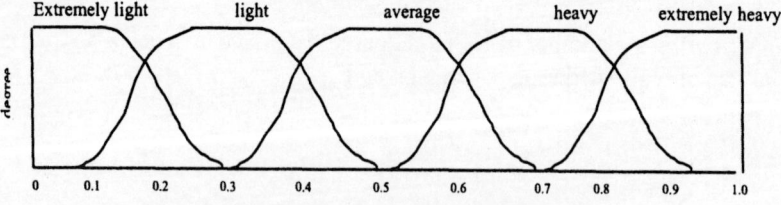

Fig. 1. Membership functions for different attributes

3.2 Fuzzy rules

The utilisation states is determines by a set of *If-Then* rule statements where the conditional statements that comprise fuzzy logic. The nine rules are state as follows:

Rule 1: **If** (b is weak) **and** (Throughput is low)   **Then** U is extremely light
Rule 2: **If** (b is weak) **and** (Throughput is moderate) **Then** U is light

Rule 8: **If** (b is strong) **and** (Throughput is moderate)**Then** U is heavy
Rule 9: **If** (b is strong) **and** (Throughput is high)   **Then** U is extremely heavy

3.3 Defuzzification

The output is defuzzified to produce a nonfuzzy value. The chosen method is called center of area[15]. Suppose the aggregated rules result in a membership function $\mu_{agg}(z)$, $z \in [z_0, s_q]$. Let the subdivide the interval $[z_0, s_q]$ into q equal subintervals by the points $z_1, z_2, \ldots z_{q-1}$. The output value z* is defined as :

$$z_c^* = \frac{\sum_{k=1}^{q-1} z_k \mu_{agg}(z_k)}{\sum_{k=1}^{q-1} \mu_{agg}(z_k)} \tag{3}$$

Based on the fuzzy system, the two Web performance metrics namely the throughput and the burstiness of a Web server LOG file is analyzed as follow in Table 1:

Table 1. Web Server utilization for every interval of 300 seconds.

Time interval		Metric		FIS
Unit	Time	throughput	Burstiness factor	Utilization index
1	0-300	4.193	0.533	0.465
2	301-600	7.544	0.800	0.659
3	601-900	7.192	0.800	0.619
.	.	.	.	
20	5701-6000	4.605	0.500	0.748
21	6001-6300	5.677	0.300	0.589

4   First approach : fuzzy Markovian utilization states

The state transition of the web server utilization is uncertain due to the users unprecedented surfing behavior and fluctuating network condition. Therefore, the web server utilization states transition is more sensible represented by a fuzzy set. Assume the discrete-parameter Markov chain $\{X_n, n \geq 0\}$ with a discrete fuzzy state space $S_i = \{i=0, 1, 2, \ldots\}$, where this set may be finite or infinite and

$\forall x \in X, \sum_{i=1}^{k} \mu_i(x) = 1$, where k is much smaller than n. The procedure of utiliz-

ing markov chain to predict the $n+1$ possible state: Let $\widetilde{MF_i}$ represents $x(1)$, $x(2)$, ..., $x(n-1)$ membership function in the fuzzy sets S_i where

$$\widetilde{MF_i} = \sum_{l=1}^{n-1} \mu_i(x_i), i = 1,2,...,k \tag{4}$$

Obviously, when $S(i)$ is crisp set, $\widetilde{MF_i}$ will become a crisps function. The occurrence of S_i initial probability, $\widetilde{P_i}$:

$$\widetilde{P_i} = \frac{\widetilde{MF_i}}{n-1}, i = 1,2,...,k \tag{5}$$

$$\widetilde{MF}_{ij} = \sum_{l=1}^{n-1} \mu_{S_i}(x_l) * \mu_{S_j}(x_{l+1}) \tag{6}$$

and the fuzzy transition from S_i to S_j, the state transition vector P_{ij},

$$P_{ij} = \frac{\widetilde{MF}_{ij}}{\widetilde{MF}_i}, \qquad i,j = 1,2,...,k \tag{7}$$

For n-state x_n and the fuzzy sets, $\mu_{Si}(x_n)$, $i=1,2,...j$. Then

$$\widetilde{P(x_n)} = [\mu_{S_1}(x_n), \mu_{S_2}(x_n),...,\mu_{S_k}(x_n)] \tag{8}$$

For the prediction of $(n+1)$ state is:

$$\widetilde{P(x_n)} * \widetilde{P^T}[\mu_{S_1}(x_n),\mu_{S_2}(x_n),...,\mu_{S_k}(x_k)] * \begin{bmatrix} P_{11} & P_{21} & P_{k1} \\ P_{12} & P_{22} & P_{k2} \\ P_{1k} & P_{2k} & P_{kk} \end{bmatrix} \tag{9}$$

4.1 Higher Transition and Steady State Determination

The higher order of the transitions probabilities are obtained as follows:

Let $\widetilde{p_i}(n) = \widetilde{P}(X_n=i)$ and $\widetilde{p}(n) = [\widetilde{p_0}(n) \quad \widetilde{p_1}(n) \quad ....]$, where $\Sigma p_k(n)=1$. Consider the initial-state probability vector,

$$\widetilde{p}(0) = [\widetilde{p_0}(0) \quad \widetilde{p_1}(0) \quad \widetilde{p_2}(0) \quad ....] \tag{10}$$

Let $\widetilde{p}(n)$ as the state probability vector after n transitions of X_n. And the $p(n)$ is defined as:

$$\widetilde{p}(n) = \widetilde{p}(0) \, P^n \tag{11}$$

The obtained transition matrix $\tilde{P}$ is converges to a steady state if the following criteria is fulfilled. Let $\{X_n, n \geq 0\}$ be a regular finite-state Markov chain with transition matrix P. Then

$$\lim_{n \to \infty} \tilde{P}^n = \hat{P} \tag{12}$$

where $\tilde{P}$ is a matrix whose rows are identical.

4.2 Numerical illustration

Assume the fuzzy states of the Web server utilization index are categorized into $Low(S_1)$, $Moderate(S_2)$, $High(S_3)$, $Extremely\ High(S_4)$. The predefined trapezoidal fuzzy sets are shown in (13) and **Figure 2**:

$$S_1 = \begin{cases} 1, & x \leq 0.30 \\ \dfrac{0.40 - x}{0.10}, & 0.30 \leq x \leq 0.40 \end{cases} \qquad S_4 = \begin{cases} \dfrac{x - 0.60}{0.15}, & 0.60 \leq x \leq 0.75 \\ 1, & x \geq 0.75 \end{cases}$$

$$S_3 = \begin{cases} \dfrac{x - 0.45}{0.10}, & 0.45 \leq x \leq 0.55 \\ 1, & 0.55 \leq x \leq 0.60 \\ \dfrac{0.75 - x}{0.15}, & 0.60 \leq x \leq 0.75 \end{cases} \qquad S_2 = \begin{cases} \dfrac{x - 0.30}{0.10}, & 0.30 \leq x \leq 0.40 \\ 1, & 0.40 \leq x \leq 0.45 \\ \dfrac{0.55 - x}{0.10}, & 0.45 \leq x \leq 0.55 \end{cases} \tag{13}$$

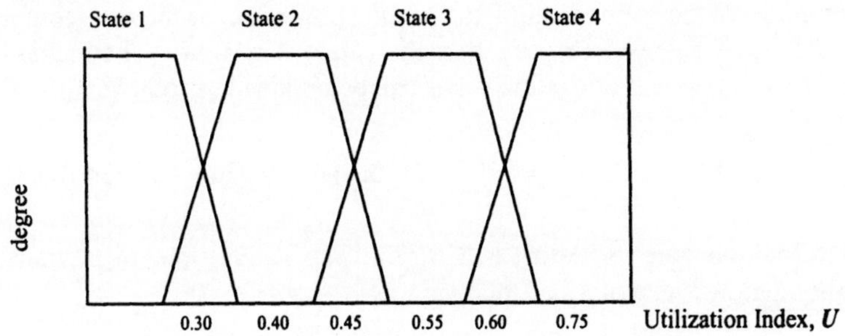

Fig. 2. Fuzzy Memberships function for different states

Based on the above membership function, the state transition of each utilization states can be obtained as follows in **Table 2**:

Table 2. Grade of membership for different states

Unit	U	S_1	S_2	S_3	S_4
1	0.465	0	0.85	0.15	0
2	0.659	0	0	0.61	0.39
3	0.619	0	0	0.87	0.13
.					
.					
20	0.748	0	0	0.01	0.99
21	0.589	0	0	1	0

According to **Table 2**, the next coming state transition is obtained from equation (7) as follows:

$$\tilde{MF}_1 = \sum_{l=1}^{20} \mu_{S_1}(x_l) = 0.55 \qquad \tilde{MF}_2 = 7.80 \quad \tilde{MF}_3 = 9.37 \quad \tilde{MF}_4 = 2.28$$

According to the equation(14), a transition probability matrix is obtained as follow:

$$\tilde{MF}_{ij} = \sum_{l=1}^{n-1} \mu_{S_i}(x_l) * \mu_{S_j}(x_{l+1}) \quad \text{and} \quad \tilde{P}_{ij} = \frac{\tilde{MF}_{ij}}{\tilde{MF}_i}, i,j = 1,2,...,k \tag{14}$$

$$\tilde{P} = \begin{bmatrix} 0.000 & 0.070 & 0.000 & 0 \\ 0.780 & 0.427 & 0.324 & 0.069 \\ 0.220 & 0.366 & 0.558 & 0.895 \\ 0.000 & 0.140 & 0.118 & 0.036 \end{bmatrix}$$

From the data of $x(21)$, the initial state probabilities vector is defined as $p(21)=[\ 0\ 0\ 1\ 0]$. Therefore the next state of the utilization is obtained by the following calculation:

$$P(x_{nj}) * P = [0.000 \quad 0.324 \quad 0.558 \quad 0.118\ ]^T$$

According to the membership function, the prediction of the next coming 300 seconds(state of x_{23}) web server's utilization state is 3(High). The higher order, says the $P(23)$ web server utilization states can be obtained as follows:

$$\tilde{P}(23) = p(0)\tilde{P}^{22} = [\ 0.0239 \quad 0.3411 \quad 0.5216 \quad 0.1134\ ]$$

Therefore the state for the x_{23} is $3(High)$. The steady state of the transition probability matrix is examines as follows:

$$\hat{p} = \tilde{P}\hat{p} \tag{15}$$

where $\hat{p}$ is the stationary distribution. By solving equation (15), the stationary distribution p is obtained:

$$\hat{p} = [p_1 \quad p_2 \quad p_3 \quad p_4] = [\ 0.0239 \quad 0.3411 \quad 0.5216 \quad 0.1134\ ]$$

5   Second approach: fuzzy Markov chain

Memoryless is one of the major unique property of Markov Chain. However, the prediction of the chosen data set seems not to fully follow the memoryless property[18]. Based on the states transition in **table 2**, the degree of visitation back to the same state(says $Si \rightarrow Si$) seems to be smaller if knowing that previously the initial state is *Si*. The longer duration of visitation in the same state will observe a lower possibility to remain in the same state.

Therefore, the memoryless property of Markov Chain is unable to capture some of the transition states or events that's involve the states before *n* are also affected the state transition from time $n \rightarrow n+1$. The state transition of time $n \rightarrow n+1$ is 'nearly' not related to the states before *n*. The term 'nearly' means the further the states before time *n*, the influence of the state transition to time $n+1$ is less.

5.1 Fuzzy markovian property

For the sake of simplicity, only the state for time *n-1* is consider that will influence the state transition from time $n \rightarrow n+1$ and assume the states before *n-1* are memoryless. Let

$$P_{k,ij} = P(X_{n+1} = k \mid X_n = j, X_{n-1} = i)$$
$$\cong P(X_i \rightarrow X_j \rightarrow X_k) \tag{16}$$

Xi(n) represents the state, X_i for time *n*. If the transition process is 'nearly' memoryless, the $\Delta P_{k,ij}$ is defined as:

$$\Delta P_{k,ij} \cong P(X_{n+1} = k \mid X_n = j, X_{n-1} = i) - P(X_{n+1} = j \mid X_n = i)$$
$$= P(X_i \rightarrow X_j \rightarrow X_k) - P(X_i \rightarrow X_j) = P_{k,ij} - P_{ij} \tag{17}$$

It is clear that if the transition process is fully memoryless, the $\Delta P_{k,ij}$ is equivalent to zero. Based on the $\Delta P_{k,ij}$ the fuzzy Markov process property memoryless level can be determined. The memoryless level property fuzzy sets, *A*, and membership function $\mu_{A(k, i, j)}$ must fulfil the following requirements:

- if $\Delta P_{k,ij} = 0$, indicates that the process is fully memoryless, with $\mu_{A(k, ij)} = 1$.
- If $\Delta P_{k,ij} \neq 0$, and the greater the $\Delta P_{k,ij}$, the smaller the transition process's memoryless. Hence $\mu_{A(k, i, j)}$ will become smaller.
- If given *i, k*, when $P(E_k \mid E_j, E_i) = 1$ or 0 and $P(E_j \mid E_i) = 1/i$ (where *i* is the number of states), therefore the transition process's are fully non-memoryless and $\mu_{A(k, ij)} = 0$.

Based on the three principles, the grade of membership function can be obtained as follows:

$$\mu_{A(k,i,\forall j)} = \begin{cases} 0, & if \ \dfrac{k}{k+1}\sum_{j=1}^{k}(\Delta P_{kij})^2 \geq 1 \\ \dfrac{k}{k+1}\sum_{j=1}^{k}(\Delta P_{kij})^2, & others \end{cases} \tag{18}$$

Assume $\underline{P_{jk}}$ is the fuzzy transition possibility. For time-*n* and time-*n+1*, the states are *Xi* and *Xj* respectively. Obviously, the process with the higher level of memoryless will yield $\underline{P_{jk}} \cong P_{jk}$, else $\underline{P_{jk}} \cong \underline{P_{kij}}$. Assume for time *n-1*, the state is define as X_{k0}, therefore:

$$P_{jk} = \mu_A(i_o, j, \forall k) * P_{jk} + [1 - \mu_A(i_o, j, \forall k)]P_{k_0, ij} \quad for\ \mu_A(i_o, j, \forall k) \approx \mu_{i_0, j}$$

$$
\begin{aligned}
P_{jk} &= \mu_{i_0, j} \bullet P_{jk} + (1 - \mu_{i_0, j})P_{k_0, ij} \\
&= P_{jk} + (1 - \mu_{i_0, j})(P_{k_0, ij} - P_{jk}) = P_{jk} + (1 - \mu_{i_0, j})\Delta P_{k_0, ij}
\end{aligned}
\tag{19}
$$

5.3 Numerical illustration

Based on the data of Table 1, the Web server utilisation states are categorised into four interval such as: *Low*=[0, 0.35], *Moderate*=[0.35, 0.50], *High*=[0.50, 0.66] and *Extremely High*=[0.66, 0.75]. The states transitions of time *n-1*, *n*, *n+1* are show as follows in **Table 3**:

Table 3. Web server utilization state transition

X	U	State	State K→I	State K→I→j
1	0.465	2	2→3	2→3→3
2	0.659	3	3→3	3→3→2
3	0.619	3	3→2	3→2→2
.			.	
.			.	
.			.	
19	0.502	3	3→4	3→4→3
20	0.748	4	4→3	
21	0.589	3		

The prediction of *x(22)* can be obtained by the following procedure:

Determine the probabilities of the transition from $i \rightarrow j \rightarrow k$, where *i*=4, *j*=3 and *k*=1,2,3,4, where P_{31}=0, P_{32}=2/10, P_{33}=6/10, P_{34}=2/10, P_{431}=0, P_{432}=0, P_{433}=1 and P_{434}=0. Based on the equation (19), μ_{43}=0.7845. Therefore, the fuzzy transition probabilities are: P_{31}= P_{31} + (1- μ_{43})ΔP_{431} = 0, P_{32}=0.1569, **P_{33}=0.6862** and P_{34}=0.1569. Due to the highest degree of membership value, the predicted following utilization state is *High*.

6 Comparison with non-fuzzy markov chain

According to the same analysed data with the interval value: *Low*=[0, 0.35], *Moderate*=[0.35, 0.50], *High*=[0.50, 0.66] and *Extremely High*=[0.66, 0.75]. The transition probability matrix is obtained as follow:

$$
P = \begin{bmatrix}
0 & 1 & 0 & 0 \\
1/7 & 3/7 & 3/7 & 0 \\
0 & 1/5 & 3/5 & 1/5 \\
0 & 0 & 1 & 0
\end{bmatrix}
$$

The state of x_{22} is obtained as $P(22)*P=[\ 0\ \ 1/5\ \ 3/5\ \ 1/5]$. From the above matrix, the x_{22} is in the state High. For the higher state, says the x_{23}, the states determination is obtained as follows:

$$P(23) = p(0) \bullet P^{22} = [\ 0.0385\ \ 0.2692\ \ 0.5769\ \ 0.1154\ ]$$

The highest probability is shows in the 3rd element of the vector. Hence the state of x_{23}, is consider as *High*.

The prediction using non-fuzzy markov chain shows some limitation comparing to the fuzzy approaches as show in the previous sessions as stated below:

- The non-fuzzy markov chain is mainly used in nearly crisp environment.
- The non-fuzzy markov chain is not able to capture the uncertainty of the states.
- Some of the data is not thoroughly follow the memoryless property of markov chain, therefore it is more appropriate include the fuzzy concept in the markov chain.

7 Conclusion

In this paper, a fuzzy inference system is established to derive server utilization index based on two workload parameters. Fuzzy Markovian model has been used to predict the possibility of the server's resource utilization states after some transition periods and its steady states are also derived. Further, the fuzzy prediction also includes the situations where the consequences states transitions are not fully independent. These fuzzy Markov models presents another alternatives which will provide a useful reference especially for future WWW accessibility and planning.

References

[1] Michele Colajanni, Philip S. Yu and Valeria Cardellini, " Dynamic Load Balancing in Geographically Distributed Heterogeneous Web Servers", Proceeding of 18th International Conference on Distributed Computing Systems, 1998 , Page(s): 295 –302

[2] Michele Colajanni, Philip S. Yu, "Scheduling Algorithms for Distributed Web Servers", Proc. ICDCS'97, Baltimore, MD, May 1997, pp. 169-176.

[3] Crovella M., and Bestavros. A., "Explaining World Wide Web Traffic SelfSimilarity", Tech. Rep.BUCS-TR-95F-015, Boston University, CD Dept, Boston MA 02215, 1995.

[4] W. Leland and M.Taqqu, "On the Self-Similar Nature of Ethernet Traffic", In Proceedings of SIGCOMM'93, 1993.

[5] Chiraphadhanakul, S., Dangprasert, P. and Avatchanakorn, V.,"Genetic Forecasting Algorithm with Financial Applications", Intelligent Information Systems, 1997. ISS '97. Proceedings, 1997 , Page(s): 174 –178, 1997.

[6] L.Feng and X.X.Guang, "A Forecasting Model of Fuzzy Self Regression", Fuzzy Sets and Systems, 38, 239-242, 1993.

[7] J.Nie, "A Fuzzy-Neural Approach To Time-Series Prediction", in Proceeding of IEEE International Conference on Neural Network (Piscataway, NJ, IEEE Service Center, 1994), pp.3164-3169, 1994.

[8] L.X.Wang and J.M.Mendel, "Generating Fuzzy Rules By Learning From Example", IEEE Transaction Systems, Man and Cybernetics, 22, 1414-1427,1992.

[9] Chin Wen Cheong, Chong Chee Way & V. Ramachandran, "Web Server Workload Forecasting in Fuzzy Environment-Linguistic Approach", *Proceeding of IEEE TENCON: Intelligent Systems and Technologies for the New Millennium, Vol.1, pp.* 361-366.

[10] Arun Iyengar, et.al. "An analysis of Web Server Performance", Proceedings of the IEEE 1997 Global Telecommunications Conference (GLOBECOM '97), Phoeniz, AZ, November 1997.

[11] M. Arlitt & C. Williamson, "Web server workload characterisation: the search for invariants", Proc. 1996 SIGMETRICS Conf. Measurement Comput. Syst.,ACM, Philadelphia, PA, May 1996, vol.37, no. 8, pp. 76-82.

[12] Daniel A. Menasce, "Capacity Planning for Web Performance: Metrics, Models & Methods", Prenctice Hall, Inc., 1998.

[13] G.Banga & P.Druschel, "measuring the Capacity of a Web Server", Usenix Symp. Internet Technol. Syst., Monterey, CA, Dec. 1997.

[14] URL http://ita.ee.lbl.gov/html/contrib/ClarkNet-HTTP.html

[15] Timothy J.Ross, Fuzzy Logic With Engineering Applications. McGraw-Hill International Edition, 1995.

[16] I.B. Turksen. Measurement of membership functions and their acquisition. *Fuzzy Sets and Systems*, 40:5--38, 1991.

[17] H. Takagi and I. Hayashi. NN-Driven Fuzzy Reasoning. *Int. J. Approximate Reasoning*, 5:191--212, 1991.

[18] C.S. Quan, et.al., "Fuzzy Prediction", Kui Chow Technology pub.1991.

Part VI

Theoretical Advances and New Paradigms

Efficient and Precise Handling of a Piecewise-Linear Fuzzy Set on Computers in an Extended Knot Form

Ayumi Yoshikawa

Okayama University, Faculty of Education, 3-1-1 Tsushima, Naka, Okayama 700-8530, JAPAN

Abstract. In this paper, an eXtended Knot Form, XKF, is introduced to handle a piecewise-linear fuzzy set, PLFS, on computers efficiently and precisely. The XKF consists of an element and three membership grades at the element and both neighbors. The redundancy of the membership grades improves the efficiency and precision of handling of PLFSs. It is also shown that the XKF is implemented with an object-oriented program language, the JAVA, and a doubly linked list.

1   Introduction

Varieties of Fuzzy Sets in Form. Many kinds of fuzzy sets in shape are introduced and used in fuzzy-set based applications. Therefore, we need to express and handle these fuzzy sets on computers. To implement them, several expressing forms have already been proposed, e.g., pointwise, vector representation, alpha level sets, function, paired representation of an element and its membership grade and so on [1]. Also, the methods for handling these forms on computers have already been proposed [1-5].

Importance of a Piecewise-Linear Fuzzy Set in KANSEI Information Processing. In this paper, I will deal with a piecewise-linear fuzzy set, PLFS, such as a trapezoidal or triangular fuzzy set, in various fuzzy sets. The reasons are as follows; The PLFS is widely used in many fields and real applications, e.g., control systems, inference systems and evaluation. Especially, in KANSEI information processing, the PLFS are used to describe KANSEI information, such as meanings of verbal hedges and subjective degree judged by human. Therefore, we need to handle the PLFS on computers adequately.

Notice of Handling of PLFSs. To describe and manipulate a PLFS precisely need to mark up all break points, which connect two neighboring linear segments. These points are called "knots" hereafter. However, some knots cannot be represented adequately in ordinary forms, e.g., the vector form, as discussed in section 2. Therefore, the adequate handling of all knots requires an improved form. To fulfill the requirement, Yoshikawa has already proposed the knot form, pairs of an element and its membership grade to express a membership function, as described in section 2 [6]. Hereafter, we call the knot form the ordinary knot form, OKF, to separate from a new knot form proposed in this paper.

Problems of the OKF. The OKF can represent a stepwise transition of membership grades at an element, i.e., a crisp transition, using triplex knots precisely. However, the triplex knots differ from an ordinary knot in their handling methodology. Therefore, the difference makes the algorithm for processing the OKF more complicate.

Aims of This Work. This paper aims to propose an improved expressing form, an eXtended Knot Form, XKF, which extends the OKF, to cancel and improve the issue related to the triplex knots. In the XKF, every knot has three membership grades per the element despite the difference of an ordinary knot and a crisp transition. As the result, the redundancy of the membership grades can not only cancel the triplex knots, but also integrate several other forms proposed already into the XKF. Also, we show an implementation of the XKF on computers with the JAVA language. The data structures and characteristics are described.

2 Characteristics of a PLFS and Problems of Some Ordinary Expressing Forms

Characteristics of a PLFS. A PLFS is a fuzzy set that is expressed in connections of linear segments. The result calculated between PLFSs becomes a PLFS, except for operations with product of their membership grades, such as the algebraic product and sum. Accordingly, we can calculate only membership grades at break points, if the all break points correspond between the two PLFSs.

Essential Attributes of Handling of a PLFS. The following items are essential to express and handle a PLFS on computers adequately.

- ☐ All break points that are necessary to express the shape are included.
- ☐ All break points that are necessary to calculate are added, if they do not exist before the calculation.
- ☐ All redundant break points are removed.

Problems of the Ordinary Forms. The ordinary forms have problems on handling a PLFS more or less. The vector form is a method for expressing a fuzzy set using only a series of membership grades that correspond to elements sampled at an equal interval. It is one of the most major expressing forms, and is widely used in many applications. However, it has the following problems.

- ☐ Some break points cannot be expressed by given sampled elements.
- ☐ Also, some intersections that an operation requires cannot be expressed by the given sampled elements.
- ☐ The form have some redundant sampled elements that are interpolated linearly.
- ☐ The form never expresses a crisp transition in a singleton and a step function.

The last problem is found not only in the vector form, but also in the pointwise form and the paired representation, then it is one of the most critical issues.

Expressing a PLFS in the OKF. Yoshikawa has already proposed the OKF, which is developed from the pointwise form, to improve the above problems [6]. The OKF expresses a PLFS using a series of break points, i.e., knots, and interpolates any membership grades between the knots linearly. Therefore, it expresses an ordinary break point as a single knot. Meanwhile, it expresses a crisp transition of membership grades as triplex knots at the element and both neighbors of the element, as shown in Fig. 1. The triplex-knots form lets us express the crisp transition precisely. However, since the triplex-knots form makes a judgment of a redundant knot complex, a new problem of complexity of the algorithm emerges. Table 1 shows the algorithm for checking the kinds of knots.

Also, similar idea to express a crisp transition was proposed in the reference of [3]. However, it is more complex than the OKF, since it requires third parameter to show whether each point is a crisp transition or not.

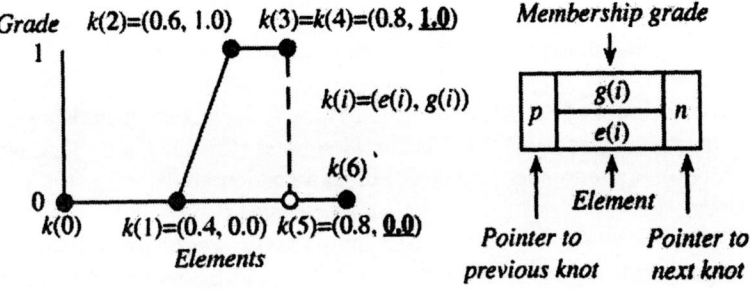

Fig. 1. Representation of a trapezoidal fuzzy set in the OKF

Table 1. The algorithm for classifying kinds of knots

Equality of grades	No. of knots	Judgment
Not equal	1	Ordinary knot
	2	Error
	3	Crisp transition
	≥ 4	Depending upon circumstances
Equal	2	Redundant knots =single ordinary knot
	3	
	≥ 4	

3   An Extended Knot Form: XKF

Expressing a PLFS in the XKF. I propose an improved form, named an eXtended Knot Form, XKF, to cancel the problem related to the triplex knots as shown above. A unit of the XKF consists of three membership grades, $g(i)$, $gl(i)$, $gr(i)$ for each element $e(i)$; $g(i)$ is the grade at $e(i)$, as same as in the OKF, and $gl(i)$ and $gr(i)$ are the grades at a left and right neighbor of $e(i)$ respectively. Therefore, a knot in the XKF is expressed as $k(i) = (e(i), gl(i) : g(i) : gr(i))$ as shown in Fig. 2. For examples, an ordinary knot has three equal membership grades, such as $(0.4, 0.0 : 0.0 : 0.0)$ at $k(1)$. On the other hand, a crisp transition has different membership grades, such as $(0.8, 1.0 : 1.0 : 0.0)$ at $k(3)$. Any membership grades between two knots, $k(i)$ and $k(i+1)$, are interpolated linearly using $gr(i)$ and $gl(i+1)$.

572

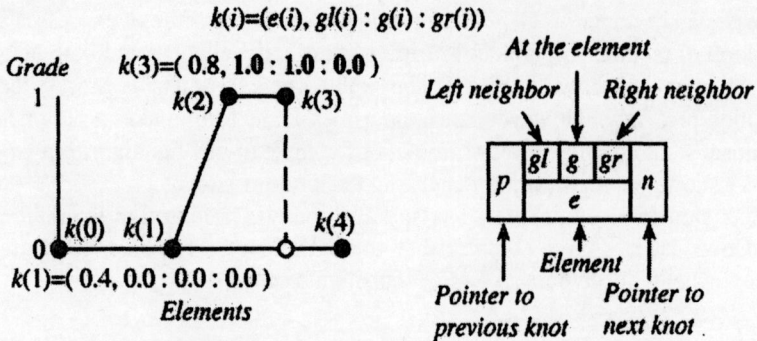

Fig. 2. Representation of a trapezoidal fuzzy set in the XKF

Characteristics of the XKF. A XKF has the following characteristics.
- The algorithm for the XKF becomes simpler than the OKF. The reason is why both an ordinary knot and a crisp transition are expressed in one same form. In other words, we never need the triplex knots.
- It can express a singleton as only single knot. Therefore, we can integrate other major forms, such as the vector form and the paired representation of elements and membership grades, into the XKF.
- It can handle a discrete fuzzy set whose elements are defined on the nominal scale. For an example, we can represent information that tallness of Taro is 0.8 as ("$Taro$", 0.0 : 0.8 : 0.0).
- It can express a type-2 fuzzy set as ($e(i)$, 0.0 : $m_i(t)$: 0.0) without no change. Where $m_i(t)$ is a fuzzy set on $T = [0, 1]$ and t means a membership grade of a type-1 fuzzy.

Problems of the XKF. The only problem of the XKF is that data per a PLFS increase slightly than the OKF due to the redundancy of expressing membership grades. Someone claims that the XKF can only approximately compute values of the algebraic product and sum. However, notice that this issue is not inherent in the XKF. Since the vector form or the paired representation can compute only the membership grades at the discrete elements, we do not notice the issue, but it exists indeed. Therefore, there is no essential difference among these forms.

A Flow of Computation. A flow of operation of PLFSs in the XKF is same as the OKF as shown in Fig. 3. Each step in the flow is as follows:
- **Step 1.** Check whether all knots of two PLFSs correspond to each other. If it is not satisfied, the redundant knots are added to satisfy.
- **Step 2.** Check whether all intersections between the two PLFSs are expressed in the corresponding knots. If it is not satisfied, the redundant knots are added to satisfy.
- **Step 3.** Execute the given operation between the membership grades over all knots.
- **Step 4.** Remove redundant knots from the result, if the knots exist after the computation.

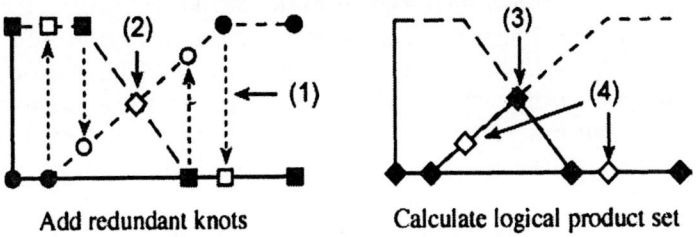

Open: redundant knots, filled: break points

Add redundant knots | Calculate logical product set

Fig. 3. An example of a flow of computation using in the XKF

In case of the algebraic product and sum, redundant knots generated in equation (1) are added in the step 1 of the above flow to approximate a parabola with broken lines.

$$dx = 2\sqrt{err/|a|}$$

(1)

Where dx, err, and a mean an interval of the added knots, a limit of the approximation error, and a second-order coefficient of the parabola respectively.

4 An Implementation of a XKF with Object-Oriented Programming

A FuzzyKnot Class. A PLFS is closed to fuzzy-set operations except for the algebraic product and sum. So we can obtain all membership grades on a universal set, calculating only the membership grades of the knots with the given operators. Accordingly, we define a FuzzyKnot, FK, class that has an element and three membership grades as private instance variables of type double, and the basic operators, such as the logical sum and product and so on, as its instance or static methods.

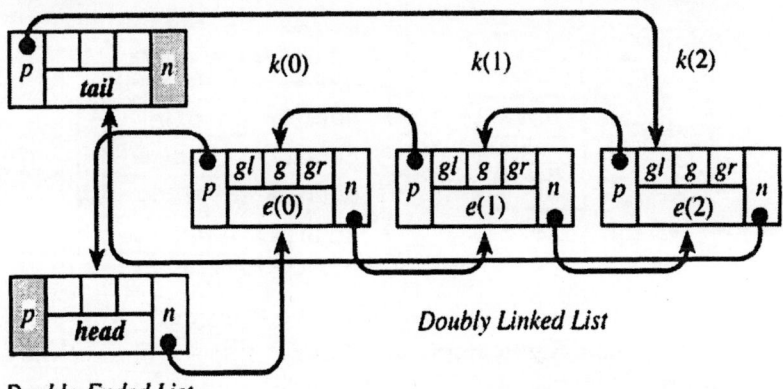

Fig. 4. Representation and manipulation of knots in the XKF on computers

Implementation of a PLFS with a Doubly Linked List. When we process PLFSs in the XKF, many redundant knots are inserted and deleted. Therefore, to

574

insert and delete those knots on memories in a computer physically requires a large amount of sort and shift operations of the knots, then it lowers the performance on the calculation. To improve it, we use a doubly linked list, DLL, despite the physical shift and sort operations. A node of the DLL, a basic unit, has two link pointers to indicate the previous and next nodes, p and n, as shown in Fig. 4. In the DLL, we can easily and simply execute to insert and delete any nodes, redirecting the link pointers. Also, we can go forward and backward through all nodes in the list, following the pointers.

Definie ordinary doubly linked list

```
┌─────────────────────┐        ┌──────────┐
│  DoublyLinkedList    │◇──────●│   Link   │
└─────────────────────┘        └──────────┘
         │ Inherit
- - - - -△- - - - - - - - - - - - - - - - ┐
┌─────────────────────┐        ┌──────────────┐
│ FuzzyDoublyLinkedList│◇──────●│  FuzzyKnot   │
└─────────────────────┘        └──────────────┘
```

Express PLFSs and | Basic structure of XKF,
   operations between knots | operations betw. grades

Class available for user directly | *Define structure of XKF*

Fig. 5. Relationship among the four classes

In this paper, we define a FuzzyDoublyLinkedList, FDLL, class, as a subclass of the DLL class. Figure 5 shows relationships among these classes in notation of the Objective Modeling Technique, OMT [7]. Then, we assign one knot of a PLFS to one node in the FDLL class. In other words, an instance variable of type Object in the FDLL class refers to an object of type FK in the FK class as shown in Fig. 6. For examples, in the FDLL class, a node assigned to a knot is constructed and the method for calculating a gravity center of a PLFS is defined.

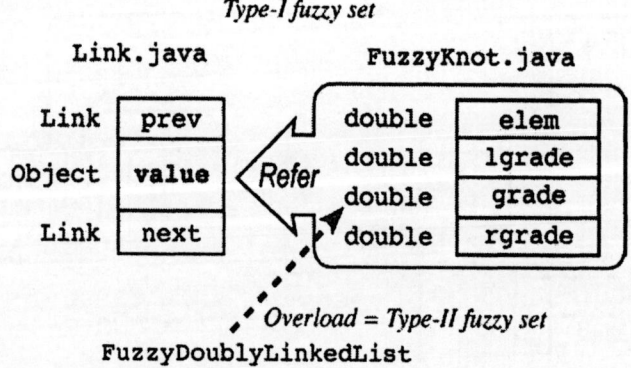

FuzzyDoublyLinkedList

Fig. 6. Relationship between a variable of type Object in Link class and an object of type FK in FK class

Advantages of Implementing the XKF with the JAVA language. We implement the XKF into a computer with the JAVA language, one of the object-oriented programming languages. Its advantages are as follows;

☐ In the JAVA, a garbage collection is automatically executed by a JAVA virtual machine, JVM. Therefore, a user never needs to free the removed knots explicitly.

☐ We can easily expand the XKF into a type-2 fuzzy set, which is defined on memberships grades of [0, 1], overloading the constructor in the FK class, as shown in Fig. 6. Also, we can expand it into a discrete fuzzy set defined on a discrete universal set in similar ways.

☐ We can use programs written in the XKF on Web pages through a WWW browser. Moreover, it does not depend on the OS of user computers.

5 Conclusion

The achievements of this work. In this paper, I have proposed a new expressing form of a PLFS, a XKF, which extends an ordinary knot form. It cancels the problem of the triplex knots for a crisp transition. Moreover, the XKF can combine several major ordinary forms, such as the vector form, the paired representation and so on, into it. Also, the capability of expressing a singleton as single knot can handle a discrete fuzzy set on the nominal universal set, such as "Taro's tallness." Then I have shown an example of implementation of the XKF on computers with a doubly linked list and the JAVA language. Therefore, we can use programs for handling PLFSs through internet browsers, writing them in the XKF.

Acknowledgement

This work is supported by Grant-in-Aid for Encouragement of Young Scientist (13750392) in Japan Society for the Promotion of Science.

Reference

1. Yamamoto, S. (1994) Software representation of fuzzy sets and logic, Tzafestas et al. (eds.) Fuzzy Reasoning in Information, Decision and Control Systems, 51-68, Kluwer Academic Publishers
2. Umano, M., et al. (1978) FSTDS System: A Fuzzy-Set Manipulation System, Information Sciences, 14, 115-159
3. Inoue, Y., et al. (1990) Fuzzy sets representation based on object-oriented paradigm (in Japanese), Proceedings of 6th Fuzzy System Symposium, 493-496

4. Fujii, T., et al. (1993) Development of practical Fuzzy C Language (in Japanese), Proceedings of 9th Fuzzy System Symposium, 533-536
5. Umano, M., et al. (1993) Data Structures and Manipulation for Fuzzy Sets on Digital Computers (in Japanese), Proceedings of 9th Fuzzy System Symposium, 77-80
6. Yoshikawa, A., et al. (1993) Improvement of Method for Computing Set Operations on a Piecewise-Linear Fuzzy Set (in Japanese), Proceedings of 9th Fuzzy System Symposium, 705-708
7. Rumbaugh, J., et al. (1991) Object-Oriented Modeling and Design, Prentice Hall

Kansei Factor Analysis Using C4.5

Kaori YOSHIDA

Kyushu Institute of Technology
680-4 Kawazu, Iizuka city, Fukuoka 820-8502, JAPAN

Abstract. Recently people come to demand more "personalized" something. We call our taste, subjectivity, such like source of individuality as "Kansei". It's necessary to summarize the needs on Kansei database system and to be clear what Kansei factor is for constructing Kansei database system reasonably. In this paper, I summarize the needs and design factors of database for Kansei information processing. Especially I focus on core data, that is database contents data, and carry out Kansei factor analysis by constructing classifiers. The result of analysis using C4.5 gave us easy understanding of users' personal view model.

1 Introduction

Conventional database systems have focused on simple structuring and normalization to deal with large bodies of information, and do not deal well with the complexities of structures needed to represent user's subjectivity. We should to using concepts from database research to expand database system for Kansei information processing. In this paper, I summarize the needs and design factors of database for Kansei information processing. Especially I focus on core data, that is database contents data, and carry out Kansei factor analysis by constructing classifiers. Because if Kansei factor would be clear, we can design more reasonable Kansei database.

2 Kansei Database System

Considering a database system for Kansei information processing, it's needed possibility of expansion especially. Here, contributed data is restricted only images. The needs of Kansei database system are summarized as follows.

- *Data Independency:* Revising data and revising program should be available independently to materialize each user's subjective requests.
- *Integrity, Consistency and Non-redundancy:* Processing and reducing the data so that the number of alternative choices to be decided among is small, and the parameters for each choice are aggregated to a high conceptual level.
- *Security and Privacy:* Each user's profile should be protected and secured from other user's access.

- *Data Sharing and Concurrent Access:* Accessing a data may happen at a same time, thus the system has to provide the data to any applications concurrently.

And also the design factors of such database system are summarized as follows.

- *Real Time Accessibility:* Mechanisms for Kansei information processing need to infer user's internal state from observed data, but such kind of data usually has various cognitive factors and irregular noises. Thus the system is required real time accessibility on variable data.
- *Adaptability:* Kansei information involves user diversity. It's important that the system has a high adaptability to satisfy user's individual requests.
- *Data Migration:* It is desired that the system can provide the user model by easy representation to users. For example, the e-mail filtering system "InfoLens [1]" and the service of personal news group auto-selected from network news groups "InfoScope [2]" represent user's profile by template style. It's easy to understand for users on the point that user can make and edit their profiles.
- *Search Capability:* Multiple search strategies and user's subjective requests are sometimes very complicated. It's required robust and conform response on the system.

This area concerns research to improve the techniques of database design and modeling. The structural model captures those semantics that are of importance in the design of the database's physical structure [3][4]. The concepts of the structural model have been expanded and applied in other research aimed at facilitating various stages of the database design process. They provide the basis for the data definition facility of critias, which implements a schema language based on the structural model [5]. They are used in the task of integrating diverse user views, whether in a centralized or distributed setting. And finally, they are used to aid the design of physical databases from the integrated user model. Further avenues of design-oriented investigation have been distributed database design and the design of database management systems.

3   Kansei Factor Analysis

There are well-known techniques for Kansei factor analysis such multi variable analysis like canonical correlation analysis, factor analysis, principal component analysis, and so on. Otherwise, learning mechanisms like neural networks and genetic algorithm are useful. But these techniques are not enough to explain the learning process represented by black box. Therefore I have used Quinlan's C4.5 [6] to build the classifiers in this work. Also, we need suitable color system that can treat Kansei information. I have used hue and tone of color as essential data for analysis.

3.1   C4.5

C4.5 is an algorithm introduced by Quinlan for inducing "classification models" also called "decision trees" from data. C4.5 starts with large sets of cases belonging to known classes. The cases, described by any mixture of nominal and numeric properties, are scrutinized for patterns that allow the classes to be reliably discriminated. These patterns are then expressed as models, in the form of decision trees or sets of if-then rules, which can be used to classify new cases, with emphasis on making the models understandable as well as accurate. The system has been applied successfully to tasks involving tens of thousands of cases described by hundreds of properties. We are given a set of records and each record has the same structure, consisting of a number of attribute/value pairs. One of these attributes represents the category of the record.

3.2   Color System

There are several color systems all over the world. In this paper, I focused on "Munsell color system [7][8]" and "Practical Color Co-ordinate System (PCCS) [9][10] for arrange values on C4.5. Here, I introduce these color systems briefly. Munsell color system is recognized as a standard system of color specification in standard Z138.2 of the American National Standards Institute, Japanese Industrial Standard for Color JIS Z 8721 [11], the German Standard Color System, DIN 6164 and several British national standards. The Munsell color-order system has been widely used in many fields of color science, most notably as a model of uniformity for colorimetric spaces and has, itself, been the subject of many scientific studies. PCCS is color system made by public in 1964 by Japan Color Research Institute, it is developed primarily to provide a systematic method for resolving color harmony problems. PCCS is constructed using the three attributes of color as a framework, and is characterized by its capacity to be used as a hue/tone, two-dimensional system.

3.3   Hue & Tone

At first, HSV (Hue, Saturation and Value) should be shown. HSV is a method of describing any color as a triplet of real values. Hue represents the color or wavelength of the color. It is sometimes called tone and is what most people think of as color. Hue is taken from the standard color wheel and is thus calibrated in degrees about the wheel. Saturation is the depth of the color. It states how gray the color is. It is a real valued parameter from 0.0 to 1.0 with 0.0 indicating full gray and 1.0 representing pure hue. The value is how black or white a color is. It also ranges from 0.0 to 1.0 but with 0.0 representing black and 1.0 white. A value of 0.5 is pure hue.

Hue is describes the distinct characteristic of color that distinguishes red from yellow from blue. These hues are largely dependent on the dominant wavelength of light that is emitted or reflected from an object. According to Japanese Industrial Standard (JIS) system, ten hue names are fixed as follows; red (R), yellow-red

(YR), yellow (Y), green-yellow (GY), green (G), blue-green (BG), blue (B), purple-blue (PB), purple (P), red-purple (RP). I used these 10 hues in the analysis.

Tone is calculated from saturation and value. 5 tones for achromatic colors are chosen as follows; black (BK), dark grayish (DkGy), medium grayish (mGy), light grayish (ltGy), white (W). I put together these 5 tones to monochrome (MN) as an element of hues later. Additionally, 12 tones for chromatic colors are chosen as follows; dark grayish (dkg), grayish (g), light grayish (ltg), pale (p), dark (dk), dull (d), soft (sf), light (lt), deep (dp), strong (s) bright (b), vivid (v).

4 Results

For Kansei factor analysis, the users selected impression words with their degrees, like "very", "slightly", on 50 full colored images shown in Figure 1. There were 10 impression words as follows; "warm", "soft", "natural", "clear", "elegant", "chic", "authentic", "classic", "gorgeous" and "dynamic". I carry out C4.5 with the users' answer. I show the extracted rules by C4.5 from Table 1 to Table 20. We can use these rules in SQL request directly if they were extracted reasonably. Note that "-" in the tables means C4.5 could not extract proper rules.

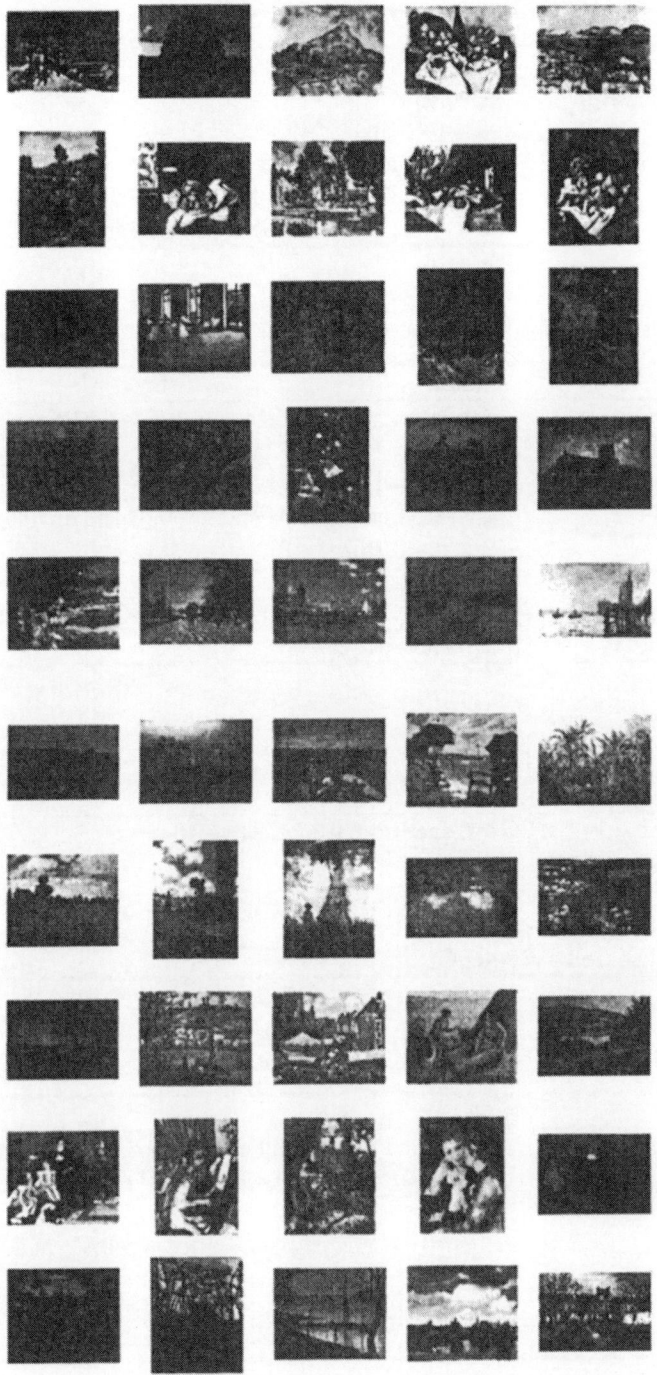

Figure 1: 50 full colored images for Kansei factor analysis

Table 1: Rules derived from values of hue on the word "warm"

	Rules
very warm	-
slightly warm	R>0.65
neutral	R<=0.65 and G<=0.15 and RP=0
not warm	(R<=0.2 and Y<=0.08 and B>0.08) or G>0.15
not warm entirely	(YR<=0.41 and Y>0.04 and B<=0.08 and PB=0 and RP>0) or (R>0.2 and B>0.08) or (Y>0.08 and B>0.08)

Table 2: Rules derived from values of tone on the word "warm"

	Rules
very warm	(g>0.09 and ltg<=0.26 and sf <=0.02) or (p<=0.11 and lt>0.03 and s<=0.26 and b>0.04)
slightly warm	g<0.09 and dp<=0.04
neutral	(p<=0.11 and sf>0.04 and lt<=0.03 and dp>0.04) or (g>0.09 and ltg>0.26)
not warm	g>0.09 and sf>0.02
not warm entirely	(g>0.01 and g<=0.09 and sf<=0.04 and lt<=0.03 and dp>0.04) or s>0.26

Table 3: Rules derived from values of hue on the word "soft"

	Rules
very soft	BG<=0 and RP>0.01 and RP<=0.04
slightly soft	-
neutral	YR>0.16 and YR<=0.48 and BG<=0.19 and RP<=0.01
not soft	-
not soft entirely	RP>0.04

Table 4: Rules derived from values of tone on the word "soft"

	Rules
very soft	(g>0.01 and p>0.01 and d<=0.12 and sf<=0.06 and v<=0.15) or (ltg>0.01 and sf>0.14 and b>0.01)
slightly soft	-
neutral	(dkg<=0.01 and ltg<=0.01) or (g>0.01 and d>0.12 and sf<0.14)
not soft	sf>0.14 and b<=0.01
not soft entirely	

Table 5: Rules derived from values of hue on the word "natural"

	Rules
very natural	(G>0.02 and G<=0.07 and B<=0.05) or (MN>0.01)
slightly natural	(YR<=0.03) or (G>0.07 and B<=0.05)
Neutral	-
not natural	-
not natural entirely	Y>0.02 and B>0.05 and RP>0.01

Table 6: Rules derived from values of tone on the word "natural"

	Rules
very natural	v<=0.18
slightly natural	v>0.18
Neutral	dkg>0.04 and s<=0.07
not natural	-
not natural entirely	dk<=0 and s>0.1 and v<=0.18

Table 7: Rules derived from values of hue on the word "clear"

	Rules
very clear	(Y>0.2 and BG>0.06) or (BG>0.06 and PB>0.02)
slightly clear	Y<=0.2 and G<=0.07 and BG>0.06 and PB<=0.02
neutral	BG<=0.06
not clear	-
not clear entirely	(YR<=0.45 and BG<=0.06 and B>0 and P<=0.01) or (R>0.8 and R<=0.84)

Table 8: Rules derived from values of tone on the word "clear"

	Rules
very clear	b<=0.02 and v>0.01 and v<=0.05
slightly clear	lt<=0.01 and v<=0.01
neutral	(lt>0.08 and b>0.02 and b<=0.11) or (sf<=0.04 and b>0.02) or (dkg>0.05 and v>0.05)
not clear	b>0.11
not clear entirely	-

Table 9: Rules derived from values of hue on the word "elegant"

	Rules
very elegant	BG>0.06 and P>0.01
slightly elegant	-
neutral	R<=0.8 and YR<=0.48 and B<=0
not elegant	-
not elegant entirely	-

Table 10: Rules derived from values of tone on the word "elegant"

	Rules
very elegant	s<=0.04
slightly elegant	-
neutral	(dkg<=0.05 and b<=0.03 and v>0.17) or (b>0.09 and v<=0.17)
not elegant	-
not elegant entirely	dk<=0 and s>0.04 and b<=0.09

Table 11: Rules derived from values of hue on the word "chic"

	rules
very chic	-
slightly chic	-
neutral	G>0.02 and RP<=0.01
not chic	-
not chic entirely	-

Table 12: Rules derived from values of tone on the word "chic"

	Rules
very chic	-
slightly chic	-
neutral	(b>0.05) or (dkg<=0.07 and d<=0.09) or (s<=0.03)
not chic	-
not chic entirely	-

Table 13: Rules derived from values of hue on the word "authentic"

	Rules
very authentic	-
slightly authentic	YR>0.41 and YR<=0.47 and B<=0.02
neutral	B>0.02
not authentic	-
not authentic entirely	-

Table 14: Rules derived from values of tone on the word "authentic"

	Rules
very authentic	dp<=0.01 and s>0.03
slightly authentic	dk>0.04 and dp>0.01 and b>0.03
neutral	(dk<=0.04 and dp>0.01 and b>0.03) or (ltg>0.01 and dk>0.03 and b<=0.03) or (sf<=0.02 and dp>0.01 and b<=0.03)
not authentic	-
not authentic entirely	-

Table 15: Rules derived from values of hue on the word "classic"

	Rules
very classic	-
slightly classic	(G>0.02 and BG<=0.02) or (MN>0.01) or (RP>0.03)
neutral	MN<=0.01 and RP<=0.03
not classic	-
not classic entirely	G<=0 and RP>0.04

Table 16: Rules derived from values of tone on the word "classic"

	Rules
very classic	-
slightly classic	(ltg<=0.18 and sf>0.13 and dp>0.02) or (b>0.16) or (d<=0.09 and dp>0.02 and s<=0.06)
neutral	(p<=0.06 and sf<=0.07 and s>0.06 and b<=0.16) or (p<=0.37 and dp<=0.02 and b<=0.16) or (ltg>0.18 and p<=0.37)
not classic	-
not classic entirely	-

Table 17: Rules derived from values of hue on the word "gorgeous"

	Rules
very gorgeous	-
slightly gorgeous	BG>0.01 and BG<=0.09 and RP>0.01
neutral	MN<=0 and YR<=0.54
not gorgeous	-
not gorgeous entirely	(BG>0.14 and BG<=0.19) or (R<=0.11 and YR>0.54)

Table 18: Rules derived from values of tone on the word "gorgeous"

	Rules
very gorgeous	-
slightly gorgeous	(p>0.19 and s>0.04) or (p>0 and dp>0 and s>0.11)
neutral	(dkg>0.02 and ltg>0.05) or (p<=0.19 and dp>0 and s>0.04 and s<=0.11)
not gorgeous	lt>0.12 and dp<=0
not gorgeous entirely	dkg<=0.02 and dp>0 and s<=0.04

Table 19: Rules derived from values of hue on the word "dynamic"

	Rules
very dynamic	-
slightly dynamic	MN>0
neutral	(MN<=0 and YR>0.37 and YR<=0.62 and G<=0.04) or (YR<=0.02) or (YR>0.12 and YR<=0.33 and Y<=0.24 and B<=0.03)
not dynamic	-
not dynamic entirely	-

Table 20: Rules derived from values of tone on the word "dunamic"

	Rules
very dynamic	sf>0.03 and s>0.19 and v>0.02
slightly dynamic	dk>0.19 and sf>0.03
neutral	(ltg<=0.18 and sf<=0.02 and v>0.02) or (ltg<=0.18 and dk<=0.19 and d<=0.16 and s<=0.19 and v>0.02) or (p<=0.02 and v<=0.02)
not dynamic	p>0.02 and dk<=0.02 and v<=0.02
not dynamic entirely	(sf>0.02 and sf<=0.03) or (ltg>0.18 and s<=0.04 and v>0.02) or (p>0.02 and dk>0.02 and v<=0.02)

5 Remarks

I summarized the needs and design factors of Kansei database. The conventional database system has not only knowledge base of image, user's personal view model and user's profile but also request logs and reference logs. They might be useful to analyze how to use personal view model according to the user's profile, what is the most reasonable graphical feature, how to adapt the user's subjective interpretation, but we could not see why or how the users have such their impression on each image. The result of analysis using C4.5 gave us easy understanding of users' personal view model. These results are very useful for database system for Kansei information. However, we have to note that C4.5 has the following weak points; precision rate is not so high relatively caused by simple calculation, it cannot correspond to complete duplicated data of same axis, it's doubt users' subjective criteria has always stability and reliability. In the future, I try to analyze another contents related Kansei information in same way and estimate the results for constructing appropriate Kansei database system.

Acknowledgment

I am grateful to Human Media Project of METI (Ministry of Economy, Trade and Industry, Japan) supported by NEDO (New Energy and Industrial Technology Development Organization, Japan) for use core samples. And I thank Mr. Y. Motomura of AIST (National Institute of Advanced Industrial Science and Technology, Japan) for helpful discussion on C4.5.

588

References

[1] Fischer G. and Stevens C., "Information Access in Complex, Poorly Structured Information Spaces," Proc. of CHI'91, pp. 63-70, 1991.

[2] Malone T., Grant K. and Turbak F. et al., "Intelligent Information Sharing Systems," Communications of the ACM, Vo.30, No.5, pp.390-402, 1987.

[3] G.G. Hendrix, "Expanding the Utility of Semantic Networks through Partitioning," SRI Artificial Intelligence Group Technical Note 105, June 1975.

[4] Gio Wiederhold, "Database Design," McGraw-Hill (in the Computer Science Series) Second edition, Jan.1983.

[5] Xiao Lei Qian and Gio Wiederhold, "Data Definition Facility of critias," Proceedings of the 4th International Conference on Entity-Relationship Approach, pp.46-55, Oct.1985.

[6] J. Ross Quinlan, "C4.5: Programs for Machine Learning (Morgan Kaufmann Series in Machine Learning)," Morgan Kaufmann Publishers, Inc., 1992.

[7] Munsell, A. H., "A Color Notation," Munsell Color Company, Baltimore, MD, 1923.

[8] Munsell, A. H., "Munsell Book of Colors", Munsell Color Company, Baltimore, MD, 1942.

[9] Japan Color Research Institute, "Practical Color Co-ordinate System," Japan Color Research Institute, 1964.

[10] Japan Color Research Institute, "PCCS Harmonic Color Chart 201-L," Japan Color Research Institute, 1999.

[11] Japanese Industrial Standards Committee, "JIS handbook 33: Colors (Colors Specification - Specification according to their three attributes)," Japanese Standard Association, 1999. (in Japanese)

An Active Search Algorithm Extending GA Based Path Planning for Mobile Robot Systems

Marcus Gemeinder and Michael Gerke

Fernuniversität Hagen, Faculty of Electrical Engineering, 58084 Hagen, Germany

Abstract. An active search algorithm is introduced extending an existing path planning software for mobile robots based on a Genetic Algorithm. This extension is executed within two different phases of the optimisation process. For each obstacle within the environment regarded a path circumventing it is computed in a preparation phase. Then, in the execution phase of the GA itself, the results of the preparation phase are used to find optimum paths. The mode of action within both phases is described in detail. Furthermore, the suitability of the approach is substantiated by an example.

1 Introduction

A research area with special importance to mobile robot systems is to devise suitable methods for planning optimum moving trajectories. Possible optimisation criteria could be path length, energy consumption, or time needed for movement.

An operational area with the crucial aspect being energy consumption are missions on other planets in our solar system. The little rover *Soujourner* has driven on the surface of Mars, and further similar missions will follow. To use the full potential of such vehicles, they should always move in an energy-saving manner. The typical task could be described as follows: for two arbitrary points in an area, find the drivable path between these points minimising the energy required.

There exist many approaches within the area of Computational Intelligence to solve the problem of finding collision-free trajectories in unknown environments (see [2], [3], [5], [6]). Most of them have in common that the aspect of energy consumption is considered only with respect to the length of the paths. Environments are modelled in a binary manner (obstacle/no obstacle), and further information about the texture of the ground is not taken into account.

In [1] a path planning software based on a Genetic Algorithm is presented, which considers changes in the texture of the ground. This is a crucial property allowing to find paths within an inhomogeneous area characterised by minimum energy consumption.

The path planning software mentioned above has shown very good results in environments of different kinds, but not in all possible environments. Therefore, the software was further developed. After a short description of the earlier software version, an active search component is presented which overcomes the drawbacks existing formerly.

2 GA Based Path Planning

In this application the genetic pool is a set of individuals, each representing a path from a given start point to a given goal. The paths are totally independent of the presence or distribution of obstacles. Information about the environment is considered later on, when the fitness of the individuals is evaluated.

Three different genetic operators are used to evolve the pool: problem specific adaptations of the standard operators *mutation* and *crossover* (see Fig. 1(a) and 1(b)), and an operator *tighten*, which is very specific for this application (see Fig. 2). For details regarding the avoidance of closed loops and the handling of exceptional situations when utilising these operators see [1].

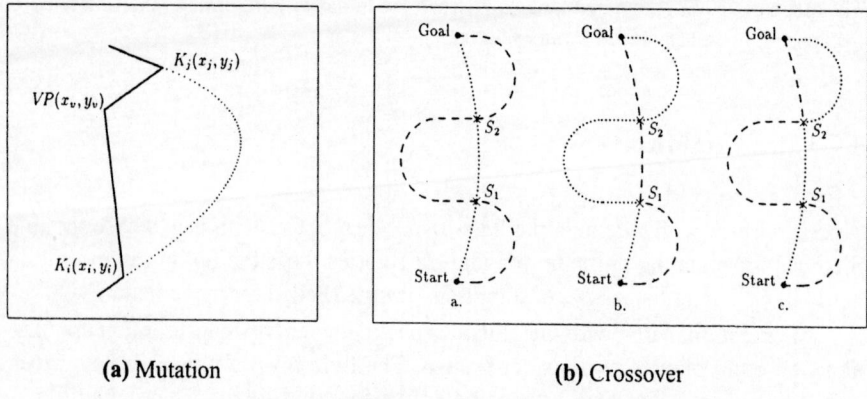

(a) Mutation **(b)** Crossover

Fig. 1. Mode of action of the genetic operators *mutation* and *crossover*. **(a)** The sub-path between K_i and K_j is replaced by two straight lines connecting K_i and K_j with an arbitrarily chosen via point VP. **(b)** The sub-paths of two paths following a randomly chosen intersection point are exchanged

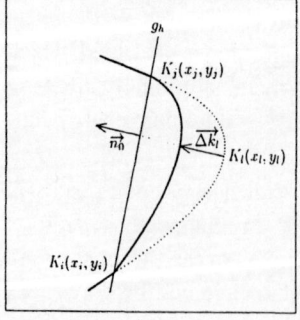

Fig. 2. Mode of action of the operator *tighten*: Every point K_l of the sub-path between two arbitrarily chosen points K_i and K_j is shifted towards the straight line g_h connecting these points. The shifting vector is parallel to the normal vector to g_h, which is directed from K_l towards g_h. The effect of *tighten* is determined by a real random number r fulfilling the inequation $0 \leq r \leq 1$. If r is equal to 0, the path is not changed, and for $r = 1$ the resulting sub-path is identical to the straight line

The regarded environment has to be subdivided into quadratic regions. Each region is well-defined by a pair of integer-numbered co-ordinates. Now, a path within this environment is a sequence of steps from the centre of one region to the centre of an adjacent region, connecting the start and the end region. Thus, a path is internally represented as an array of pairs of integer co-ordinates.

Each region corresponds to an entry of a matrix. In this matrix the environment information is stored as follows: each entry is a real value, indicating the texture-dependent amount of energy that would be needed if the robot moved through the corresponding region. This energy factor has to lie in the interval between 0 and 1. Weighting to extend this interval is possible. If the value is set equal to 1, the corresponding region cannot be used for the movement. This way obstacles are modelled.

The fitness is evaluated according to the energy consumption. This value is determined by adding up the energy costs for the single movements from one region's centre to the next region's centre, considering the corresponding entries of the environment matrix. If one or more entries are equal to 1, the path is not drivable. Such paths are ranked after the drivable ones according to the number of obstacles touched. For a more precise description see [1].

3 Active Search Extending the GA

A main disadvantage of the earlier version of the path planning software is that all modifications of paths take place without regarding the environment matrix. No information about positions of obstacles is taken into account. This is the main reason why the software shows very promising results in certain environments such as the one in Fig. 3(a), but fails completely in environments similar to the one in Fig. 3(b).

For this reason, an active search component was developed to overcome such drawbacks. This new component is executed in two different phases. The actual search process is performed in advance before the GA's execution starts. The information obtained in this first phase is utilised during the actual optimisation process – the second phase – to eliminate obstacle contacts. The mode of operation within both phases is described in detail in the following paragraphs.

3.1 Preparation Phase: Computing Circumventing Paths

Before the GA's execution starts, for each obstacle a path circumventing it is determined by the search algorithm. To begin with, a second environment map – the obstacle map – is computed disregarding the different energy factors: the only matter of interest in computing this map are the obstacles. One after another, the regions of the considered environment are investigated. If a region belonging to an obstacle is found, a recursive function is executed which determines all regions belonging to this very obstacle. All these regions are marked by setting the corresponding entries of the matrix representing the obstacle map to the same integer value. This way one obtains a matrix reflecting the distribution of obstacles in the environment, only.

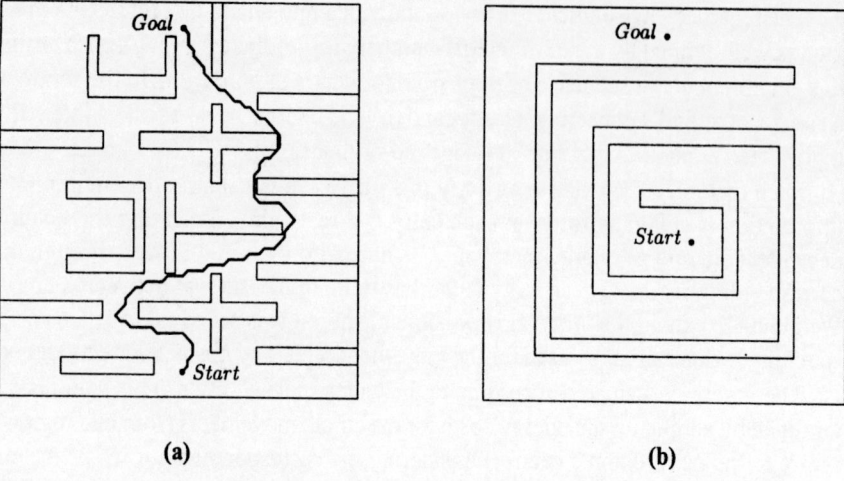

(a)   **(b)**

Fig. 3. Example environments. **(a)** Environment with a drivable path not further optimised. **(b)** Environment for which no drivable path is found by the unmodified algorithm

After that, for each obstacle a path circumventing it is determined. For each obstacle, the search process starts from an arbitrary region adjacent to this obstacle. Then, the path is developed step by step. The regions adjacent to the current one are investigated, each. The possible directions are chosen according to a fixed priority list (see Fig. 4). First, the region lying in direction 1 is investigated; this region is added to the circumventing path if it fulfils four conditions:

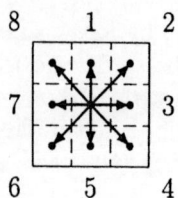

Fig. 4. The adjacent regions are investigated in ascending order. Therefore, direction 1 has the highest priority

- It lies within the environment regarded.
- It is not part of the obstacle considered[1].
- It is horizontally or vertically adjacent to a region belonging to the obstacle.
- It has not been visited before during the current search process.

If one or more of the conditions above are not fulfilled, the region is discarded and the region lying in direction 2 is investigated, and so on. A new region – or node

---

[1] It is not necessary to demand 'not part of any obstacle', because if the region is part of another obstacle, the next condition is not fulfilled, anyway.

– is appended to a path in a way avoiding superfluous sequences of regions (see Fig. 5 or, for a more detailed description, [1]).

Fig. 5. Elimination of superfluous nodes in order to avoid undesired node sequences

The process terminates if the next region to be appended is identical to the start region of the search process. The result is a closed path circumventing the obstacle. In contrast to the paths solving the optimisation problem outlined in Sect. 2, these paths can be traversed bidirectionally (see Fig. 6(a)).

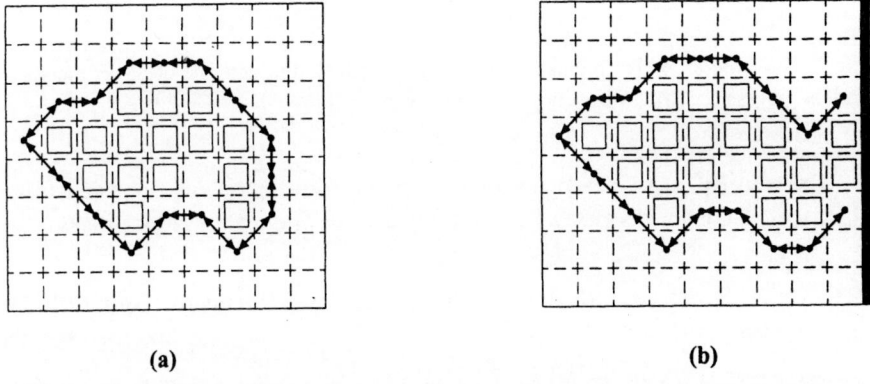

(a)   (b)

Fig. 6. Paths circumventing obstacles. (a) Obstacle located away from the environment's edges. (b) Obstacle located at an edge

At this point, two special cases have to be considered. Firstly, if the obstacle is located at the edge of the environment regarded, the search process cannot proceed if a region lying adjacent to the obstacle and the edge[2] is reached. No step in one of the eight directions is possible without breaking one of the rules above. In this situation, the search process is re-started, beginning from this very region. The result is a non-closed circumventing path beginning and ending at an edge of the environment (see Fig. 6(b)). If the obstacle is tangent to the edges of the environment at more than one place, a non-closed path is determined in an analogous way for each segment of the obstacle's edge.

---

[2] For instance, the regions in the last column and rows three, resp. six, in Fig. 6(b).

Secondly, a further step can become impossible during the search process if the path found so far leads into the obstacle. In Fig. 7(a) a situation is depicted in which no further step is allowed. In such a case, the actual region is marked in the same way as the obstacle in the obstacle map – for it has no sense to visit it later on – and the algorithm proceeds again from the antecedent region. In the example situation this leads to the same exceptional situation as before (see Fig. 7(b)). Therefore, another step backwards is taken. After that, the search process can proceed in the regular way as depicted in Fig. 7(c).

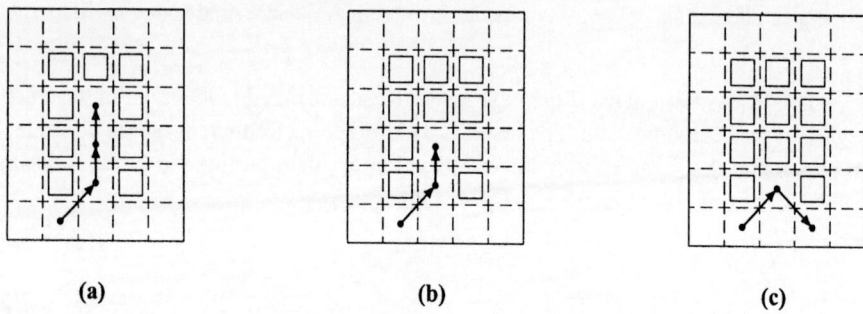

(a)   (b)   (c)

Fig. 7. Handling of an exceptional situation. **(a)** A path leading into the obstacle, blocking further processing. **(b)**, **(c)** Breaking the deadlock

The search process is executed for every single obstacle. Having completed this, the circumventing paths are available during the following optimisation process.

3.2 Execution Phase: Utilising Circumventing Paths During the Optimisation Process

As stated before, a path is a sequence of steps from one region to an adjacent region connecting the start and the end region of the given optimisation problem. In the unmodified version of the path planning GA, all paths are evolved utilising genetic operators which take no information about the environment into account. Therefore, in complex environments with many obstacles or obstacles shaped like the one depicted in Fig. 3(b), nearly every path computed contains regions belonging to obstacles.

Figure 8(a) shows an obstacle and a path passing through it. Now, the modified GA presented in this paper detects such unwanted obstacle contacts and eliminates them as follows. The sequence of regions lying within the obstacle is replaced by a segment of the circumventing path belonging to this obstacle. If the circumventing path is a non-closed one like the one depicted in Fig. 6(b), the segment is determined without ambiguity. In case of a closed loop, there are two options: there exist two segments of the circumventing path which connect the last path region preceding the obstacle with the first region following it. Therefore, each time the direction of the traversal – clockwise or not – is determined randomly as follows. Let n_1 be the number of regions belonging to the segment lying in clockwise direction and n_2, respectively, the corresponding number for the segment lying in the opposite direction[3]. Then, the first segment is chosen with probability[4]

$$p_1 = \frac{n_2}{n_1 + n_2} \, . \tag{1}$$

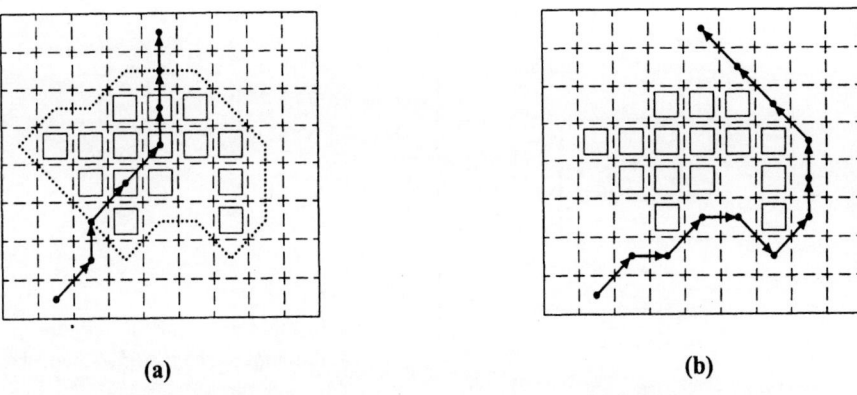

Fig. 8. Elimination of obstacle contacts. **(a)** Example path passing through an obstacle. **(b)** Final path circumventing the obstacle

Having chosen the segment, the three path sections are connected in a way to avoid superfluous node sequences as stated above. Figure 8(b) shows the final result after applying this replacement strategy. If a path passes more than one obstacle, the process is repeated for each single obstacle.

In Fig. 9 the same environment as above is depicted. Whereas the unmodified algorithm was not suitable for finding a feasible path within this environment, the modified version is able to compute the depicted optimum path.

The new operator *circumvent* is not a genuine genetic operator. In contrast to the other operators, the influence of randomness is very small or non-existent, respectively. In addition, the operator is not applied to single paths with a certain probability, but to each path after applying the other genetic operators *mutation*, *crossover* and *tighten*. Therefore, when the fitness of the individuals is evaluated, all paths represented by these individuals are feasible.

[3] Alternatively, one could compute the corresponding amounts of energy needed to traverse the two different segments.

[4] Cp. the distribution of electric currents passing through resistances placed in parallel.

4 Conclusions

A Genetic Algorithm based path planning software was extended by a search component which takes environment information into account. The two phases of the new component's modus operandi were described in detail, and the suitability of the approach was documented by reporting the result of the modified algorithm's application to a problem previously unsolved.

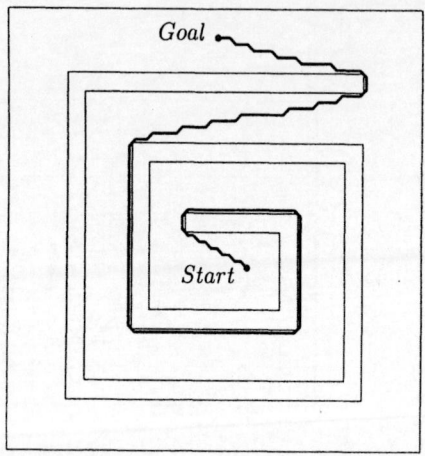

Fig. 9. Example environment with an optimum path found by the modified algorithm

Furthermore, the modified path planning algorithm is successfully being utilised in a small size experimental environment employing Khepera robots (for more information about that robot see [4]).

The extended GA works very promisingly. Utilising the active search overcomes the drawbacks of the earlier version of the path planning software. Thus, the extended GA is suitable for universal usage now.

References

1. Gemeinder M., Gerke M. (2001) GA Based Search for Paths with Minimum Energy Consumption for Mobile Robot Systems. Proc. 7th Fuzzy Days, Dortmund, (Accepted)
2. Gerke M. (1998) Genetische Bahnplanung für mobile Roboter. 8. Workshop Fuzzy-Control des GMA-FA 5.22, Dortmund
3. Lee S., Kardaras G. (1997) Collision-Free Path Planning with Neural Networks. Proc. 1997 IEEE International Conference on Robotics and Automation, Albuquerque
4. Löffler A., Mondada F., Rieckert U., editors (1999) Experiments with the Mini-Robot Khepera. Proc. 1st International Khepera Workshop, Paderborn
5. Xiao J., Michalewicz Z., Zhang L., Trojanowski K. (1997) Adaptive Evolutionary Planner/Navigator for Mobile Robots. IEEE Transactions on Evolutionary Computation 1(1)
6. Zhao M., Ansari N., Hou E.S.H. (1994) Mobile Manipulator Path Planning by a Genetic Algorithm. Journal of Robotic Systems 11(3)

Genetic Programming for Combining Neural Networks for Drug Discovery

W. B. Langdon[1], S. J. Barrett[2], and B. F. Buxton[1]

[1] Computer Science, University College, Gower Street, London, WC1E 6BT, UK
{W.Langdon, B.Buxton}@cs.ucl.ac.uk
http://www.cs.ucl.ac.uk/staff/W.Langdon
Tel: +44 (0) 20 7679 4436, Fax: +44 (0) 20 7387 1397
[2] GlaxoSmithKline Research and Development, Harlow, Essex, UK

Abstract. We have previously shown [Langdon and Buxton, 2001b] on a range of benchmarks genetic programming (GP) can automatically fuse given classifiers of diverse types to produce a combined classifier whose Receiver Operating Characteristics (ROC) are better than [Scott *et al.*, 1998]'s "Maximum Realisable Receiver Operating Characteristics" (MRROC). I.e. better than their convex hull. Here our technique is used in a blind trial where artificial neural networks are trained by Clementine on P450 pharmaceutical data. Using just the networks, GP automatically evolves a composite classifier.

1 Introduction

There are an increasing range of cases where computer based systems are able to provide huge volumes of data but rendering it intelligible is seldom attempted or left to labour intensive intervention. This has provoked interest in artificial intelligence techniques to try and extract information from the data. A common task is classification. Here we are particularly interested in data rich Cheminformatics applications where we wish to be able to predict how chemicals, particularly potential drugs, will behave. Intelligent classification techniques such as artificial neural networks (ANN) have had limited success at predicting potential drug activity. Using genetic programming many diverse classifiers can be fused to yield a superior classifier.

Any classifier makes a trade off between catching positive examples and raising false alarms. Where the costs of these are not known in advance it may be useful to be able to tune the classifier to favour one over the other. The Receiver Operating Characteristics (ROC) of a classifier provides a helpful way of illustrating this trade off.

[Scott *et al.*, 1998] has previously suggested the "Maximum Realisable Receiver Operating Characteristics" for a combination of classifiers is the convex hull of their individual ROCs. However the convex hull is not always the best that can be achieved [Yusoff *et al.*, 1998]. Previously we showed [Langdon and Buxton, 2001b] in at least some cases better classifiers, in terms of their ROCs, can be automatically produced. Here we apply our technique to a real world application namely classifying potential drug compounds as to whether they are inhibitors of a P450 enzyme.

Section 2 gives the back ground to data fusion. The collection of the P450 data at GlaxoSmithKline Pharmaceuticals is described in Sect. 3. Section 4 describes using the SPSS Clementine data mining tool to train 60 artificial neural networks on 699 chemical features. These ANN are used as the primary classifiers by the genetic programming data fusion system, which is described in Sect. 5. The results are given in Sect. 6. Finally we conclude in Sect. 7.

2 Background

There is considerable interest in automatic means of making large volumes of data intelligible to people. Arguably traditional sciences such as Astronomy, Biology and Chemistry and branches of Industry and Commerce can now generate data so cheaply that it far outstrips human resources to make sense of it. Increasingly scientists and Industry are turning to their computers not only to generate data but to try and make sense of it. Indeed the new science of Bioinformatics has arisen from the need for computer scientists and biologists to work together on tough, data rich problems such as drug discovery.

The terms Data Mining and Knowledge Discovery are commonly used for the problem of getting information out of data. In addition to traditional techniques, a large range of "intelligent" or "soft computing" techniques, such as artificial neural networks, decision tables, fuzzy logic, radial basis functions, inductive logic programming, support vector machines, are being increasingly used. Genetic programming, using Receiver Operating Characteristics (ROC), offers an automatic way of combining diverse classifiers to yield a superior composite.

More information on ROC curves can be found in our previous work [Langdon and Buxton, 2001a] [Langdon and Buxton, 2001b] [Langdon and Buxton, 2001c] and the companion web pages to this paper (http://www.cs.ucl.ac.uk/staff/W.Langdon/wsc6/). Briefly any binary classifier can be characterised by two scalars. Its "true positive" rate (TP) and its "false positive" rate (FP). I.e. the fraction of positive examples it correctly classifies and the fraction of negative examples it gets wrong (false alarms). When plotted against each other TP v. FP lie inside a unit square. An ideal classifier has TP = 1 and FP = 0. I.e. the upper left corner of the square (see Fig. 4). Many classifiers have a sensitivity parameter. This allows the user to trade off TP against FP. By varying the sensitivity the FP,TP point traces a curve. A good classifier will have a curve which lies as close to (0,1) as possible. A very poor classifier's ROC will lie near the diagonal (0,0) – (1,1). It is common to use the area under the ROC as a measure of the classifier's performance. (Although a single scalar measurement cannot capture all the possible variations between two curves it is widely used and is in most cases satisfactory).

A new classifier can always be constructed by randomly choosing between two available classifiers. [Scott *et al.*, 1998] showed that the ROC of the new classifier lies on a line between the ROC's of the two real classifiers. If we choose evenly (50%), then the new point lies halfway between them. (The element of chance may mean it cannot be used in some applications). Changing the ratio from 50% moves the point from the midpoint towards one of the existing classifiers. Since this can be done for any number of classifiers operating at any of their sensitivity parameter settings, a new classifier can be constructed at any point within the convex hull of their ROC curves. In fact only those on the convex hull are of interest, since they are bound to be better than points within the hull. Scott showed a new classifier whose ROC is the convex hull of the input classifiers ROC curves can indeed be constructed in practise. He called this the "Maximum Realisable Receiver Operating Characteristics" (MRROC). In fact it is possible in principle to do better and as we shall show, GP can do better in this application.

3   The Pharmaceutical Data

Volume inhibition data was extracted from the GlaxoSmithKline biological results database for all compounds screened against a P450 enzyme assay using high through put screening (HTS). If all data for an individual compound were initially within 15% inhibition points of each other they were retained and averaged. Otherwise noisy data was discarded. The clean data was then thresholded at an appropriate level of inhibition dividing the compounds into "actives" (enzyme inhibitors) and "inactives" (non-inhibitors).

The actives were separated from the inactives and each set hierarchically clustered separately using Ward's linkage in combination with Tanimoto similarity, computed from Daylight 2Kbit string chemical fingerprint data. Clusters were defined at 0.8tan (Tanimoto) for the (smaller) actives set and at 0.75 for the (larger) inactives set, as a first step to reduce the gross imbalance in actives vs. inactives. Then, from each cluster in each partitioning, centroid compounds were selected to reduce the internal chemical structure bias and the volume of the original data. This noise removal and clustering resulted in a "controlled diversity" dataset of 2256 compounds with a balance of approximately 4 inactives for every active. A total of 699, 2-d numerical, chemical features from a diverse array of families (electronic, structural, topological/shape, physico-chemical, etc.) were then computed for each centroid molecule, starting from a SMILES representation of it's primary chemical structure. 1500 compounds (300 actives, 1200 inactives) were selected for use as the training set, whilst the remaining 756 were retained as a separate "holdout" set.

4   Artificial Neural Networks

We used the Clementine data mining tool to train artificial neural networks (ANN) to model the training data. These models were then frozen and made available to genetic programming as a function. ANN models were generated using "train net" node using the "Quick" (BackProp) method, with options to prevent over training set (50% training data used) and to stop on 300 cycles. A fixed random seed was employed.

The ANN models were trained using Clementine on subsets of 1500 training records, containing the 699 features. These were divided by GlaxoSmithK-line into 15 groups of about 50 attributes (features). Four single layer perceptrons were trained on each group.

It is well know that this kind of neural network performs best when trained on "balanced data sets", i.e. data sets containing an equal mix of active and inactive examples. However drug discovery tasks are seldom like this. It is common, as here, for many compounds to be inactive and only a few to be active. Four data sets were prepared. Each contained the same 300 active examples and 300 different inactive examples. That is each data set was balanced. Each neural network was trained on one of the 15 groups of attributes selected from one of the four balanced data sets. Making a total of 60 networks.

A script was written which automatically manipulated the Clementine stream (see Fig. 2). The script feeds one of the four sets of balanced examples via a "type node" (which selects the FEATURE SET) into Clementine's artificial neural network trainer (TRAIN NET NODE in Fig. 1). The script loops through each of the four balanced training sets and loops through the 15 groups of chemical attributes, exporting neural network models for each combination of FEATURE SET and DATA SET.

5   Genetic Programming Configuration

The genetic programming data fusion system is deliberately almost identical to that described in [Langdon and Buxton, 2001b].

5.1   Function Set

The Function set is the collection of basic floating point operations that are available to form the individual programs in the GP population. They (and the other GP parameters) are summarised in Table 1. The functions include the four basic arithmetic operations "+", "−", "×" and "÷". Note however ÷ by zero always yields "1". This "protects" it and prevents the GP system failing with a divide-by-zero fault. Max (Min) takes two arguments and returns the value of the largest (smallest). MaxA (MinA) also takes two arguments but returns the signed value of the largest (smallest) in

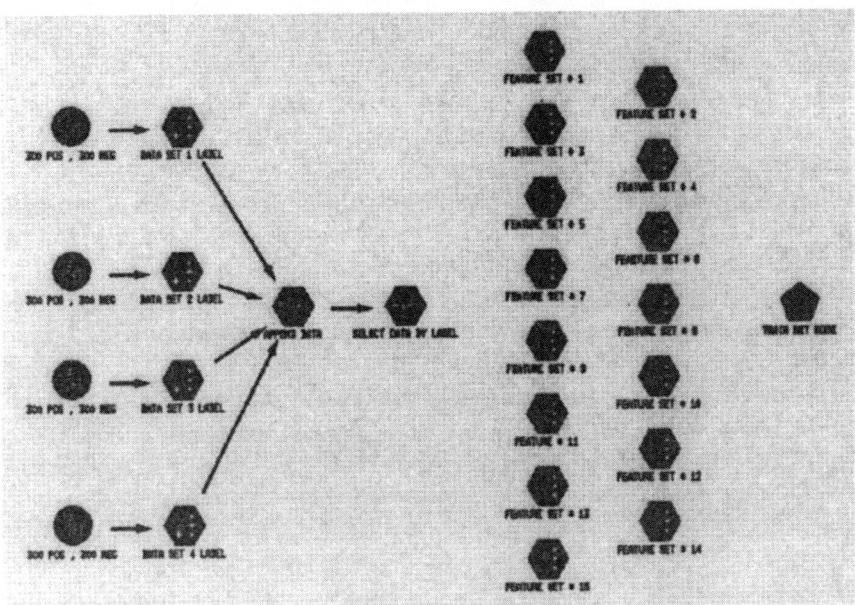

Fig. 1. Clementine streams for training multiple Artificial Neural Networks. The command script for switching the training stream between the data files is given in Fig. 2

```
For F in Feature_Set#'s 1 to 15   # selects FEATURE SET typenode name
  For [dataset] N from 1 to 4    # select DATA SET
    set m=F><N                   # concat. FEATURE SET & DATA SET names
    set :trainnet.netname = M         # rename TRAIN NET node
    connect ^F between select and ^M  # connects dataset to 'feaset'
    execute : trainnet
    disconnect ^F
    export generated ^M in [file directory] codexport
  endfor
endfor
```

Fig. 2. Clementine script used to train 15 groups of Artificial Neural Networks. The ANN in each group are trained on a group of about 50 chemical features. Each group contains four networks, each trained on the same active chemicals but different inactive examples. Cf. Fig. 1

absolute terms. E.g. MaxA(-2,1) returns -2. INT returns the integer part of its input. E.g. INT(3.23) returns 3. FRAC returns the fractional part of its input. E.g. FRAC(3.23) returns 0.23. Finally IFLTE takes four arguments. If the first is less than or equal to the second, IFLTE returns the value of its third argument. Otherwise it returns the value of its fourth argument. E.g. FRAC(0, 0.345, ANN1..., ANN2...) returns the value given to it by the subtree starting with ANN1.

In order to use the neural networks within the GP they are packaged up and presented to GP as 60 problem specific functions. (The GP is run separately from Clementine using 60 files containing the ANNs. Note the ANN are frozen and not retrained). Each returns the classification given by the corresponding neural network for the current chemical. Clementine codes its multi-layer perceptron neural networks to have one output neuron which yields a floating point value between zero and one. For a mid-point threshold, this is exactly the value returned by the function inside the GP system. Values near either zero or one mean the neural network is highly confident of its answer. While values near 0.5 suggest the classifier is less confident. Normally the output of the neural network is converted into a binary classification (i.e. the chemical is active or is inactive) by testing to see if the value is greater or less than 0.5. This gives a single point in the ROC square. I.e. one trade off between catching all positives but raising too many false alarms. However we can change this trade off. So that instead of getting a single point, we get a complete curve in the ROC square. This is easily done by replacing the fixed value of 0.5 by a tunable threshold. By continuously varying the threshold from below zero to above one the (binarized) output of any of the neural networks will be biased from saying every chemical is inactive, through the usable range, to catching all positive examples but being 100% wrong on the negative examples (by saying all chemicals are active). In fact we leave the choice of suitable operating point to the GP to automatically evolve. This is done by making the threshold point an argument to the function. These arguments are treated like any other by the GP and so can be any valid arithmetic operation, including the neural networks themselves.

5.2 Terminal Set

The terminals or leafs of the trees being evolved by the GP are either constants or the threshold (see Table 1). The threshold allows us to change the bias of the tree produced by GP. As with the neural networks, changing the bias allows each tree to sweep out an ROC curve rather than operate at a single point. This is done by running each tree with the threshold taking values 0, 0.1, 0.2, ... 1.0 (i.e. 11 values). Note, as with all the functions and terminals, it is entirely up to the evolutionary process how it uses the threshold parameter.

5.3 Representation

The GP is set up to signal its prediction of the class of each data value by returning a floating point value, whose sign indicates the class and whose magnitude indicates the "confidence".

Following earlier work [Jacobs et al., 1991,Soule, 1999,Langdon, 1998] each GP individual is composed of five trees. Each of which is capable of acting as

Table 1. GP Data Fusion Parameters

Objective:	Evolve a Non-Linear Combination of Neural Networks with Maximum ROC Convex Hull Area on P450
Function set:	INT FRAC Max Min MaxA MinA MUL ADD DIV SUB IFLTE 60 ANN trained on P450 data
Terminal set:	T 0 0.1 0.2 0.3 0.4 0.5 0.6 0.7 0.8 0.9 1 plus 100 unique random constants 0..1
Fitness:	Area under convex hull of 11 ROC points.
Selection:	generational (non elitist), tournament size 7
Wrapper:	$\geq 0 \Rightarrow$ active, inactive otherwise
Pop Size:	500
No size or depth limits	
Initial pop:	ramped half-and-half (5:8) (half terminals are constants)
Parameters:	50% size fair crossover [Langdon, 2000], 50% mutation (point 22.5%, constants 22.5%, shrink 2.5% subtree 2.5%)
Termination:	generation 50

a classifier. The use of signed numbers makes it natural to combine classifiers by adding them. I.e. the classification of the "ensemble" is the sum of the answers given by the five trees. Should a single classifier be very confident about its answer this allows it to "out vote" all the others. Note that although this has some similarity with some neural network "ensembles", the GP can combine the supplied classifiers in an almost totally arbitrary non-linear way. It is not constrained to a weighted linear sum of all or even a subset of them.

5.4 Fitness Function

The fitness function is crucial to any optimisation process, including evolutionary computation techniques. By giving intermediate results a score (in GP known as the individual program's fitness) it guides the optimiser. Often fitness functions are not given sufficient thought. For example when classifying, one scheme is simply to assign an individual's fitness equal to the number of correct predictions it makes. In some cases this is sufficient. But where one class is much more common than the other, this can lead to a trap where the optimiser creates a classifier which always says which ever class occurs most often. This classifier has a high score but no predictive ability. By using ROC curves we avoid this trap.

Since we may not know in advance the trade off between the costs of misclassification of the two classes, we use GP to produce a *tunable* classifier. We asses the usefulness of each candidate classifier produced by GP from its Receiver Operating Characteristics (ROC) curve.

To assign a fitness to each evolved classifier. The tuning parameter is set to values 0.1 a part, starting at 0 and increasing to 1. For each setting it is used to predict the activity of each chemical in the training set. These predictions are compared with measured activity. The proportions of active

604

chemicals correctly predicted (TP) and the proportion of inactive one incorrectly predicted (FP) are calculated. Each TP,FP pair gives a point on a curve. The fitness of the classifier is the area under the convex hull of these (plus the fixed points 0,0 and 1,1).

5.5 Genetic Operations and other Parameters

As mentioned above, we used the same mix of genetic operators as had proved themselves in earlier experiments. Other work (for example [Angeline, 1998]) has suggested that a high mutation rate and a mixture of different mutation operators for evolving pattern matching functions. As described in Table 1, 50% of new classifiers were created by random changes of a single parent from the previous generation. On average 112.5 programs per generation were created by point mutation on the functions and 112.5 by random changes to the constants. Point mutation was applied on average to 10% of functions within the program. Point mutation on a function randomly replaces it with another (which takes the same number of arguments). Mutating a constant, replaces it with another randomly constant. However the new constant is not chosen uniformly, instead a new value is chosen from an approximately Normal distribution centred on the current value and standard deviation of 5% of the mean. There are 110 available constants. The one closest to the randomly generated value replaces the original constant. (The new constant will be different from the original). Note bigger programs will on average have more changes made to them.

Two other mutation operators were also used. On average 12.5 programs in each generation were created using "subtree" mutation and 12.5 by "subtree shrink" mutation. Both select a subtree within (a copy of) the parent program and replace it with another. Note unlike point mutation and constant mutation, these mutation operators change the size and shape of the program. In subtree mutation the new subtree is created using the same algorithm used to create the initial random population, i.e. "ramped-half-and-half" [Koza, 1992, pages 92–93]. This tends to increase the size of programs. (The behaviour of ramped-half-and-half is discussed in [Luke and Panait, 2001]). In contrast "subtree shrink" replaces the subtree with another chosen at random from within it. This must be smaller leading to the new program being smaller than the program in the previous generation from which it was created.

On average 250 of the new programs are created using size fair crossover [Langdon, 2000]. In genetic programming crossover selects two fit programs to be parents for a new one. The new program is composed of parts drawn from (copies of) the two parents. It thus provides an analogue of sexual reproduction in the artificial evolution used by genetic programming. There are a number of crossover operators available in GP [Langdon and Poli, 2001], size fair crossover was introduced to reduced the tendency seen in the usual GP crossover operator [Koza, 1992] for programs in successive generations to increase in size (known as bloat [Langdon et al., 1999]) but still allow size and shape to be determined by the evolutionary process rather than by hard limits or initial conditions.

Size fair, like Koza's subtree crossover, creates a new program by selecting a subtree from each parent program. In one the subtree is discarded and replaced by the subtree selected in the other parent. The difference is, in the older crossover both subtrees are selected independently at random. In size fair the subtree to be removed is selected at random but the the subtree to be inserted is chosen from those in the other parent program of a chosen size. The size is given by the size of the subtree to be deleted. To allow size fair crossover to change the size of the new program, the inserted subtree's size need not be identical to that being deleted but is randomly chosen from 1 to up to twice the size of the subtree tree being deleted. However care is taken to ensure size fair does not, of itself, lead to either an average increase or decrease in size. (Details are given in [Langdon, 2000]).

5.6 GP Training Data

The 1500 examples used to train the neural networks were randomly split into 1000 to be used to training the GP and 500 (containing 100 active examples) kept back as a verification set.

6 Results

In one run the GP evolved a combined classifier with a fitness of 0.90 on the training data and 0.86 on both the validation and holdout data. Figure 3 shows its performance (on the training set vs. the holdout set) in comparison with that of the neural networks. If performance on the training set and holdout were identical, then all the points would lie along the diagonal and there would be no over fitting. In practise, since we are dealing with finite samples, there is bound to be some statistical scatter. As scatter of points is around the diagonal this indicates there is little over fitting.

To obtain improved performance from any classifier fusion the input classifiers must be different. Figure 3 shows their performance varies considerably. Doing pairwise comparisons, most ANN are significantly different (McNemar's test at 5%) from each other on the training data. The GP selected only 17 ANN for inclusion in the evolved classifier. These are shown with double crosses in Fig. 3. Figure 3 shows GP has selected a broad range of ANNs.

Figure 4 shows the Receiver Operating Characteristics of the evolved classifier, measured on its training data and on the holdout set. There is slight reduction in performance on the holdout set, potentially indicating some over fitting. Figure 4 also shows, for comparison, the ROC of the MRROC classifier produced by taking the convex hull of the 60 ANNs. The MRROC is shown measured on the training data and on the holdout data. Of course

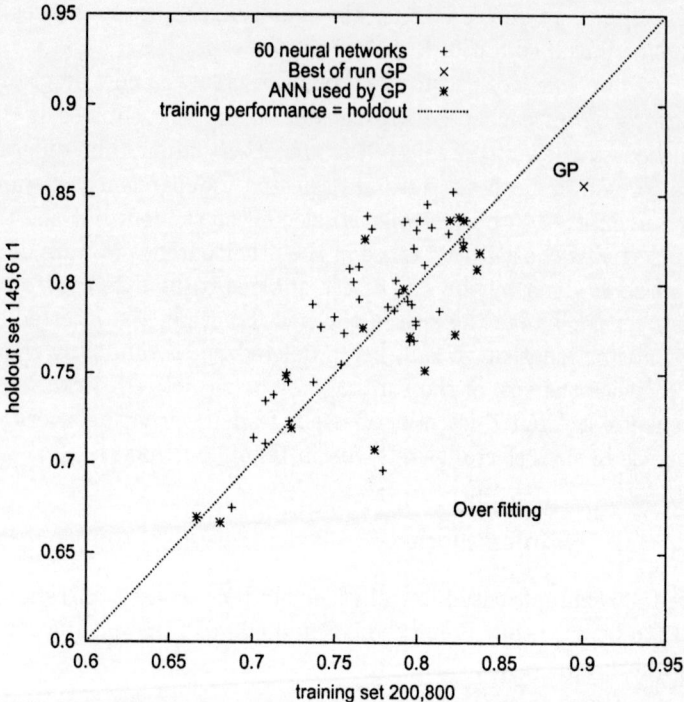

Fig. 3. Performance of 60 given neural networks and genetic programming. The area under the Receiver Operating Characteristics (ROC) is plotted on the training data (horizontal) v. holdout data (vertical). Points below the diagonal indicate a degree of over training.

the MRROC is convex on the training data, but need not be on the hold out data. On both the training and holdout data, the evolved classifier is slightly better than the MRROC combination of the 60 ANN trained by Clementine and therefore better then each of the ANN individually.

7 Conclusions

Where accuracy is paramount (and so its stochastic basis can be ignored) [Scott *et al.*, 1998]'s convex hull classifier, MRROC, offers an automatic means of combining classifiers. However it is not guaranteed to be optimal. In [Langdon and Buxton, 2001b] we showed, using Scott's own bench marks, that genetic programming can do better than the MRROC both in theory and practise. Nevertheless we cannot guarantee GP will always do better and so it is important to demonstrate it on interesting applications. Here we have shown (cf. Figs. 3 and 4) that our GP technique can be used in a large classification application related to drug discovery with just a few minutes of computer processing time.

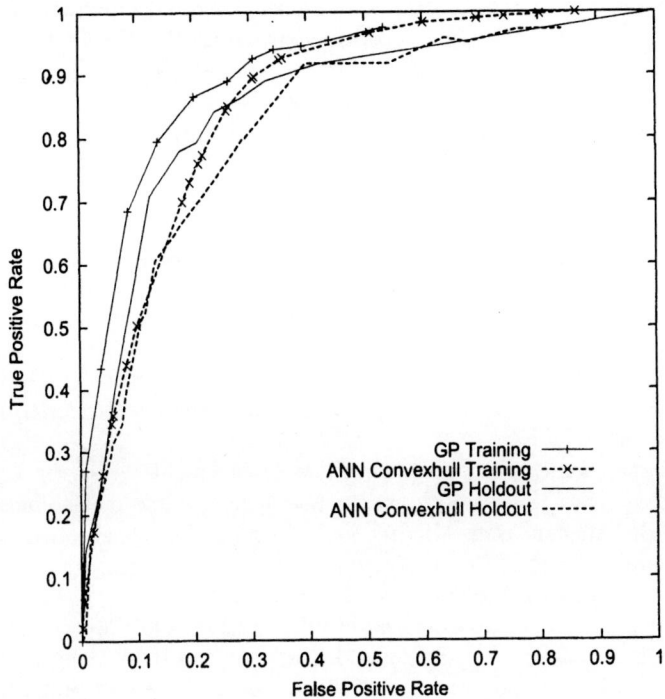

Fig. 4. Receiver Operating Characteristics of evolved composite classifier. For comparison the convex hull of the 60 given neural networks, on the training and holdout data, is given. Note the convex hull classifier is no longer convex when used to classify the holdout data.

References

[Angeline, 1998] Peter J. Angeline. Multiple interacting programs: A representation for evolving complex behaviors. *Cybernetics and Systems*, 29(8):779–806, November 1998.

[Jacobs *et al.*, 1991] Robert A. Jacobs, Michael I. Jordon, Steven J. Nowlan, and Geoffrey E. Hinton. Adaptive mixtures of local experts. *Neural Computation*, 3:79–87, 1991.

[Koza, 1992] John R. Koza. *Genetic Programming: On the Programming of Computers by Means of Natural Selection*. MIT Press, 1992.

[Langdon and Buxton, 2001a] W. B. Langdon and B. F. Buxton. Evolving receiver operating characteristics for data fusion. In Julian F. Miller, Marco Tomassini, Pier Luca Lanzi, Conor Ryan, Andrea G. B. Tettamanzi, and W. B. Langdon, editors, *Genetic Programming, Proceedings of EuroGP'2001*, volume 2038 of *LNCS*, pages 87–96, Lake Como, Italy, 18-20 April 2001. Springer-Verlag.

[Langdon and Buxton, 2001b] W. B. Langdon and B. F. Buxton. Genetic programming for combining classifiers. In Lee Spector, Erik D. Goodman, Annie Wu, W. B. Langdon, Hans-Michael Voigt, Mitsuo Gen, Sandip Sen, Marco Dorigo, Shahram Pezeshk, Max H. Garzon, and Edmund Burke, editors, *Proceedings of the Genetic and Evolutionary Computation Conference (GECCO-2001)*, pages 66–73, San Francisco, California, USA, 7-11 July 2001. Morgan Kaufmann.

[Langdon and Buxton, 2001c] W. B. Langdon and B. F. Buxton. Genetic programming for improved receiver operating characteristics. In Josef Kittler and Fabio Roli, editors, *Second International Conference on Multiple Classifier System*, volume 2096 of *LNCS*, pages 68–77, Cambridge, 2-4 July 2001. Springer Verlag.

[Langdon and Poli, 2001] W. B. Langdon and Riccardo Poli. *Foundations of Genetic Programming*. Springer, 2001.

[Langdon et al., 1999] W. B. Langdon, Terry Soule, Riccardo Poli, and James A. Foster. The evolution of size and shape. In Lee Spector, W. B. Langdon, Una-May O'Reilly, and Peter J. Angeline, editors, *Advances in Genetic Programming 3*, chapter 8, pages 163–190. MIT Press, 1999.

[Langdon, 1998] W. B. Langdon. *Data Structures and Genetic Programming*. Kluwer, 1998.

[Langdon, 2000] W. B. Langdon. Size fair and homologous tree genetic programming crossovers. *Genetic Programming and Evolvable Machines*, 1(1/2):95–119, April 2000.

[Luke and Panait, 2001] Sean Luke and Liviu Panait. A survey and comparison of tree generation algorithms. In Lee Spector, Erik D. Goodman, Annie Wu, W. B. Langdon, Hans-Michael Voigt, Mitsuo Gen, Sandip Sen, Marco Dorigo, Shahram Pezeshk, Max H. Garzon, and Edmund Burke, editors, *Proceedings of the Genetic and Evolutionary Computation Conference (GECCO-2001)*, pages 81–88, San Francisco, California, USA, 7-11 July 2001. Morgan Kaufmann.

[Scott et al., 1998] M. J. J. Scott, M. Niranjan, and R. W. Prager. Realisable classifiers: Improving operating performance on variable cost problems. In Paul H. Lewis and Mark S. Nixon, editors, *Proceedings of the Ninth British Machine Vision Conference*, volume 1, pages 304–315, University of Southampton, UK, 14-17 September 1998.

[Soule, 1999] Terence Soule. Voting teams: A cooperative approach to non-typical problems using genetic programming. In Wolfgang Banzhaf, Jason Daida, Agoston E. Eiben, Max H. Garzon, Vasant Honavar, Mark Jakiela, and Robert E. Smith, editors, *Proceedings of the Genetic and Evolutionary Computation Conference*, volume 1, pages 916–922, Orlando, Florida, USA, 13-17 July 1999. Morgan Kaufmann.

[Yusoff et al., 1998] Y. Yusoff, J. Kittler, and W. Christmas. Y. Yusoff, J. Kittler, and W. Christmas. Combining multiple experts for classifying shot changes in video sequences. In *IEEE International Conference on Multimedia Computing and Systems*, volume II, pages 700–704, Florence, Italy, 7-11 June 1998.

Training MLP Networks by the Differential Evolution Algorithm

Miika Lindfors, Jouni Lampinen

Lappeenranta University of Technology, Laboratory of Information Processing,
P.O.Box 20, FIN-53851 Lappeenranta, Finland

Abstract. In this article feasibility of training multilayer perceptron networks by applying a recently introduced evolutionary algorithm, called Differential Evolution, was investigated. The Differential Evolution algorithm was compared with four variations of the Back-Propagation algorithm on training the network. The initial results of our experiments suggested that the Back-Propagation typically finds a good solution relatively fast, but cannot improve the solution further on, when more iterations are performed. Consequently, the Differential Evolution algorithm did not, in the early stages of the training process, provide solutions as good as those obtained with the Back-Propagation. However, after the Back-Propagation got stuck on a locally optimal solution, the Differential Evolution finally overtook it due to its global optimization capabilities. Our results suggest that it depend on the time available for training, which of the compared algorithms provide the best training result. However, the conclusions are preliminary and limited by the problems studied so far. For example, until now, only relatively small networks have been used for experimentation.

1. Introduction

1.1 Background

Recently, various general-purpose optimization methods for seeking approximate solutions to nonlinear global optimization problems have been under continuous development. Especially, various evolutionary algorithms, e.g. Genetic Algorithms and Evolution Strategies, have been found to be promising stochastic population based methods for solving problems of this important class. Due to already demonstrated advantages of evolutionary algorithms as global optimization methods, it is motivated to inspect such a stochastic methods as well in the context of training artificial neural networks. Namely, training neural networks can also be viewed as a difficult nonlinear global optimization problem. Traditionally, mainly local optimizers like the Back-Propagation have been applied to solving this problem. However, their usage is limited to relatively small networks since local optimizers are prone to get stuck in locally optimal solutions. Finding a solution to these training problems is still remaining as one of the most important open problems in this field.

In this article applicability of a novel evolutionary algorithm, the Differential Evolution (DE), was initially investigated and the DE was compared with four variants of the Back-Propagation algorithm in training weights of a multilayer perceptron network (MLP) [1]. The primary motivation behind this initial investigation was finding out, whether or not the DE provides good enough development potential for justifying further investigations, and whether the DE is capable of providing a potential alternative for the widely applied Back-Propagation algorithms, perhaps by providing effectiveness, efficiency and robustness advantages, or capabilities for training larger networks.

1.2 Training MLP Networks

In general, training MLP Networks can be considered as a nonlinear global optimization problem, where the objective is to minimize the error between the training data set and the corresponding MLP network output by finding optimum values for the weights assigned to the network. In the following, the structure of MLP network and its training by the Differential Evolution algorithm will be briefly outlined.

The basic processing unit of MLP Networks, Perceptron node or neuron, consists of inputs, outputs, weights and biases. Inputs x_1, x_2, ... , x_n multiplied by their respective real-valued weights w_1, w_2,..., w_n are summed by the neuron. Bias x_0 has the constant value 1.

$$y = \sum_{i=0}^{n} w_i x_i \qquad (1)$$

Then, the sum is transformed by an activation function. The structure of a perceptron is illustrated in Fig. 1. [7].

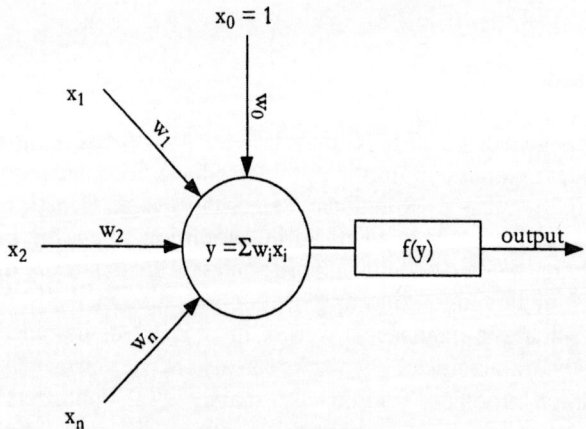

Fig. 1. Perceptron

A Multilayer Perceptron (MLP) network consists of interconnected layers of neurons (see Fig. 2). The layers between the input and output layers are called hidden layers. A linear function is usually used as an activation function in the output layer. Common activation functions used in the hidden layers are the sigmoid function,

$$y(x) = \frac{1}{1 + e^{-ax}} \tag{2}$$

and hyperbolic tangent function,

$$y(x) = \tanh(x) \tag{3}$$

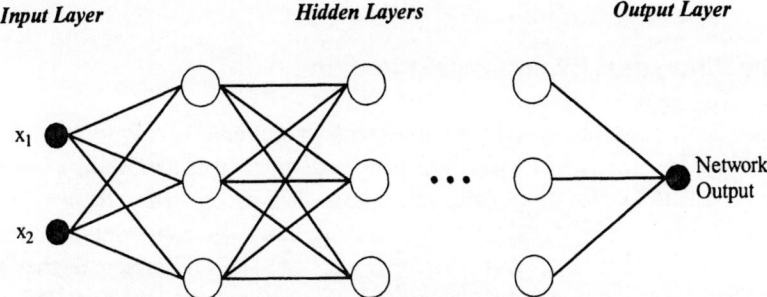

Fig. 2. Structure of a MLP network

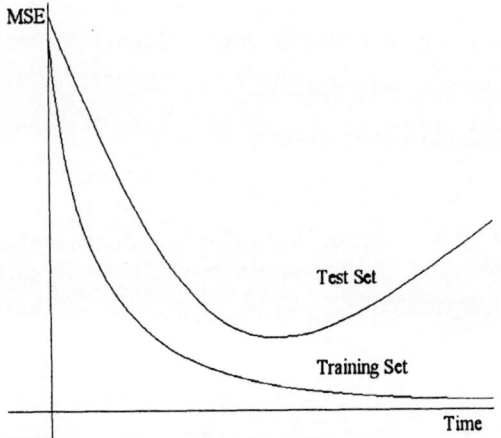

Fig. 3. Overfitting phenomenon

At first, inputs to the network x_1, x_2, ... , x_n are given as inputs to all neurons in the first hidden layer. After that, outputs from all neurons in the first hidden layer are calculated. Consequently, those outputs are inputs to the next hidden layer.

A training set is used for training the MLP network. It consists of a finite number of input and output vector pairs (i_i, d_i). Input vectors are inputs to the network. Outputs o_i are calculated for each input i_i by the network. After that actual outputs o_i are compared to desired outputs d_i. Performance of the network is determined by some error function, for example the mean square error (MSE) between o_i and d_i. After that the weights are adapted by some training algorithm, so that the error function is minimized. An independent test set is used for testing the network by approximating its generalization capability outside of the training set. During the training process, the test set error may start to increase while the training set error is still improving. This phenomena is called overfitting (see Fig. 3) and if it happens network training is stopped in order to avoid generalization capability loss [7].

1.3 The Differential Evolution (DE) algorithm

Price and Storn first introduced the Differential Evolution (DE) algorithm a few years ago [16,18]. The DE can be classified as an *evolutionary optimization algorithm*. At present, the best known representatives of this class are *genetic algorithms* [5] and *evolution strategies* [15], while due to its novelty, the DE cannot yet considered as a widely known method. However, currently, there are several variants of the DE [14]. The particular version used throughout this investigation is the *DE/rand/1/bin* scheme. This particular scheme will be mentioned only briefly, since more detailed descriptions are provided in [14,16,17,18].

Generally, the function to be optimized, f, is of the form:

$$f(X): R^D \to R \tag{4}$$

The optimization goal is to minimize the value of this *objective function f(X)*,

$$\min(f(X)) \tag{5}$$

by optimizing the values of its parameters:

$$X = (x_1, ..., x_D), \quad X \in R^D \tag{6}$$

where X denotes a vector composed of D objective function parameters. Usually, the parameters of the objective function are also subject to lower and upper boundary constraints, $x^{(L)}$ and $x^{(U)}$, respectively:

$$x_j^{(L)} \leq x_j \leq x_j^{(U)} \qquad j = 1, ..., D \tag{7}$$

As with all evolutionary optimization algorithms, the DE operates on a *population*, P_G, of candidate solutions, not just a single solution. These candidate solutions are the *individuals* of the population. In particular, the DE maintains a population of constant size that consists of *NP*, real-valued vectors, X_{iG}, where i indexes the population and G is the *generation* to which the population belongs.

$$P_G = \left(X_{1,G},...,X_{NP,G}\right) \qquad G = 0,...,G_{max} \qquad (8)$$

Additionally, each vector contains D real parameters (*chromosomes* of individuals):

$$X_{i,G} = \left(x_{1,i,G},...,x_{D,i,G}\right) \qquad i = 1,..., NP, \ G = 0,...,G_{max} \qquad (9)$$

In order to establish a starting point for optimum seeking, the population must be initialized. A natural way to seed the initial population, $P_{G=0}$, is with random values chosen from within the given boundary constraints:

$$x_{j,i,0} = rand_j[0,1] \cdot \left(x_j^{(U)} - x_j^{(L)}\right) + x_j^{(L)} \qquad i = 1,..., NP, \ \ j = 1,..., D \qquad (10)$$

where $rand_j[0,1]$ denotes a uniformly distributed random value within the range: [0.0,1.0] that is chosen anew for each j.

From the 1st generation forward, vectors in the current population, P_G, are randomly sampled and combined to create candidate vectors for the subsequent generation, P_{G+1}. The population of candidate, or "trial" vectors $P'_{G+1} = U_{i,G+1} = u_{j,i,G+1}$ (where $i = 1,...,NP$, $j = 1,...,D$), is generated as follows:

$$u_{j,i,G+1} = \begin{cases} v_{j,i,G+1} & \text{if } rand_j[0,1) \le CR \vee j = k \\ x_{j,i,G} & \text{otherwise} \end{cases} \qquad (11)$$

where

$$v_{j,i,G+1} = x_{j,r3,G} + F \cdot \left(x_{j,r1,G} - x_{j,r2,G}\right)$$

$$i = 1,..., NP, \ \ j = 1,..., D$$

$$k_i \in \{1,..., D\}, \text{random parameter index}$$

$$r_1, r_2, r_3 \in \{1,..., NP\}, \text{randomly selected,}$$

$$\text{except} : r_1 \ne r_2 \ne r_3 \ne i$$

$$CR \in [0,1], \ \ F \in (0,1+]$$

The randomly chosen indexes, r_1, r_2 and r_3 are different from each other and also different from the running index, i. New, random, integer values for r_1, r_2 and r_3 are chosen for each value of the index i, i.e., for each individual. The index, k, refers to a randomly chosen chromosome which is used to ensure that each individual trial vector, $U_{i,G+1}$, differs from its counterpart in the previous generation, X_{iG}, by at least one parameter. A new, random, integer value is assigned to k prior to the construction of each trial vector, i.e., for each value of the index i.

F and CR are DE's control parameters. Like *NP*, both values remain constant during the search process. F is a real-valued factor in the range (0.0,1.0+] that scales the differential variations. The upper limit on F has been empirically determined. So far, it is considered that values of F greater than unity do not appear to be productive.

614

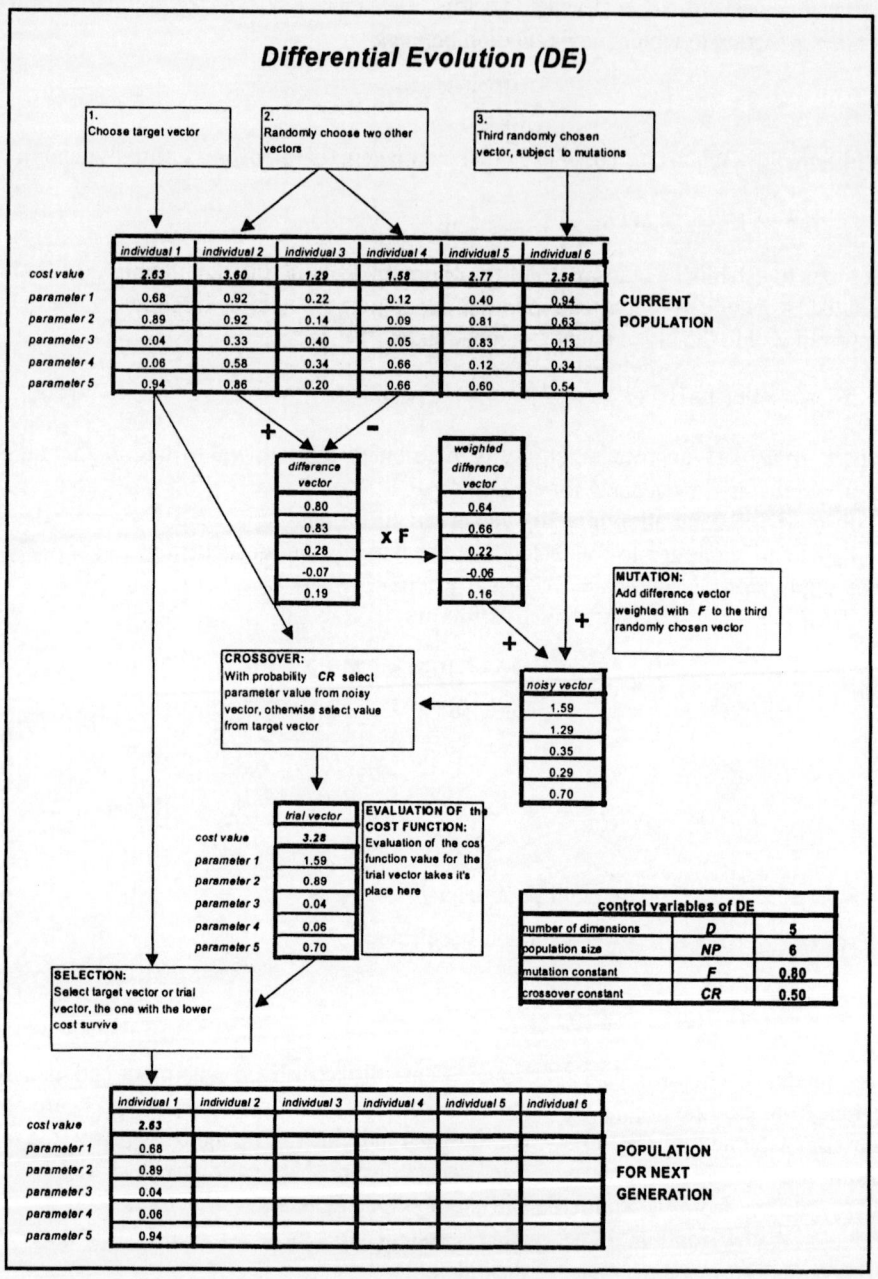

Fig. 4. Differential Evolution works directly with the floating-point valued variables of the objective function, not with their (binary) encoding. The functioning of DE is illustrated here for the case of a simple objective function $f(X) = x_1 + x_2 + x_3 + x_4 + x_5$ [9].

CR is a real-valued crossover factor in the range [0.0,1.0] that controls the probability that a trial vector parameter will come from the randomly chosen, mutated vector, $v_{j,i,G+1}$, instead of the current vector, $x_{j,i,G}$. Usually, suitable values for F, CR and NP can be found by trial-and-error after a few tests using different values.

DE's selection scheme also differs from other evolutionary algorithms. The population for the next generation, P_{G+1}, is selected from the current population, P_G, and the child population, according to the following rule:

$$X_{i,G+1} = \begin{cases} U_{i,G+1} & \text{if } f\left(U_{i,G+1}\right) \leq f\left(X_{i,G}\right) \\ X_{i,G} & \text{otherwise} \end{cases} \tag{12}$$

Thus, each individual of the temporary population is compared with its counterpart in the current population. Assuming that the objective function is to be minimized, the vector with the lower objective function value wins a place in the next generation's population. As a result, all the individuals of the next generation are as good or better than their counterparts in the current generation. The interesting point concerning DE's selection scheme is that a trial vector is only compared to one individual, not to all the individuals in the current population.

2. Experiments

The experimentation was performed with a MLP network having two inputs, one hidden layer containing 5 or 10 nodes, and one output. A sigmoid function was applied as the activation function in the hidden layer. Consequently, the activation function in the output layer was linear. The weights of the MLP network were initialized randomly between −1 and 1 for the DE.

As a test problem, the MLP network was trained to approximate four different functions by applying each of the compared five training algorithms in turn. In each of the training cases the same training and test data sets, depending on the function to be approximated, were applied with all compared training algorithms.

The training set consisted of 50 input points, $x_{1(i)}$ and $x_{2(i)}$, and the corresponding output points, $y_i = (x_{1(i)}, x_{2(i)})$, where $i = 1,...,50$. Respectively, the test set consisted of 100 input points $x_{1(i)}, x_{2(i)}$ and the corresponding output points $y_i = (x_{1(i)}, x_{2(i)})$, where $i = 1,...,100$.

The mentioned four functions to be approximated by the network, denoted by f from now on, were used for generating the datasets used for experimentation and comparisons. The training and test sets were created for each approximated function, f, by generating random input values from the specified domains of the functions (see Eq. 13-16) and then computing the corresponding output value by applying the function f. The following functions were subject to approximation and therefore used for generating the training and test data sets:

Rosenbrock's function

$$f(x_1,x_2) = 100(x_1^2 - x_2)^2 + (1 - x_1)^2, \text{ subject to } x_1, x_2 \in [-2.048, 2.047] \tag{13}$$

Rana's function

$$f(x_1, x_2) = x_1 \sin(\sqrt{|x_2 + 1 - x_1|}) \cos(\sqrt{|x_1 + x_2 + 1|})$$
$$+ (x_2 + 1) \cos(\sqrt{|x_2 + 1 - x_1|}) \cdot \sin(\sqrt{|x_1 + x_2 + 1|}) \qquad (14)$$
$$\text{, subject to } x_1, x_2 \in [-512, 511]$$

Stretched V Sine Wave function

$$f(x_1, x_2) = (x_1^2 + x_2^2)^{0.25}(\sin^2(50(x_1^2 + x_2^2)^{0.1}) + 1) \text{, subject to } x_1, x_2 \in [-100, 100] \qquad (15)$$

Schwefel's function

$$f(x_1, x_2) = -x_1 \sin(\sqrt{|x_1|}) - x_2 \sin(\sqrt{|x_2|}) \text{, subject to } x_1, x_2 \in [-512, 512] \qquad (16)$$

Thus, the experiments were performed simply by training a three-layered MLP approximation of these functions by applying each of the compared five training algorithms: Differential Evolution, BFGS quasi-Newton Back-Propagation, Gradient Descent Back-Propagation, Levenberg-Marquardt Back-Propagation and RPROP Back-Propagation. The mean square error (MSE) between the actual values y_i and the network outputs o_i was used as the objective to be minimized.

Two sizes of the hidden layer, 5 and 10 neurons, were used. Experiments with the DE algorithm were performed applying two different population sizes, $NP = 2D$ and $NP = 3D$, where D is the number of parameters in the objective function. The other DE's search parameters, F and CR, were both set to 0.90.

Each experiment was repeated five times. Thus, 240 experiments were performed in total. Time (processor time) was used as the stopping criterion for all training algorithms: 20 000 seconds with 5 hidden neurons and 80 000 seconds with 10 hidden neurons. Since the definition of a training epoch differs from a training algorithm to another, this stopping criterion was used to guarantee fair comparisons. Furthermore, by applying this stopping criteria, varying computational overheads introduced by each training algorithm were considered. All experiments were carried out on Sun platform under Solaris 2.6 operating system and by using Matlab 5.3.0.10183 .

3. Results

Some results of the experiments are summarized in Table 1. Note here, that in order to avoid overfitted results, the results reported in Table 1 (training time and MSE) were extracted at the moment, when the independent test set error was of its minimum. Generally, concerning the experiments performed, the following trends were observed. Typically, the Back-Propagation algorithms found a fair solution relatively fast, but failed improving the solution when more time was used for training. Rather expectedly, as local optimizers these methods found local solutions quickly, but get stuck on it, failing to continue towards better solutions, or even to the global optimum.

Respectively, the solution with the DE algorithm was not as good as with the Back-Propagation algorithms during the early stages of training, but usually the DE achieved a significantly better solution during the later stages of training. This trend is illustrated in Fig.5. The results suggest that the choice of the training algorithm depend on the time available for training. In the cases studied here, the Back-Propagation algorithms were better at the beginning of the training, but after the Back-Propagation got stuck on a local solution the DE soon overtook it, and trained the network better than the Back-Propagation.

Table 1. Results of the experiments. In this table only the DE results are presented, which were done with a population size 3D (D is the number of weights in the net). MLP layer configurations (input-hidden-output layers) 2-5-1 and 2-10-1 were used in experiments.

Function approximated by the NN	Number of nodes in hidden layer		Training algorithm [1) 2) 3)]				
			DE	BFG	GD	LM	RP
Rana's function (Eq. 14)	5	MSE [4)]	37 442	45 678	55 608	56 679	60 860
	5	Time [5)]	9 752	1 197	5 550	20	197
			DE	BFG	GD	LM	RP
	10	MSE [4)]	35 376	46 375	53 675	48 007	57 678
	10	Time [5)]	56 232	160	170	12	813
Rosenbrock's function (Eq. 13)	5		DE	BFG	GD	LM	RP
	5	MSE [4)]	10 823	179 768	239 455	203 433	290 251
	5	Time [5)]	12 039	370	1 695	2 725	6 527
			DE	BFG	GD	LM	RP
	10	MSE [4)]	18 271	108 961	129 190	1 161	198 041
	10	Time [5)]	69 655	25 541	4 416	2 317	3 930
Schwefel's function (Eq. 16)	5		DE	BFG	GD	LM	RP
	5	MSE [4)]	59 582	78 758	80 049	78 064	79 718
	5	Time [5)]	7 537	201	262	7	467
			DE	BFG	GD	LM	RP
	10	MSE [4)]	73 102	77 792	73 611	80 600	78 126
	10	Time [5)]	4 623	19	1 440	2	310
Stretched V Sine Wave function (Eq. 15)	5		DE	BFG	GD	LM	RP
	5	MSE [4)]	10	17	13	12	17
	5	Time [5)]	10 950	5 645	1 902	1 260	57
			DE	BFG	GD	LM	RP
	10	MSE [4)]	16	17	9	10	13
	10	Time [5)]	54 332	1 778	2 002	369	31

[1)] Training algorithms: **DE**: Differential Evolution, **BFG**: Quasi-Newton Back-Propagation, **GD**: Gradient Descent Back-Propagation, **LM**: Levenberg-Marguardt Back-Propagation, **RP**: RPROP Back-Propagation.

[2)] MATLAB Neural Networks Toolbox were used for experimentation, except the MATLAB implementation of the DE-algorithm, which was programmed by Miika Lindfors.

[3)] With the DE-algorithm the population size NP=3D was used here. The D is the number of weights to be trained. For the other tuning parameters of DE, F=0.9 and CR=0.9 were applied.

[4)] Mean Squared Error, averaged over five independent training attempts for each training algorithm.

[5)] Time that was needed to get the value on the MSE-column. Note that this is not the time that was used for the whole training process.

It appears that the DE is a rather slow training method, but still provides a progress rate comparable to the Back-Propagation algorithms. Another important aspect is the quality of the training result. In the case of approximating the Rosenbrock's function (Eq. 13), it was possible to find a better solution using the Levenberg-Marquardt algorithm than with the DE algorithm. Another exception from the general trend was the difference between the DE and the Back-Propagation, in the case of the Stretched V Sine Wave function and 10 hidden neurons, where the Gradient Descent Back-Propagation obtained a significantly better result (see Table 1). In all the other cases, the DE was the best alternative from the achieved mean squared error (MSE) point of view. Thus, assuming that the longer training time required by the DE is available, the DE appeared to be at least a reasonable alternative to the Back-Propagation. For finding a good suboptimal solution as fast as possible, the Back-Propagation is likely to outperform the DE.

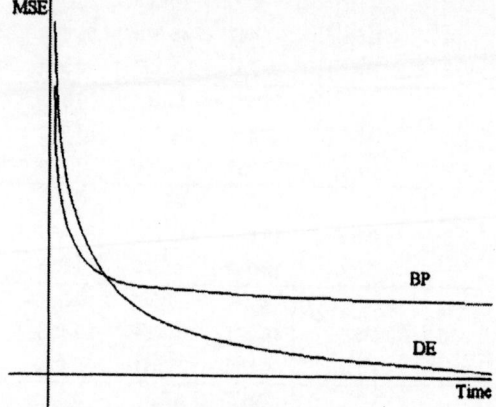

Fig. 5. Comparison of DE and Back-Propagation algorithms. An illustration of the general trend observed during the experimentation. Back-Propagation algorithms typically train the network better at the beginning of the training, but after some time, depending on the training problem, the DE overtakes Back-Propagation.

4. Conclusions

The preliminary results of our experiments suggest that the DE-algorithm is a reasonable alternative for the widely applied Back-Propagation algorithms for training MLP networks. Especially if there is enough time available for performing the learning by the DE, a training result with a significantly lower MSE can be often achieved. If the training time available is relatively short, the Back-Propagation algorithms are likely to result in a lower MSE.

For setting the tuning parameters of the DE for training MLPs, on the basis of the experiments, our preliminary recommendation is to set F to 0.9 and CR to 0.9. See [9,14] for more details on how to set the internal parameters of the DE. For the population size, NP, our preliminary recommendation is setting NP to 2-3 times the number of weights in the network to be trained. An interesting point was the fact that the MLP network can be trained well although the size of population is relatively small. A population size 2-3 times the number of weights in the network appeared to be large enough.

The conclusions drawn here are based on a rather limited set of experiments and should be considered as preliminary results only. Further research is required to fully explain and understand the DE's properties in this context. So far, only some empirical evidences exist about the efficiency and effectiveness of our approach. Another limitation of our experiments and conclusions is that only moderate size networks have been trained until now. Larger networks should be subject to future experimentation.

However, our main conclusion concerning the promising applicability of the DE-algorithm for training NNs is in line with the conclusions of the recent investigations performed by other researchers who have investigated the training of NNs by applying DE [2,3,4,6,8,10,11,12,13].

Based on the results of this limited feasibility study, further investigations in this field are required. On the basis of the results obtained so far, further studies are also well justified. In our future work, the possibilities for learning large networks will be one of the most important subjects. On the basis of the current result we expect, that the main limitation of the DE's applicability for training large networks will be the time required for training. For this reason, in order to speed up the convergence rate of the DE, various modifications, specific to neural network training, are under development at the moment. Another interesting subject for future investigations is the DE's capabilities to optimize also the topological parameters of neural network during the training process. In principle at least, as a global optimizer based on a direct search, the DE can be applied for that purpose.

Acknowledgements

The authors would like to thank researcher Joni Kämäräinen from Lappeenranta University of Technology for sharing his knowledge, including his constructive comments during this work. Joni have also programmed a refined MATLAB implementation of our DE code for training MLPs. This MATLAB Neural Networks Toolbox compatible code, accompanied with further information about our project, can be found via the Internet http://www.it.lut.fi/project/nngenetic/.

References

1. Alander J.T. (1999). An Indexed Bibliography of Genetic Algorithms and Neural Networks. Available from: ftp://ftp.uwasa.fi/cs/report94-1/gaNNbib.ps.Z.
2. Fischer M., Hlavackova-Schindler K., Reismann M. (1999). A Global Search Procedure for Parameter Estimation in Neural Spatial Interaction Modelling. Papers in Regional Science 78(2):119–134, 1999. ISSN 1056–8190. Available via the Internet: http://wigeo.wu-wien.ac.at/~reismann/13.pdf.
3. Fischer M.M., Hlavackova-Schindler K., Reismann M. (1999). An Evolutionary Mutation-Based Algorithm for Weight Training in Neural Networks for Telecommunication Flow Modelling. In: Mohammadian M. (ed.), Computational Intelligence for Model-

ling, Control and Automation. Evolutionary Computation and Fuzzy Logic for Intelligent Control, Knowledge Acquisition and Information Retrieval, 17.-19. February 1999; Vienna, Austria. Concurrent Systems Engineering Series Vol.55, pp. 54–59. IOS Press, Amsterdam, Netherlands.

4. Fischer M.M., Reismann M, Hlavackova-Schindler K. (1999). Parameter Estimation in Neural Spatial Interaction Modelling by a Derivative Free Global Optimization Method. The IV International Conference on GeoComputation, Mary Washington College, Fredericksburg, VA, USA, 25-28 July 1999. Available via the Internet: http://www.geovista.psu.edu/sites/geocomp99/Gc99/007/gc_007.htm.

5. Goldberg D.E. (1989). Genetic Algorithms in Search, Optimization and Machine Learning. Addison-Wesley, Reading (MA). ISBN 0-201-15767-5.

6. Schmitz G.P.J., Aldrich C. (1999). Combinatorial Evolution of Regression Nodes in Feedforward Neural Networks. Neural Networks 12(1):175-189, 1999.

7. Koikkalainen P. Neuraalilaskennan mahdollisuudet. Teknologian kehittämiskeskus, TEKES 43/94, 1994. In Finnish.

8. Lampinen J., Zelinka I. (1999). DELA – an Evolutionary Learning Algorithms for Neural Networks. In: Ošmera, P. (ed.). Proc. of MENDEL'99, 5th Int. Conf. on Soft Computing, June 9.–12. 1999, Brno University of Technology, Brno (Czech Republic), pp. 410–414. ISBN 80-214-1131-7.

9. Lampinen J., Zelinka I. (2000). On Stagnation of the Differential Evolution Algorithm. Proc. of MENDEL 2000, 6th Int. Conf. on Soft Computing, June 7-9, 2000. Brno University of Technology, Brno (Czech Republic), pp.76–83. ISBN 80-214-1609-2. Available via the Internet: http://www.lut.fi/~jlampine/debiblio.htm.

10. Masters T., Land W. (1997). A New Training Algorithm for the General Regression Neural Network. 1997 IEEE International Conference on Systems, Man, and Cybernetics, Computational Cybernetics and Simulation., Volume: 3, pp. 1990–1994.

11. Li Gang, Tu Yiqing, Tong Fu (1999). A Fast Evolutionary Algorithm for Neural Network Training Using Differential Evolution. In: Luo J., Xu B., Wang Y., Li X., Lu J. (eds.). Proc. of ICYCS'99, Fifth Int. Conf. for Young Computer Scientists. vol.1. 17-20 Aug. 1999; Nanjing, China. Vol. 1, pp. 507–511. Int. Acad. Publishers, Beijing, China.

12. Plagianakos V.P, Vrahatis M.N. (1999). Neural Network Training with Constrained Integer Weights. Proceedings of the 1999 Congress on Evolutionary Computation, CEC'99, Vol. 3, pp. 2007–2013. IEEE, Piscataway, NJ, USA. ISBN 0-7803-5536-9.

13. Plagianakos V.P., Vrahatis M.N. (1999). Training Neural Networks with 3-bit Integer Weights. In: Banzhaf W., et.al. (eds.). Proc. of the Genetic and Evolutionary Computation Conf. GECCO-99, Orlando, USA, 13-17 July 1999, Vol. 1, pp. 910–915. Morgan Kaufmann Publishers, San Francisco, CA, USA.

14. Price K. (1999). An Introduction to Differential Evolution. In: Corne D., Dorigo M., Glover, F. (eds.) (1999). New Ideas in Optimization. McGraw-Hill, London (UK), pp. 79–108. ISBN 007-709506-5.

15. Schwefel H-P (1995). Evolution and Optimum Seeking. John Wiley & Sons Inc., New York.

16. Storn R., Price K. (1995). Differential Evolution - a Simple and Efficient Adaptive Scheme for Global Optimization Over Continuous Spaces. Technical Report TR-95-012, ICSI, March 1995. Available from: ftp://ftp.icsi.berkeley.edu/pub/techreports/1995/tr-95-012.ps.Z.

17. Storn R., Price K. (1997). Differential Evolution – A simple evolution strategy for fast optimization. Dr. Dobb's Journal, April 97, pp. 18–24 and p. 78.

18. Storn R., Price, K. (1997). Differential Evolution – a Simple and Efficient Heuristic for Global Optimization over Continuous Spaces. Journal of Global Optimization, 11(4):341–359, December 1997. Kluwer Academic Publishers.

A Genetic Algorithm Based on Cell Loss for Dynamic Routing in ATM Networks

P Cortes[1], J Muñuzuri[1], J Larrañeta[1] and L Onieva[1]

[1] Seville University, Escuela Superior Ingenieros. Grupo Ingeniería Organización. Camino de los Descubrimientos s/n. Sevilla 41092. Spain.

Abstract. New B-ISDN will have to ensure a high standard of quality of service measured as delay in the communications, guarantee of safety communications and of course no loss of information. Although the routing optimization problem has been previously dealt with in the bibliography, not many works have been presented according to the routing in ATM networks using cell loss as the overall criterion. Here a new model based on cell loss is presented for an ATM network with matrix switches and queues at the end. A quickly evaluated genetic algorithm is proposed to solve the model.

1 Introduction

In the near future telecommunication networks with higher capacity and speed will be needed to give solution to the set of transport and traffic volume requirements. In the eighties of the past century, the ITU-T recommended the adoption of Asynchronous Transfer Mode (ATM) as switching mode and the Synchronous Digital Hierarchy (SDH) as transmission mode for the future broadband network. In this context, the routing algorithms will have to ensure a high standard of quality of service measured as delay in the communications, guarantee of safety communications and of course no loss of information.

Several references based on soft computing techniques can be found dealing with the routing operation, most of them are based on circuit or packet switched networks. So, diverse authors (e.g. [5] or [12]) have shown the use of neural approaches to solve the real time traffic routing problem. A fuzzy multiobjective optimization model to develop a routing algorithm to guarantee the various quality of service characteristics requested by the wide range of applications supported by B-ISDN has been presented in [1]. The use of cutting planes has been analyzed to deal with the routing problem with capacity limitations, [4]. A hybrid neural-genetic procedure to deal with the bandwidth allocation of virtual paths in ATM networks has been developed in [3]. Here we propose the use of a genetic algorithm to route the communications in an ATM operation mode ensuring no relevant loss of information through a new model based on cell loss.

The rest of the paper deal with the ATM mode description in section 2, including the different time scales to be considered in ATM operation and the loss models in ATM switches with queues at the end. A new model based on cell loss is presented in section 3. Next, section 4 introduces a genetic algorithm to solve the problem. Section 5 shows the results in randomly generated and real life problems. Finally, we review the main aspects in the latest section.

2 The Asynchronous Transfer Mode

The Asynchronous Transfer Mode (ATM) was created to provide fastest communications in both public and private networks. ATM can be characterized as a hybrid between circuit and packet switched networks, [11]. ATM combines circuit and packet switched networks maintaining the channel structures, by means of timeslot division, as in circuit switched, and the packet structure, by means of ATM cells, as in packet switched.

ATM cells are 53 bytes sized, 5 bytes for the header and 48 bytes for the payload. Into the header, the VCI (Virtual Channel Identifier) and VPI (Virtual Path Identifier) fields specify the circuit and path that each cell should flow.

The VPI and VCI identifiers are only valid into each connection link, and acts identifying a logic circuit into the physical link, see figure 1. When the cell arrives to the switch, the VPI/VCI are found into the routing tables and the cell is routed trough the output port changing the identifier when necessary. The virtual circuit (or channel) concept is an abstraction representing the cell unidirectional transport associated to an identifier (VCI). The VCI as the VPI indicate an ordered cell flow associated to a concrete connection. The virtual connection channel (VCC) concept means the connection of the different virtual paths from the origin of communication until the destination of communication, as next figure 1 depicts.

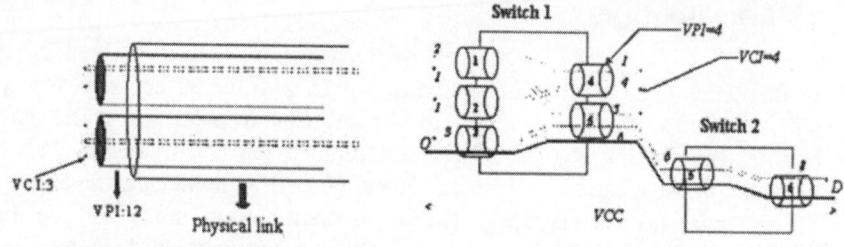

Fig. 1. Virtual circuits, virtual paths and virtual connection channel

Into the switches, the communication is transported from an inflow logic channel into an outflow logic channel. This procedure is known as routing.

Cell loss in ATM switches appears when very much cells are routed onto the same link, being higher the cell number than the switch buffer size. Habitual values for cell loss probability are from 10^{-8} to 10^{-11}.

The switches can be classified in accordance with the structure of the switching fabric architecture. The matrix switch is one of the most recommended switches, where a connection matrix establishes the interconnection of any input point with any output point. One of the better delay-performance relations is obtained for matrix switches with queues at the end. This option allows transmit more than one cell in the same timeslot. To ensure no cell loss the transference must be N (number of input ports) times faster than the cell homing rate. Moreover, the queue location at the end avoids the line header blocking.

The ATM service categories are four. Constant Bit Rate service (CBR) to maintain real time communications as video or audio with very strict delay requirement. The CBR service provides a connection with large bandwidth and very low cell loss probability. Variable Bit Rate service (VBR) indicated for frame relay traffic. Available Bit Rate service (ABR) used for unknown characteristics traffic and moderately restrictive attending to cell loss and without delay requirement as real time applications. Finally, Unspecified Bit Rate service (UBR) to use the remaining available bandwidth.

2.1 Time scales in ATM networks

The traditional Erlang model considers one unique time scale: the connection scale or call scale. In bandwidth traffic this only consideration is not possible. At this respect, a detailed study is done in [10], considering three time scale levels: the call level, the burst level and the cell level proposing a control procedure acting at the burst and call level. Here, we consider the cell level necessary to deal with the switch cell loss analysis, so the flow variables must be referred to the cell level scale. In addition, we consider the call level to ensure the call routing. In ATM, once the first cell is routed onto a virtual circuit all the remaining cells follow the first one through the same circuit, so the path variables must be referred to the call level scale. Figure 2 reveals the different levels and their interrelations.

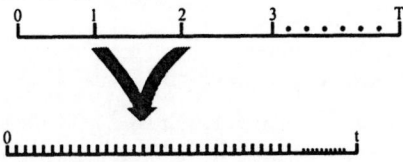

Fig. 2. Timescales in ATM networks

The cell loss function depends on the switch internal structure. In our case, we are considering matrix switches with queues at the end. In ATM switches, the transmission time for one cell is measured as a slot, approximately 2.8 μseg at 155.52 Mbps. The ATM networks are synchronized, so when a cell leaves the buffer at the end of the slot, it is synchronized and served at the beginning of the following slot. The cell arrivals can take place in any instant during a slot. The precise arrival instant is no relevant because the cells arriving during the slot n will not leave the buffer before the slot $n+1$.

We use the term port to designate each of the inputs and outputs at the switch, so we deal with input ports and output ports relative to the input links and the output links. We will use independently the terms port or link. The following figure 6 depicts the previous description.

624

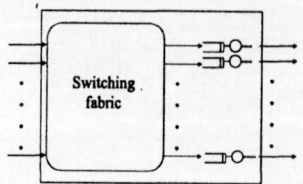

Fig. 3. Ports in an ATM switch

2.2 Cell loss model in ATM networks

In ATM networks, each switch is composed by a set of buffers and a switching fabric. The buffers are finite capacity queues where the cells wait to be served; each buffer can be well modeled as a Markov chain. The switching fabric can be implemented by means of an interconnection matrix, central memory, bus or interconnection ring. Therefore, the problem can be assimilated to a queuing system. In this line, the cell loss rate, L, is described in [11] as

$$L = E(k) - \rho \tag{1}$$

Being ρ the average number of served cells per timeslot and $E(K)$ the expected cell number arriving to the system.

According to these expressions we can obtain the total number of lost cells in each switch output port, supposing size-b buffers in a timeslot and measured as cells/slot. We note this cell loss function as $L_{ij}^{\tau}(\lambda)$ meaning the cell loss in the output port j, relative to the switch i in the instant τ when the arrival rate is λ.

It is habitual to use the cell loss probability (CLP) calculated as CLP=$L/E(k)$ or the cell loss rate (CLR) calculated as the total lost cells with respect to the total transmitted cells during a period. In this paper, we make use of the total cell loss as the objective function for evaluating the routing options, as well as the CLR.

Finally, we have to remark that the buffer size is a critical parameter. Each ATM switch must have a size of queue enough to ensure low cell loss and low delay conditions. These two conditions are opposing. Next figure 8 represents the CLP versus the queue capacity. It can be appraised how a not very large queue size guarantees acceptable CLP. Anyhow, we implement a buffer size based on standard sizes for control admission call, [6], as can be seen in the appendix.

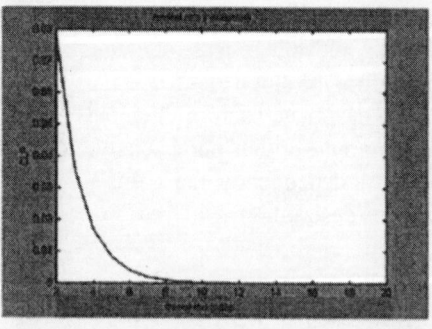

Fig. 4. Cell loss probability versus queue size

3 Model based on cell loss

ATM networks can be well suited by graphs $G = (N, E)$ being N the set of nodes or switches and E the set of links. Many research has been done under the perspective of minimizing the delay [2], in such cases the only delay considered is the delay into the switch due to the queue and the real switching process because the rest of sources of delay, electric-optic conversion or signal propagation, are second order factors.

ATM cells can be lost through four manners: due to line transmission failures, Usage Parameter Control (UPC) in the switching network, buffer sizes or switching fabric limitation. We consider as cell loss only those inside the switch. It is not a strong limitation because it represents the largest part of the cell loss in ATM networks. Moreover, in our model, we consider only CBR traffic, also this is not a strong restriction because of VBR, ABR and UBR traffic are much less restrictive with the quality conditions according to cell loss (our objective function).

Under these conditions, the input data are the network topology and the demand of communication under CBR service conditions and the maximum allowed cell loss. Moreover, we assume links of extra capacity. It is a realistic supposition because telecommunication links are over dimensioned due to the prevision of future traffic increments. So here, we do not consider the capacity problem.

□ **Parameters:**
- **N** Set of nodes (switches).
- **E** Set of links.
- **M** Set of communication origin-destination pairs ($O(m)$ origin of communication m and $D(m)$ destination of communication m).
- **H(m)** Set of feasible paths to connect each origin-destination pair m.
- **T** Temporal horizon at call scale. This horizon is composed by call time division $t \in T$. Subsequently, the call time scale is divided into the cell time scale, $\tau \subset t$.
- **N(h)** Set of nodes belonging to path h.
- **E(h)** Set of links belonging to path h.
- **B(i)** Set of nodes located before node i.
- **A(i)** Set of nodes located after node i.

□ **Variables:**
- $P_h^{m,t}$ Binary variable indicating if the connection between the origin and destination of communication $O(m)$ and $D(m)$ is established by the path h, referred at the call scale $t \in T$.
- $X_{h,ij}^{m,\tau}$ Continuous flow variable (cell/slot) associated to the pair m over the link (i,j) in the expressed direction (from i to j), being (i,j) a link of the path h, referred to the cell scale.
- $l_{h,ij}^{m,\tau}$ Continuous variable (cell/slot) determining the cell loss at the link (i,j) due to the port j into the switch i in the connection path h used by the pair of communication m, in the timeslot $\tau \subset t$ (so it is referred to the cell scale).
- F_{ij}^{τ} Total cell flow variable (cell/slot) that should be routed from node i to node j. It is referred to the cell scale $\tau \subset t$.

It is important to note the difference between the variables F_{ij}^{τ} and $X_{h,ij}^{m,\tau}$. The first of them refers to the total flow of cells that ideally should go through the link (i,j), supposing no cell loss, meanwhile the second variable has in account the cell loss possibility. Next figure 5 depicts an example. The switches v, w, y, z are previously located to the node i, and the nodes j, k are located after the node i. For an instant τ, there is a flow of cells from nodes v, w, y, z to the node i, the flow from nodes v, z must be routed to the node j meanwhile the flow from nodes w, y must be routed to the switch k.

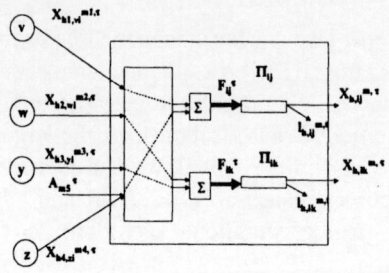

Fig. 5. Total flow and cell loss into the switch

☐ **Data:**

- A_m^{τ} Demand of the origin-destination pair m in the timeslot $\tau \in t$. It is referred to the cell scale and expressed in cell/slot.
- GoS^m Grade of service. Maximum number of lost cells for each connection m, measured as cells/slot.

Once has been introduced the notation the model can be written as follows:

$$MIN \quad \sum_{t \in T} \sum_{\tau \in t} \sum_{(i,j) \in E} \sum_{m \in M} \sum_{h \in H(m)} l_{h,ij}^{m,t} \tag{2}$$

$s.t.$

$$\sum_{h \in H(m)} P_h^{m,t} = 1 \quad \forall m \in M, \forall t \in T \tag{3}$$

$$X_{h,ij}^{m,\tau} + l_{h,ij}^{m,\tau} = \begin{cases} A_m^{\tau} P_h^{m,t} & \text{if } i = O(m) \\ \\ X_{h,ki}^{m,\tau-1} & \text{if } i \neq O(m) \end{cases} \tag{4}$$

$\forall \tau \subset t, \forall t \in T, \forall h \in H(m), \forall m \in M, \forall i \in N : \forall (i,j) \in E(h), \forall (k,i) \in E(h)/i,j,k \in N(h)$

$$F_{ij}^{\tau} = \sum_{m \in M} \sum_{\substack{h \in H(m), \\ (q,i) \in E(h): j \in N(h)}} \sum_{q \in B(i)} X_{h,qi}^{m,\tau-1} + \sum_{m \in M/i=O(m)} A_m^{\tau} \quad \forall (i,j) \in E, \forall \tau \subset t, \forall t \in T \tag{5}$$

$$\sum_{m \in M} \sum_{h \in H(m)} l_{h,ij}^{m,\tau} \geq L_{ij}^{\tau}(F_{ij}^{\tau}) \quad \forall i \in N, \forall (i,j) \in E, \forall \tau \subset t, \forall t \in T \tag{6}$$

$$\sum_{\tau \in t} \sum_{h \in H(m)} \sum_{(i,j) \in E(h): j \in A(i)} l_{h,ij}^{m,\tau} \leq GoS^m \quad \forall m \in M, \forall t \in T \tag{7}$$

$$l_{h,ij}^{m,\tau} \geq 0 \quad \forall (i,j) \in E, \forall m \in M, \forall \tau \subset t, \forall t \in T \tag{8}$$

$$X_{h,ij}^{m,\tau} \geq 0 \quad \forall (i,j) \in E, \forall m \in M, \forall \tau \subset t, \forall t \in T \tag{9}$$

$$P_h^{m,t} \in (0,1) \quad \forall m \in M, h \in H(m), t \in T \tag{10}$$

The constraint (3) is referred to the call scale and it determines how the communication for pair m must be done trough one only path, h, according to the ATM mode. The constraint (4) is the flow balance equation; the term on the left shows the flow through the link (i,j) belonging to path h, that is used to connect the pair m, plus the loss in the buffer j of the switch i. The term on the right shows two situations: in the first of them, the node i is the origin of communication, so all the created traffic is sent over the path h; in the second one, the node i is an intermediate node of the path h, so the expression must include the flow sent by the node before the switch i in the path h, i.e. the link (k,i), this traffic will carry a timeslot delay due to the switching process and includes the losses into the previous switches. This balance equation is imposed at the cell scale. The constraint (5) determines the total flow including all the flows homing from the nodes located previously to the node i and being routed to the switch j. In constraint (6), the cell loss is characterized as $L_{ij}{}^\tau(F_{ij}{}^\tau)$ for the output port j relative to the switch i in the instant τ when the arrival rate is the total flow homing to the switch i and out flowing to port j, i.e. $F_{ij}{}^\tau$. The term on the left reflects the sum for all the origin-destination pairs and their feasible paths for the connection including the link (i,j) and the term on the right sets the losses at the output port j. Finally, the constraint (7) is a bound on the network grade of service, so a maximum cell loss is imposed for each path of communication. The objective function (2) has been discussed in section 2.2 and it includes the total amount of losses for all the pairs of communication, for each path, for each link and for the entire horizon.

4   The genetic algorithm

To solve the problem, our proposal is a genetic algorithm. We state as individuals of the population the routing structure for each connection. So, the genetic encoding, figure 6, is determined by a binary matrix with the next considerations:
- Each individual must contain so many lines as periods of connection. We consider t as the time for an ATM communication (call scale).
- Each of the lines is composed by a number of fields, M, representing the different origin-destination pairs of communication.
- The field associated to each pair is shaped by so many elements as feasible paths exist between that pair of communication. Only one route is feasible for each communication, having the value of 1.

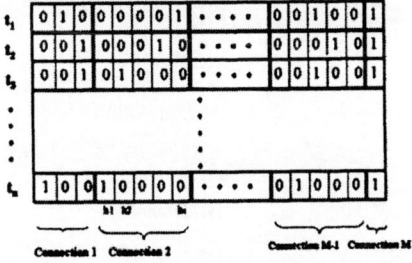

Fig. 6. Individual chromosome

The population is randomly generated and is composed by N individuals as defined in the previous figure 6. With these conditions, we propose a genetic algorithm with the following characteristics:

- *Random parents selection*. All the individuals from the size N population are randomly selected. This will enrich the population genetic variety.
- *Uniform crossover operators*. If both parents use the same route for the same call scale time, t, and the same pair m, the offspring maintains this connection in the same time t. Otherwise, it is elected the alternative of each of the parents with the same ½ probability. Crossovers are done with preventing incest from happening and duplicate generation control to enrich the genetic variety of the population.
- *Mutation operator*. A mutation process acts changing an active route per other.
- *Ranking based replacement*. We use a hypergeometric function to let more probability of replacement to the individuals with worse fitness. So, the individual in ranking position-i, have a replacement probability equal to $q(1-q)^i$, being q the replacement probability of the worst individual.
- *Stop criterion*. We propose a dual control for the stop criterion. First, we set a maximum number of iterations, N_MAX. Second, we follow the suggestion in [7] to control the population entropy level. We calculate the entropy of each pair of communication, S_m, as function of the activation frequency, $fr^t_{h(m)}$, of each route h of the pair m in instant t (11). After that the population entropy, S, can be calculated as the sum of the origin-destination pair entropies weighted with respect to the number of alternative routes for each pair (12).

$$S_m = -\sum_{t \in T} \sum_{h \in H(m)} fr^t_{h(m)} \cdot \ln fr^t_{h(m)} \tag{11}$$

$$S = \sum_{m \in M} \left(\frac{|H(m)|}{\sum_{m \in M} |H(m)|} \cdot S_m \right) \tag{12}$$

- *Fitness evaluation*. Each individual of the population is evaluated according to the algorithm described in section 6.1.

Once the variable $P_h^{m,t}$ has been fixed, and grouping constraints (5) and (6) into one, the resultant model is given by (13), (14), (15) and (16):

$$X_{ij}^{m,\tau} + l_{ij}^{m,\tau} = \begin{cases} A_m^\tau & \text{if } i = O(m) \\ \\ X_{ki}^{m,\tau} & \text{if } i \neq O(m) \end{cases} \tag{13}$$

$$\forall \tau \subset t, \forall m \in M, \forall i \in N : \forall (i,j) \in E(h), \forall (k,i) \in E(h), /i,j,k \in N(h)$$

$$\sum_{m \in M} l_{ij}^{m,\tau} \geq L_{ij}^\tau (\sum_{m \in M} \sum_{q \in B(i)} X_{qi}^{m,\tau-1} + \sum_{m \in M/i=O(m)} A_m^\tau) \quad \forall i \in N, \forall (i,j) \in E(h): q \in A(i), \forall \tau \subset t \tag{14}$$

$$l_{ij}^{m,\tau} \geq 0 \qquad \forall (i,j) \in E, \forall m \in M, \forall \tau \subset t \tag{15}$$

$$X_{ij}^{m,\tau} \geq 0 \qquad \forall (i,j) \in E, \forall m \in M, \forall \tau \subset t \tag{16}$$

The grade of service constraint is erased from the model and it will be had in account as a feasibility control of the individuals, discarding those individuals not satisfying the grade of service control.

4.1 Fitness evaluation algorithm

1. For each τ in t:

   1.1. Calculate:
   $$F_{ij}^{\tau} = \sum_{m \in M} \sum_{\substack{q \in B(i) \\ q,i,j \in h}} X_{qi}^{m,\tau-1} + \sum_{m \in M / i = O(m)} A_m^{\tau} \tag{17}$$

   1.2. Calculate:
   $$l_{ij}^{\tau} = \sum_{m \in M} l_{ij}^{m,\tau} = L_{ij}^{\tau}(F_{ij}^{\tau}) \tag{18}$$

   1.3. The total l_{ij}^{τ} is demand proportionality distributed into each
   connection as:
   $$l_{ij}^{m,\tau} = l_{ij}^{\tau}\left(\frac{A_m^{\tau}}{\sum_{m:(i,j) \in h(m)} A_m^{\tau}} \right) \tag{19}$$

   1.4. $X_{ij}^{m,\tau}$ is calculated substituting $l_{ij}^{m,\tau}$ into the flow balance equation as:
   $$X_{ij}^{m,\tau} + l_{ij}^{m,\tau} = \begin{cases} A_m^{\tau} & \text{if } i = O(m) \\ \\ X_{ki}^{m,\tau-1} & \text{if } i \neq O(m) \end{cases} \tag{20}$$
   $$\forall i \in N : \forall (i,j) \in E(h), \forall (k,i) \in E(h), / i,j,k \in N(h)$$

2. The fitness is calculated as:
   $$\sum_{t \in T} \sum_{\tau \in t} \sum_{(i,j) \in E(h)} \sum_{m \in M} l_{ij}^{m,\tau} \tag{21}$$

3. Test the grade of service feasibility:
   $$\sum_{\tau \in t} \sum_{(i,j) \in E(h)} l_{ij}^{m,\tau} \geq GoS^m \tag{22}$$

5 Computational results

All the tests have been run on a 200 MHz PC Pentium MMX workstation. We note that ATM routing is done among nodes at the same hierarchical level, so the number of nodes has not to be too much large for realistic problems, [9]. The test networks were randomly generated and one additional real life network has been considered. The parameters are given bellow and are shown in figure 7:

- Network A: 7 nodes and 18 links.   Network B: 12 nodes and 46 links.
- Network C: 15 nodes and 62 links.   Network D: 18 nodes and 52 links.
- Network E: Real life transport network with 9 nodes and 24.

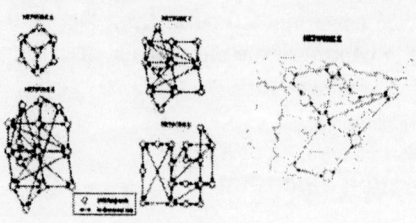

Fig. 7. Test networks

We consider three different CBR demand scenarios representing three peak levels of demand. The lowest level (scenario I) is equivalent to a demand of 0.265 Mbps between every origin-destination pair, the medium level (scenario II) is equivalent to 0.8 Mbps and the highest level (scenario III) is equal to 1.55 Mbps.

The five networks and the three demand levels imply a total amount of 15 different scenarios being analyzed. In table 1, we have used CLR parameter as an appropriated indication of the routing quality.

Table 1. CLR and CPU time results.

Network	Demand scenario	CLR ($\times 10^{-14}$)	Average CPU time (seconds)
A	I	4.76	
	II	2.24	125.1
	III	1.4	
B	I	3.37	
	II	1.64	375.2
	III	1.67×10^4	
C	I	2.82	
	II	80.3	585.6
	III	2.11×10^7	
D	I	2.05	
	II	1.21×10^3	618.5
	III	7.2×10^7	
E	I	3.89	
	II	1.82	179.67
	III	1.05	

The table shows the increment of CLR when the number of nodes grows up because of the new connections with the rest of nodes. On the other hand, the different demand scenario affects directly to CLR excepting in *network A* case. For such case, the volume of traffic is very low and CLR does not grow up due to an upper increment of the total transmitted cells than the lost cells. The situation turns to critical only for very high density of nodes and traffic (cases C and D in scenario III). All of them are cases with a very large number of switches, even higher than realistic problems. Figure 8 reveals CLR ($\times 10^{-14}$) behavior versus the number of nodes represented in a logarithmic scale: $CLR^* = ln\,(CLR \times 10^{-14})$.

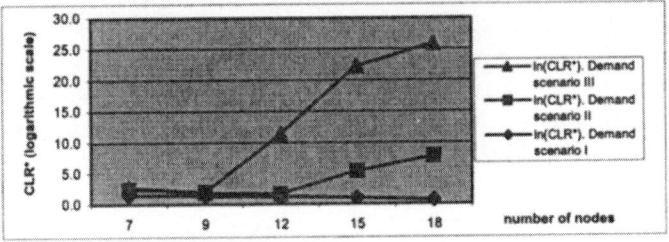

Fig. 8. CLR in demand scenario I

Finally, with respect to the CPU time consumption we have to state that solutions are calculated in efficient time when we are dealing with real life dynamic routing in ATM networks, solving them in the order of seconds.

We have considered also for the real life *network E* the total number of lost cells for comparison with parameter CLR. Moreover, we have considered two additional demand scenarios: scenario IV with a demand of 3.11 Mbps and scenario V with a demand of 3.9 Mbps. The next table 2 summarizes the results.

Table 2. CLR *versus* total lost cells for network E

Network	Demand scenario	Total lost cells ($\times 10^{-6}$)	CLR ($\times 10^{-14}$)
E	I	1.58	3.89
	II	2.16	1.82
	III	2.5	1.05
	IV	1.75×10^{3}	3.68×10^{2}
	V	1.4×10^{8}	2.25×10^{4}

The global losses increase in line with the traffic demand. While the network is not very saturated the total lost cells behavior shows bounded levels, however when the increment of traffic leads to the network congestion (critical point) the total lost cells grow strongly up. The figure 9 reveals the critical points for both total lost cells $\times 10^{-6}$ and CLR$\times 10^{-14}$ parameters.

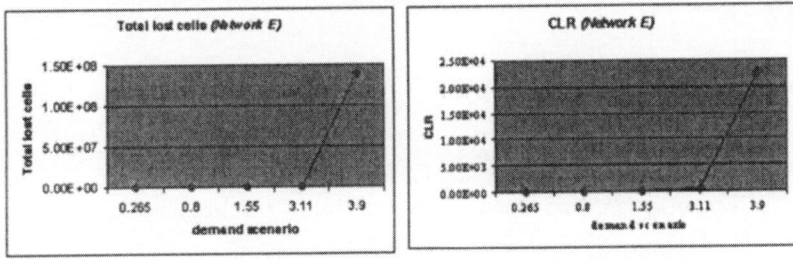

Fig. 9. Total lost cells and CLR as function of the demand scenario. Case network E

6 Conclusions

In this paper, we have dealt with the routing problem in ATM networks considering matrix switches and queues at the end. We have presented a new model based on cell loss and a genetic algorithm to solve it with random parent selection, replacement based on fitness ranking, uniform crossover operators and entropy control. The fitness estimation is quickly evaluated by means of the individual chromosome encoding. The procedure showed an adequate behavior related to the cell loss rate (CLR), the total lost cells and the CPU time consumption. The tests were evaluated in both randomly generated and real life networks.

Appendix. Algorithm pseudo-code and parameters

1. Randomly generation of an initial sized-N population.
2. Population individuals fitness evaluation.
3. *Do while* {number of iterations < N_MAX}&{entropy > $S_{minimum}$}
   - 3.1. Crossover/mutation randomly selection.
   - 3.2. Parent(s) randomly selection.
   - 3.3. Calculate the new individual fitness.
   - 3.4. Calculate the replacement probability of the individuals according to the hypergeometric rule.
   - 3.5. Replace the selected individual by the new generated individual.

Table 3. Genetic algorithm parameters

Population size, N	50
Crossover probability, p_c (mutation probability, $1-p_c$)	0.8 (0.2)
Ranking based replacement probability, q	0.2/0.7[a]
Maximum number of iterations, N_MAX	50
Minimum entropy level, $S_{minimum}$	0.25
Buffer size, b	10 cells

[a] Replacement probability: 0.2 for first 20 iterations and 0.7 for following iterations

References

1. Aboelela, E. and Douligeris, C. (1999) Fuzzy generalized network approach for solving an optimization model for routing in B-ISDN. Telecommunication Systems **12**, 237-263.
2. Amiri, A. and Pirkul, H. (1999) Routing and capacity assignment in backbone communication networks under time varying traffic conditions. European Journal of Operational Research **117**, 15-29.
3. Chou, L-D and Wu, J.L. (1998) Bandwidth allocation of virtual paths using neural-networks based genetic algorithms. IEE Proceedings on Communications Vol 145, No.1, 33-39.
4. Dahl, G., Martin, A. and Stoer, M. (1999) Routing through virtual paths in layered telecommunication networks. Operations Research **47**, 693-702.

5. Ghanwani, A. (1998) Neural and delay based heuristics for the Steiner problem in networks. European Journal of Operational Research **108**, 241-265.
6. Giroux, N. and Ganti, S. (1999) Quality of Service in ATM Networks (Prentice Hall).
7. Grefenstette, J.J. (1987) Incorporating problem-specific knowledge into genetic algorithms, in: L. Davis, ed., Genetic Algorithms and their Applications (Morgan Kaufmann, Los Angeles).
9. Lee, M-J and Yee, J-R (1994) An algorithm for optimal minimax routing in ATM networks. Annals of Operations Research **49**, 185-206.
10. Medova, E. (1998) Chance-constrained stochastic programming for integrated services network management. Annals of Operations Research **81**, 213-229.
11. Pitts, J.M. and Schormans, J.A. (1996) Introduction to ATM Design and Performance (Wiley&Sons).
12. Wang, C. and Weissler, P.N. (1995) The use of artificial neural networks for optimal message routing. IEEE Network, March-April,16-24.

Gene Expression Programming in Problem Solving

Cândida Ferreira

Departamento de Ciências Agrárias
Universidade dos Açores
9701-851, Angra do Heroísmo, Portugal
candidaf@gene-expression-programming.com
http://www.gene-expression-programming.com

Gene expression programming is a full fledged genotype/phenotype system that evolves computer programs encoded in linear chromosomes of fixed length. The structural organization of the linear chromosomes allows the unconstrained and fruitful (in the sense that no invalid phenotypes will follow) operation of important genetic operators such as mutation, transposition, and recombination as the expression of each gene results always in valid programs. Although simple, the genotype/phenotype system of gene expression programming is the first artificial genotype/phenotype system with a complex and sounding translation mechanism. Indeed, the interplay between genotype (chromosomes) and phenotype (expression trees) is at the core of the tremendous increase in performance observed in gene expression programming. Furthermore, gene expression programming shares with genetic programming the same kind of tree representation and, therefore, with GEP it is possible, for one thing, to retrace easily the steps undertaken by genetic programming and, for another, to explore easily new frontiers opened up by the crossing of the phenotype threshold. In this tutorial, the fundamental differences between gene expression programming and its predecessors, genetic algorithms and genetic programming, are briefly summarized so that the evolutionary advantages of gene expression programming could be better understood. The work proceeds with a detailed description of the main players in this new algorithm, focusing mainly on the interactions between them and how the simple yet revolutionary structure of the chromosomes allows the efficient, unconstrained exploration of the search space.

1. Genetic algorithms at large

The aim of this introduction is to bring into focus the basic differences between gene expression programming (GEP) and its predecessors, genetic algorithms (GAs) and genetic programming (GP). According to Mitchell (1996), gene expression programming is, like GAs and GP, a genetic algorithm as it uses populations of individuals, selects them according to fitness, and introduces genetic variation using one or more genetic operators. The fundamental difference between the three algorithms resides in the nature of the individuals: in GAs the individuals are

636

symbolic strings of fixed length (chromosomes); in GP the individuals are non-linear entities of different sizes and shapes (parse trees); and in GEP the individuals are encoded as symbolic strings of fixed length (chromosomes) which are then expressed as non-linear entities of different sizes and shapes (expression trees).

1.1. Genetic algorithms

Genetic algorithms, invented by J. Holland in the 1960s, applied biological evolution theory to computer systems (Holland 1975). Like all evolutionary computer systems, GAs are an oversimplification of biological evolution. In this case, solutions to a problem are encoded in character strings (usually 0's and 1's), and a population of these solutions is left to evolve in order to find a solution to the problem at hand. Populations, and therefore solutions, evolve because individual solutions (chromosomes) reproduce with modification. This is obviously the prerequisite for evolution to occur. Modification in the original GA was introduced by mutation, crossover, and inversion. In addition, for evolution to occur, individuals must pass the sieve of selection. They are selected according to fitness, being the fitness rigorously determined and its value used to reproduce them proportionately. The higher the fitness, the higher the probability of leaving more offspring.

The chromosomes of GAs are simple replicators (e.g., Dawkins 1995), and therefore they survive by virtue of their properties alone. This is equivalent to say that they function simultaneously as genome and phenome. So, the chromosomes are not only keepers of the genetic information that is replicated and transmitted with modification to the next generation, but are also the object of selection. The variety of functions GAs' chromosomes are able to play is severely limited by this dual role and by their structural organization, specially the simple language of chromosomes and their fixed length. This very much resembles a simple RNA World, where the linear RNA genome is also capable of exhibiting structural diversity. In this case, the whole structure of the RNA molecule determines the functionality and, therefore, the fitness of the individual. For instance, it wouldn't be possible in such systems to use only a particular region of the genome as a solution to the problem: the whole genome is always the solution. Obviously these systems are severely constrained.

1.2. Genetic programming

Genetic programming, invented by Cramer in 1985 and further developed by Koza (1992), solved the problem of fixed length solutions by creating non-linear entities with different sizes and shapes. The alphabet used to create these entities was also more varied, creating a richer, more versatile system of representation. However, the created individuals lacked a simple, autonomous genome, functioning simultaneously both as genome and phenome. Again, in the jargon of evolutionary theory, the entities of GP are simple replicators that survive by virtue of their own properties. The non-linear entities (parse trees) of GP resemble protein molecules in their use of a richer alphabet and in their complex, hierarchical representation. Thus, GP entities are capable of exhibiting a great variety of functionalities. But these entities are very difficult to reproduce with modification because the genetic modifications are done directly on the parse tree itself. Consequently, most modifications generate structural impossibilities. As a comparison, it is worth noticing that, in nature, the expression of any protein gene results always in a valid protein structure (in nature, there is no such thing as a structurally incorrect protein).

So, in GP, the genetic operators act directly on the parse tree and, although at first sight this might appear advantageous, it greatly limits this technique (it is impossible to make an orange tree produce mangos only by grafting and pruning). Furthermore, the pallet of genetic operators available to GP is very limited, because most of them would result in invalid parse trees. Consequently, GP uses almost exclusively a special kind of recombination that operates at the level of parse trees. In this GP-specific crossover, selected branches are exchanged between two parent parse trees to create offspring. The idea behind its implementation was to exchange smaller, mathematically concise blocks in order to evolve more complex, hierarchical solutions composed of smaller building blocks.

The mutation operator in GP also differs from point mutations in nature in order to guarantee the creation of syntactically correct programs. The mutation operator selects a node in the parse tree and replaces the branch beneath that node by a randomly generated branch. Again, the overall shape of the tree is not greatly changed by this kind of mutation.

Permutation is the third operator used in GP and, like recombination and mutation, is greatly constrained: the arguments to a chosen function are selected and exchanged. In this case the overall shape of the tree remains unchanged.

Although J. Koza described these three operators as the basic GP operators, crossover is practically the only genetic operator used in most GP implementations. Not surprisingly, in GP, huge populations of parse trees are used with the aim of creating all the necessary building blocks with the inception of the initial population in order to guarantee the discovery of a solution only by moving these initial building blocks around.

Finally, due to the dual function of the parse trees (genome and phenome), and like GAs, GP is incapable of a simple, rudimentary expression: in all cases, the entire parse tree is the solution.

1.3. Gene expression programming

Gene expression programming was invented by myself in 1999 (Ferreira 2001), and is the natural development of GAs and GP. GEP uses the same kind of diagram representation of GP, but the entities evolved by GEP (expression trees) are the expression of a genome. Therefore, with GEP, the second evolutionary threshold - the Phenotype Threshold - is crossed, providing new and efficient solutions to evolutionary computation.

The great insight of GEP consisted in the invention of chromosomes capable of representing any expression tree. For that a new language was created so that the information of GEP chromosomes could be read and expressed. Also important is that the structure of GEP chromosomes allows the easy implementation of multiple genes, each encoding a sub-program. On the other hand, the structural and functional organization of GEP genes and their interplay with expression

638

trees, always guarantees the production of valid programs, no matter how much or how profoundly we modify the chromosomes.

In contrast to its analogous cellular gene expression, the expression of the genetic information in GEP is rather simple. The main players in GEP are only two: the chromosomes and the expression trees (ETs), being the latter the expression of the genetic information encoded in the former. As in nature, the process of information decoding is called translation. And this translation implies obviously a kind of code and a set of rules. The genetic code is very simple: a one-to-one relationship between the symbols of the chromosome and the functions or terminals they represent. The rules are also very simple: they determine the spatial organization of the functions and terminals in the ETs and the type of interaction between sub-ETs.

In GEP there are therefore two languages: the language of genes and the language of ETs, and knowing the sequence or structure of one, is knowing the other. In nature, despite being possible to infer the sequence of proteins given the sequence of genes and vice versa, we practically know nothing about the rules that determine the three-dimensional structure of proteins. But in GEP, thanks to the simple rules that determine the structure of ETs and their interactions, it is possible to infer immediately the phenotype given the sequence of a gene, and vice versa. This bilingual and unequivocal system is called *Karva* language.

The tutorial proceeds with the presentation of the structural and functional organization of GEP chromosomes; how the chromosomes are translated into expression trees; how the chromosomes function as genotype and the expression trees as phenotype; and how an individual program is created, matured, and reproduced, leaving offspring with new properties, thus, capable of adaptation.

2. The genome of GEP individuals

In GEP, the genome or chromosome consists of a linear, symbolic string of fixed length composed of one or more genes. Despite their fixed length, we will see that GEP chromosomes code for ETs with different sizes and shapes.

2.1. Open reading frames and genes

The structural organization of GEP genes is better understood in terms of open reading frames (ORFs). In biology, an ORF or coding sequence of a gene begins with the 'start' codon, continues with the amino acid codons, and ends at a termination codon. However, a gene is more than the respective ORF, with sequences upstream of the start codon and sequences downstream of the stop codon. Although in GEP the start site is always the first position of a gene, the termination point not always coincides with the last position of a gene. It is common for GEP genes to have non-coding regions downstream of the termination point. For now we won't consider these non-coding regions, because they don't interfere with the product of expression.

Consider, for example, the algebraic expression:

$$\frac{a \cdot b}{c} + \sqrt{d-e}$$

(2.1)

It can also be represented as a diagram:

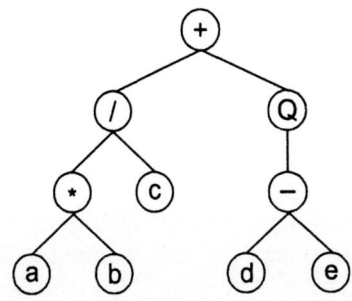

where 'Q' represents the square root function.

This kind of diagram representation is in fact the phenotype of GEP chromosomes, being the genotype easily inferred from the phenotype as follows:

```
0123456789
+/Q*c-abde
```
(2.2)

which is the straightforward reading of the ET from left to right and from top to bottom (exactly as we read a page of text). The expression (2.2) is an ORF, starting at '+' (position 0) and terminating at 'e' (position 9). I named these ORFs K-expressions (from Karva notation).

Consider another ORF, the following K-expression:

```
012345678901
*-/Qb+b+aaab
```
(2.3)

Its expression as an ET is also very simple and straightforward. To correctly express the ORF, we must follow the rules governing the spatial distribution of functions and terminals. The start position (position 0) in the ORF corresponds to the root of the ET. Then, below each function are attached as many branches as there are arguments to that function. The assemblage is complete when a baseline composed only of terminals (the variables or constants used in a problem) is formed. So, for the K-expression (2.3) above, the following ET is formed:

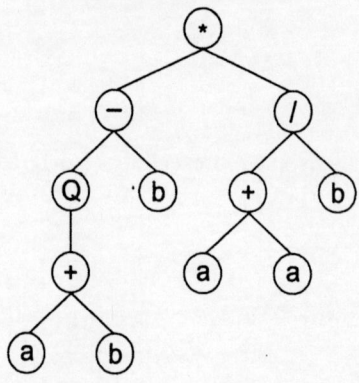

Looking at the structure of GEP ORFs only, it is difficult or even impossible to see the advantages of such a representation, except perhaps for its simplicity and elegance. However, when ORFs are analyzed in the context of a gene, the advantages of this representation become obvious. As I said, GEP chromosomes have fixed length, and they are composed of one or more genes of equal length. Therefore the length of a gene is also fixed. Thus, in GEP, what varies is not the length of genes which is constant, but the length of the ORFs. Indeed, the length of an ORF may be equal or less than the length of the gene. In the first case, the termination point coincides with the end of the gene, and in the last case, the termination point is somewhere upstream of the end of the gene.

So, what is the role of these non-coding regions in GEP genes? They are in fact the essence of GEP and evolvability, for they allow the modification of the genome using any genetic operator without restrictions, producing always syntactically correct programs without the need for a complicated editing process or highly constrained ways of implementing genetic operators. Indeed, this is the paramount difference between GEP and previous GP implementations, with or without linear genomes.

Let's analyze then the structural organization of GEP genes in order to understand how they invariably code for syntactically correct programs and why they allow the unconstrained application of any genetic operator.

2.2. GEP genes

GEP genes are composed of a head and a tail. The head contains symbols that represent both functions and terminals, whereas the tail contains only terminals. For each problem, the length of the head h is chosen, whereas the length of the tail t is a function of h and the number of arguments of the function with more arguments n, and is evaluated by the equation:

$$t = h \ (n - 1) + 1$$

(2.4)

Consider a gene for which the set of functions F = {Q, *, /, -, +} and the set of terminals T = {a, b}. In this case, $n = 2$; and if we chose an $h = 15$, then $t = 16$. Thus, the length of the gene g is 15+16=31. One such gene is shown below (the tail is shown in bold):

```
01234567890123456789012345674890
/aQ/b*ab/Qa*b*-ababaababbabbbba
```
(2.5)

It codes for the following ET with eight nodes:

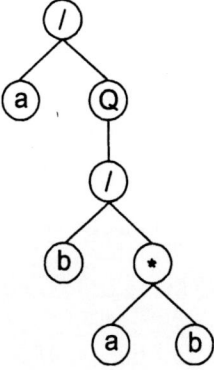

In this case, the ORF ends at position 7, whereas the gene ends at position 30.

Suppose now a mutation occurred at position 2, changing the 'Q' into '+'. Then the following gene is obtained:

```
01234567890123456789012344567890
/a+/b*ab/Qa*b*-ababaababbabbbba
```
(2.6)

And its expression gives an ET with 18 nodes. In this case, the termination point shifts 10 positions to the right (position 17).

Obviously the opposite might also happen, and the ORF can be shortened. For example, consider again gene (2.5) above, and suppose a mutation occurred at position 5, changing the '*' into 'b':

```
01234567890123456789012344567890
/aQ/bbab/Qa*b*-ababaababbabbbba
```
(2.7)

Its expression results in an ET with six nodes. In this case, the ORF ends at position 5, shortening the parental ET in two nodes.

So, despite its fixed length, each gene has the potential to code for ETs of different sizes and shapes, being the simplest composed of only one node (when the first element of a gene is a terminal) and the biggest composed of as many nodes as the length of the gene (when all the elements of the head are functions with maximum arity).

It is evident from the examples above, that any modification made in the genome, no matter how profound, results always in a structurally correct ET. The only thing we must be careful about, is in not disrupting the structural organization of genes, maintaining always the boundaries between head and tail and not allowing symbols representing functions on the tail. We will pursue these matters further in section 3 where the mechanisms and effects of different genetic operators are thoroughly analyzed.

2.3. Multigenic chromosomes

GEP chromosomes are usually composed of more than one gene of equal length. For each problem or run, the number of genes, as well as the length of the head, are *a priori* chosen. Each gene codes for a sub-ET and the sub-ETs interact with one another forming a more complex multi-subunit ET.

Consider, for example, the following chromosome with length 45, composed of three genes (the tails are shown in bold):

```
012345678901234012345678901234012345678901234
Q/*b+Qababaabaa-abQ/*+bababbab**-*bb/babaaaab
```
(2.8)

It has three ORFs, and each ORF codes for a sub-ET (Figure 1). Position zero marks the start of each gene. The end of each ORF, though, is only evident upon construction of the respective sub-ET. As shown in Figure 1, the first ORF ends at position 8 (sub-ET$_1$); the second ORF ends at position 2 (sub-ET$_2$); and the last ORF ends at position 10 (sub-ET$_3$). Thus, GEP chromosomes contain several ORFs, each ORF coding for a structurally and functionally unique sub-ET. Depending on the problem at hand, these sub-ETs may be selected individually according to their respective fitness (for example, in problems with multiple outputs), or they may form a more complex, multi-subunit ET where individual sub-ETs interact with one another by a particular kind of posttranslational interaction or linking. For instance, algebraic sub-ETs can be linked by addition or multiplication whereas Boolean sub-ETs can be linked by OR, AND or if(x,y,z).

The linking of three sub-ETs by addition is illustrated in Figure 1, c. Note that the final ET could be linearly encoded as the following K-expression:

```
012345678901234567890123456
++*Q-*-/ab*bb/*bbaba+b+Qaba
```
(2.9)

However, to evolve solutions to complex problems, it is more effective the use of multigenic chromosomes, for they permit the modular construction of complex, hierarchical structures, where each gene codes for a small building block. These small building blocks are separated from each other, and thus can evolve independently. Furthermore, these multigenic systems are much more efficient than unigenic ones. Indeed, GEP is a highly efficient hierarchical invention system capable of discovering simple blocks and using them to form more complex structures.

a) 012345678901234012345678901234012345678901234
Q/*b+Qa**babaabaa**-abQ/*+**bababbab*** *-*bb/**babaaaab**

b)

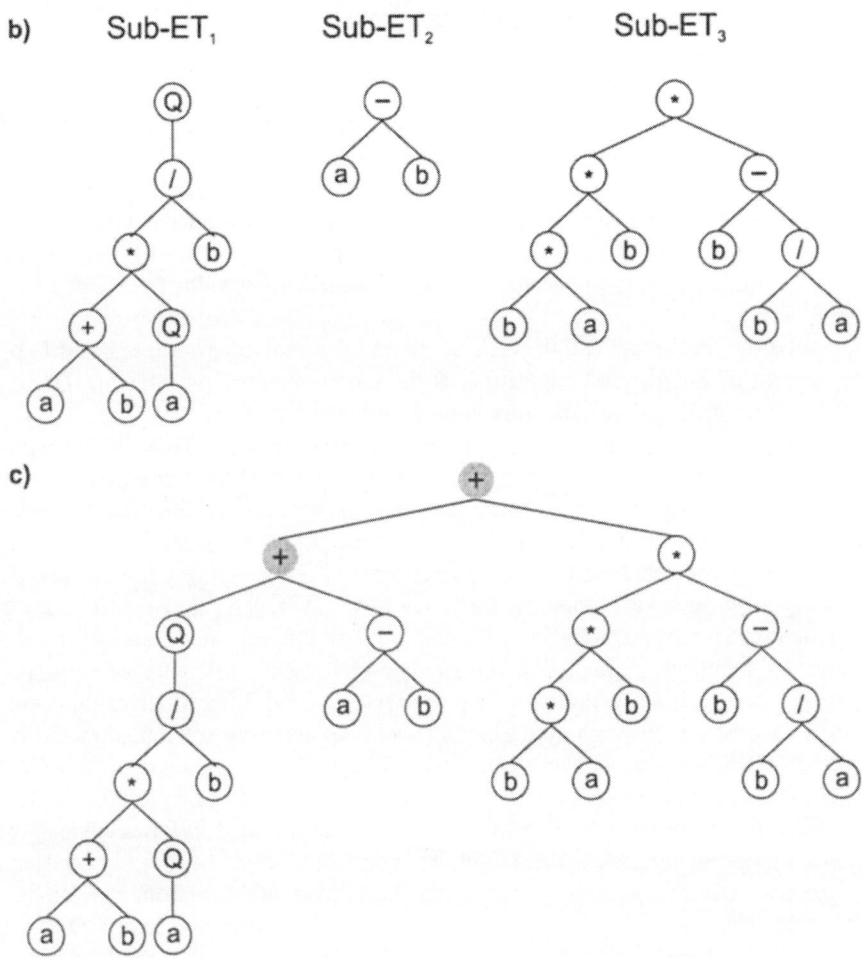

Fig. 1. Expression of GEP genes as sub-ETs. **a)** A three-genic chromosome with the tails shown in bold. Position zero marks the start of each gene. **b)** The sub-ETs codified by each gene. **c)** The result of posttranlational linking with addition. The linking functions are shown in gray.

So, for each problem, the type of linking function, as well as the number of genes and the length of each gene, are *a priori* chosen. While attempting to solve a problem, we can always start by using a single-gene chromosome and then proceed by increasing the length of the head. If it becomes very large, we can increase the number of genes and obviously choose a function to link the sub-ETs. We can start with addition for algebraic expressions or OR for Boolean expressions, but in some cases another linking function might be more appropriate (like multiplication or IF, for instance). The idea, of course, is to find a good solution, and GEP provides the means of finding one very efficiently.

3. Genetic operators and evolution

Genetic operators are the core of all genetic algorithms, and two of them are common to all evolutionary systems: selection and replication. Indeed, all artificial systems use a scheme to select individuals more or less according to fitness. Some schemes are totally deterministic, whereas others include a touch of unpredictability. For GEP, I chose one of the latter, namely, a fitness proportionate roulette-wheel scheme (Goldberg 1989) coupled with the cloning of the best individual (simple elitism) as it pretty accurately mimics nature and produces very good results.

According to fitness and the luck of the roulette, individuals are selected to be replicated. Although vital, replication is the most uninteresting operator. During replication, chromosomes are dully copied into the next generation. The fitter the individual the higher the probability of leaving more offspring. Thus, during replication, the genomes of the selected individuals are copied as many times as the outcome of the roulette. The roulette is spun as many times as there are individuals in the population, maintaining always the same population size.

Although the center of the storm, selection and replication, by themselves, do nothing in terms of evolution. In fact, they can only cause genetic drift, making populations less and less diverse with time until all the individuals are exactly the same. So, the touch stone of all evolutionary systems is modification, or more specifically, the genetic operators that cause variation. And different algorithms create this modification differently. For instance, GAs normally use mutation and recombination; GP uses almost exclusively GP-specific recombination; and GEP uses mutation, recombination and transposition.

With the exception of GP, which is severely constrained in terms of tools of genetic modification, in GAs and GEP, it is possible to implement easily a vast set of genetic operators capable of causing genetic diversification (from now on, unless otherwise stated, I will use the designation 'genetic operators' to refer to those with intrinsic transforming power, putting selection and replication aside) because the chromosomes of both algorithms allow their easy implementation. In fact, I implemented several genetic operators in GEP in order to shed some light on the dynamics of evolutionary systems, but what is important is to provide for the necessary degree of genetic diversification to allow an efficient evolution. Mutation alone (by far the most important operator) is capable of wonders. However, the interplay of mutation and the other genetic operators not only allows an effective evolution but also allows the duplication of building blocks, their circulation in the genetic pool, the creation of repetitive sequences, etc., making things really interesting.

In the remainder of this section we will see how genetic operators work and how their implementation in GEP is a child's play due to the genotype/phenotype mapping.

3.1. Mutation

Mutations can occur anywhere in the chromosome. However, the structural organization of chromosomes must remain intact. In the heads, any symbol can change into another (function or terminal); in the tails, terminals can only change into terminals. This way, the structural organization of chromosomes is maintained, and all the new individuals produced by mutation are structurally correct programs.

Consider the following three-genic chromosome:

```
01234567890012345678900123456789 0
Q+bb*bbbaba-**--abbbaaQ*a*Qbbbaab
```

Suppose a mutation changed the '*' at position 4 in gene 1 to '/'; the '-' at position 0 in gene 2 to 'Q'; and the 'a' at position 2 in gene 3 to '+', obtaining:

```
01234567890012345678900123456789 0
Q+bb/bbbabaQ**--abbbaaQ*+*Qbbbaab
```

Note that if a function is mutated into a terminal or vice versa, or a function of one argument is mutated into a function of two arguments or vice versa, the ET is modified drastically. Note also that the mutation on gene 1 is an example of a neutral mutation, as it occurred in the non-coding region of the gene. It is worth emphasizing that the non-coding regions of GEP chromosomes are ideal places for the accumulation of neutral mutations. In summary, in GEP there are no constraints neither in the kind of mutation nor the number of mutations in a chromosome: in all cases the newly created individuals are syntactically correct programs.

3.2. Transposition and insertion sequence elements

The transposable elements of GEP are fragments of the genome that can be activated and jump to another place in the chromosome. In GEP there are three kinds of transposable elements: i) short fragments with a function or terminal in the first position that transpose to the head of genes except the root (insertion sequence elements or IS elements); ii) short fragments with a function in the first position that transpose to the root of genes (root IS elements or RIS elements); iii) and entire genes that transpose to the beginning of chromosomes.

3.2.1. Transposition of IS elements

Any sequence in the genome might become an IS element, being therefore these elements randomly selected throughout the chromosome. A copy of the transposon is made and inserted at any position in the head of a gene, except the first position. The transposition operator randomly chooses the chromosome, the start of the IS element, the target site, and the length of the transposon.

Consider the following two-genic chromosome:

```
012345678901234560123456789 0123456
-aba+Q-baabaabaabQ*+*+-/aababbaaaa
```

Suppose that the sequence 'a+Q' in gene 1 (positions 3-5) was randomly chosen to become an IS element and transpose between positions 2-3 in gene 2, obtaining:

```
0123456789012345601234567890123456
-aba+Q-baabaabaabQ*+a+Q*+ababbaaaa
```

Note that, on the one hand, the sequence of the transposon becomes duplicated but, on the other, a sequence with as many symbols as the IS element was deleted at the end of the head of the target gene (in this case the sequence '-/a' was deleted. Thus, despite the insertion, the structural organization of chromosomes is maintained, and therefore all the new individuals created by transposition are syntactically correct programs.

3.2.2. Root transposition

All RIS elements start with a function, and thus are chosen among the sequences of the heads. For that, a point is randomly chosen in the head and the gene is scanned downstream until a function is found. This function becomes the start position of the RIS element. If no functions are found, the operator does nothing.

This operator randomly chooses the chromosome, the gene to be modified, the start of the RIS element, and its length. Consider the two-genic chromosome below:

```
0123456789012345601234567890123456
*-bQ/++/babbabbba//Q*baa+bbbabbbbb
```

Suppose that the sequence 'Q/+' in gene 1 was randomly chosen to become an RIS element. Then, a copy of the transposon is made into the root of the gene, obtaining:

```
0123456789012345601234567890123456
Q/+*-bQ/babbabbba//Q*baa+bbbabbbbb
```

Note that during transposition, the whole head shifts to accommodate the RIS element, losing, at the same time, the last symbols of the head (as many as the transposon length). In this case, the sequence '++/' was deleted, and the transposon became only partially duplicated. As with IS elements, the tail of the gene subjected to transposition and all nearby genes stay unchanged. Note, again, that the newly created programs are syntactically correct because the structural organization of the chromosome is maintained.

3.2.3. Gene transposition

In gene transposition an entire gene functions as a transposon and transposes itself to the beginning of the chromosome. In contrast to the other forms of transposition, in gene transposition, the transposon (the gene) is deleted at the place of origin.

Apparently, gene transposition is only capable of shuffling genes and, for ETs linked by commutative functions, this contributes nothing to adaptation in the short run. However, gene transposition is very important when coupled with recombination (see below), for it allows not only the duplication of genes but also a more generalized recombination of genes or smaller building blocks.

This operator randomly chooses the chromosome to undergo gene transposition, and randomly chooses one of its genes (except the first, obviously) to transpose. Consider the following chromosome composed of 3 genes:

```
0123456789012012345678901201234567 89012
/+Qa*bbaaabaa*a*/Qbbbbbabb/Q-aabbaaabbb
```

Suppose gene 3 was chosen to undergo gene transposition. Then the following chromosome is obtained:

```
0123456789012012345678901201234567 89012
/Q-aabbaaabbb/+Qa*bbaaabaa*a*/Qbbbbbabb
```

Note that for numerical applications where the function chosen to link the genes is commutative, the expression evaluated by the chromosome is not modified. But the situation differs in other applications where the linking function is not commutative, for instance, the IF function chosen to link sub-ETs in Boolean problems. Note that, in those cases, gene transposition has a very drastic effect, generating most of the times nonviable individuals.

3.3. Recombination

In GEP there are three kinds of recombination: one-point recombination, two-point recombination and gene recombination. In all types of recombination, two chromosomes are randomly chosen and paired to exchange some material between them, creating two new daughter chromosomes. Usually the daughter chromosomes are as different from each other as they are from their mothers.

3.3.1. One-point recombination

In one-point recombination the chromosomes are paired and split in the same point. The material downstream of the recombination point is afterwards exchanged between the two chromosomes.

Consider the following parent chromosomes:

```
012345678901234560123456789 0123456
+*-b-Qa*aabbbbaaa-Q-//b/*aabbabbab
++//b//-bbbbbbbbb-*-ab/b+bbbaabbaa
```

Suppose bond 6 in gene 1 (between positions 5 and 6) was randomly chosen as the crossover point. Then, the paired chromosomes are cut at this bond, and exchange between them the material downstream of the crossover point, forming the offspring below:

```
012345678901234560123456789 0123456
+*-b-Q/-bbbbbbbbb-*-ab/b+bbbaabbaa
```

++//b/a*aabbbbaaa-Q-//b/*aabbabbab

It is worth noticing that with this kind of recombination, most of the times, the offspring created exhibits different traits from those of the parents.

3.3.2. Two-point recombination

In two-point recombination the chromosomes are paired and two points are randomly chosen as crossover points. The material between the recombination points is afterwards exchanged between the two parent chromosomes, forming two new daughter chromosomes.

Consider the following parent chromosomes:

```
0123456789012345601234567890123456
*-+Q/Q*QaaabbbbabQQab*++-aabbabaab
Q/-b-+/abaabbbaab/*-aQa*babbabbabb
```

Suppose bond 5 in gene 1 (between positions 4 and 5) and bond 7 in gene 2 (between positions 6 and 7) were chosen as crossover points. Then, the following chromosomes are created:

```
0123456789012345601234567890123456
*-+Q/+/abaabbbaab/*-aQa*-aabbabaab
Q/-b-Q*QaaabbbbabQQab*++babbabbabb
```

It's worth noticing that the non-coding regions of GEP chromosomes are ideal regions where chromosomes can be split to cross over without interfering with the ORFs and, in fact, during search, these regions are most favored by crossover.

3.3.3. Gene recombination

In the third kind of GEP recombination, gene recombination, entire genes are exchanged between two parent chromosomes, forming two daughter chromosomes containing genes form both mothers. The exchanged genes are randomly chosen and occupy the same position in the parent chromosomes. Consider the following parent chromosomes:

```
0123456789012012345678901201234567789012
/+/ab-aabbbbb-aa**+aaabaaa-+--babbbbaab
+baQaaaabaaba*-+a-aabbabbb/ab/+bbbabaaa
```

Suppose gene 2 was chosen to be exchanged. In this case the following offspring is formed:

```
0123456789012012345678901201234567789012
/+/ab-aabbbbb*-+a-aabbabbb-+--babbbbaab
+baQaaaabaaba-aa**+aaabaaa/ab/+bbbabaaa
```

Note that, with this kind of recombination, similar genes can be exchanged but, most of the times, the exchanged genes are very different and new material is introduced in the population.

It is worth noticing that this operator is unable to create new genes: the individuals created are different arrangements of existing genes. Understandingly, when gene recombination is used as the unique source of genetic variation, more complex problems can only be solved using very large initial populations in order to provide for the necessary diversity of genes. However, the creative power of GEP is based not only in the shuffling of genes or building blocks, but also in the constant creation of new genetic material.

4. Solving a simple problem with GEP

The aim of this section is to study a successful run in its entirety in order to understand how populations of GEP individuals evolve towards a perfect or good solution.

In symbolic regression or function finding the goal is to find an expression that satisfactorily explains the dependent variable. The input into the system is a set of fitness cases in the form $(a_{(i,0)}, a_{(i,1)}, ..., a_{(i,n-1)}, y_i)$ where $a_{(i,0)}$ - $a_{(i,n-1)}$ are the independent variables and y_i is the dependent variable. The set of fitness cases consists of the adaptation environment where solutions adapt, discovering, in the process, solutions to problems.

In the example of this section, a simple test function was chosen, being therefore the fitness cases computer generated. Thus, in this case, we know exactly which function we are aiming at (remember, however, that in real-world problems the function is obviously unknown). So, suppose we are given a sampling of the numerical values from the curve

$$y = 3a^2 + 2a + 1$$

(4.1)

over 10 randomly chosen points in the real interval [-10, +10] and we wanted to find a function fitting those values within a certain error. In this case, we are given a sample of data in the form of 10 pairs (a_i, y_i), where a_i is the value of the independent variable in the given interval and y_i is the respective value of the dependent variable (a_i values: -4.2605, -2.0437, -9.8317, -8.6491, 0.7328, -3.6101, 2.7429, -1.8999, -4.8852, 7.3998; the corresponding y_i values can be easily evaluated). These 10 pairs are the fitness cases (the input) that will be used as the adaptation environment. The fitness of a particular program will depend on how well it performs in this environment.

There are five major steps in preparing to use gene expression programming, and the first is to choose the fitness function. For this problem we could measure the fitness f_i of an individual program i by the following expression:

$$f_i = \sum_{j=1}^{C_t} \left(M - \left| C_{(i,j)} - T_j \right| \right)$$

(4.2)

where M is the range of selection, $C_{(i,j)}$ the value returned by the individual chromosome i for fitness case j (out of C_t fitness cases) and T_j is the target value for fitness case j. If, for all j, $|C_{(i,j)} - T_j|$ (the precision) less or equal to 0.01, then the precision is equal to zero, and $f_i = f_{max} = C_t \cdot M$. For this problem, we will use an $M = 100$ and, therefore, $f_{max} = 1000$. The advantage of this kind of fitness function is that the system can find the optimal solution for itself.

The second major step consists in choosing the set of terminals T and the set of functions F to create the chromosomes. In this problem, the terminal set consists obviously of the independent variable, i.e., T = {a}. The choice of the appropriate function set is not so obvious, but a good guess can always be done in order to include all the necessary functions. In this case, to make things simple, we will use the four basic arithmetic operators. Thus, F = {+, -, *, /}.

The third major step is to choose the chromosomal architecture, i.e., the length of the head and the number of genes. In this problem we will use an $h = 6$ and three genes per chromosome.

The fourth major step in preparing to use gene expression programming is to choose the linking function. In this case we will link the sub-ETs by addition.

And finally, the fifth major step is to choose the set of genetic operators that cause variation and their rates. In this case we will use a combination of all genetic operators (mutation at $p_m = 0.051$; IS and RIS transposition at rates of 0.1 and three transposons of length 1, 2, and 3; one-point and two-point recombination at rates of 0.3; gene transposition and gene recombination both at rates of 0.1).

To solve this problem, I chose an evolutionary time of 50 generations and a small population of 20 individuals in order to simplify the analysis of the evolutionary process and not fill this text with pages of encoded individuals. However, one of the advantages of GEP is that it is capable of solving relatively complex problems using small population sizes and, thanks to the compact Karva notation, it is possible to fully analyze the evolutionary history of a run.

To show evolution at work, I chose a successful run in which a perfect solution was found in generation 3. The initial population of this run, together with the fitness of each individual, is shown below:

```
Generation N: 0
012345678901201234567890120123456789012
+**/*/aaaaaaa/+a/a*aaaaaaa/a-*a+aaaaaaa-[ 0] = 577.3946
--aa++aaaaaaa+-/a*/aaaaaaa/--a-aaaaaaa-[ 1] = 0
/***/+aaaaaaa*+/+-aaaaaaaa++aa/aaaaaaa-[ 2] = 463.6533
-/+/++aaaaaaa+-//+/aaaaaaa+-/a/*aaaaaaa-[ 3] = 546.4241
++a/*aaaaaaa+-+a*-aaaaaaa-a/-*aaaaaaa-[ 4] = 460.8625
*+*a-*aaaaaaa*a/aa/aaaaaaa//+*a/aaaaaaa-[ 5] = 353.2168
*/**+aaaaaaaa+a/**+aaaaaaa----+/aaaaaaa-[ 6] = 492.6827
*aa-+-aaaaaaa+a/-+/aaaaaaa***/-*aaaaaaa-[ 7] = 560.9289
+/-*//aaaaaaa*+*//+aaaaaaa-/**+*aaaaaaa-[ 8] = 363.4358
--a+*/aaaaaaa+a++--aaaaaaa+a+aa+aaaaaaa-[ 9] = 386.7576
+-*-**aaaaaaa*/-+**aaaaaaa*+--++aaaaaaa-[10] = 380.6484
/a-**/aaaaaaa/-a/a/aaaaaaa+/a/-*aaaaaaa-[11] = 0
+--+//aaaaaaa+*+/*-aaaaaaa/*-a-+aaaaaaa-[12] = 551.2066
-a/+a/aaaaaaa*/--/aaaaaaaa*-+/a+aaaaaaa-[13] = 308.1296
/+/-+-aaaaaaa+-a/aaaaaaaaa**+-*-aaaaaaa-[14] = 0
//-*+/aaaaaaa//*a+aaaaaaaa/a++a*aaaaaaa-[15] = 489.5392
```

```
*a-a*-aaaaaaa+*+-a/aaaaaaa*/*aa*aaaaaaa-[16]  = 399.2122
-a++*/aaaaaaa+/aa-*aaaaaaa---/**aaaaaaa-[17]  = 317.6631
--a/*aaaaaaaa++*+-aaaaaaaa+-/*+-aaaaaaa-[18]  = 597.8777
*+++-/aaaaaaa/--///aaaaaaa+-+aaaaaaaaaa-[19]  = 661.5933
```

Note that three of the 20 individuals are nonviable and thus have fitness 0. The best of generation, chromosome 19, has fitness 661.5933. Note that the second gene of this individual returns 0 and, therefore, might be considered a pseudogene. The descendants of the individuals of the initial population are shown below:

```
Generation N: 1
012345678901201234567890120123456789012
*+++-/aaaaaaa/--///aaaaaaa+-+aaaaaaaaaa-[ 0]  = 661.5933
-a++*/aaaaaaa+//a--aaaaaaa---/**aaaaaaa-[ 1]  = 0
+-*-**aaaaaaa*/-+**aaaaaaa*+--++aaaaaaa-[ 2]  = 380.6484
+-*-**aaaaaaa*/-+**aaaaaaa*/*a**aaaaaaa-[ 3]  = 356.9471
+-+aaaaaaaaaa*+++-/aaaaaaa/--///aaaaaaa-[ 4]  = 661.5933
*aa-+-aaaaaaa+a/++/aaaaaaa***+-*aaaaaaa-[ 5]  = 567.9289
*a-a*-aaaaaaa+/*-a/aaaaaaa*+-*++aaaaaaa-[ 6]  = 449.802
*aa-+-aaaaaaa+a/-+/aaaaaaa*+--++aaaaaaa-[ 7]  = 961.8512
/***/+aaaaaaa*+/+-aaaaaaaa-a/-*aaaaaaaa-[ 8]  = 470.5862
+--+//aaaaaaa+*+/*-aaaaaaa/*-a-+aaaaaaa-[ 9]  = 551.2066
*+++-/aaaaaaa-//--/aaaaaaa+--+aaaaaaaaaa-[10] = 0
--+a*-aaaaaaa++a/*aaaaaaaa-a/-*aaaaaaaa-[11]  = 487.3099
-a++*/aaaaaaa+/aa-*aaaaaaa---/**aaaaaaa-[12]  = 317.6631
++a/*aaaaaaaa+-+a*-aaaaaaa++aa/aaaaaaaa-[13]  = 451.464
+--+/-aaaaaaa+a/**+aaaaaaa----+/aaaaaaa-[14]  = 493.5336
*/-a++aaaaaaa+/aa-*aaaaaaa---/**aaaaaaa-[15]  = 356.4241
+/-*//aaaaaaa*+a//+aaaaaaa-/+*+*aaaaaaa-[16]  = 493.9218
*/**+aaaaaaaa+*+/*aaaaaaaa***/-*aaaaaaa-[17]  = 448.4805
+-*-**aaaaaaa*/-+**aaaaaaa*+--++aaaaaaa-[18]  = 380.6484
++a/*aaaaaaaa+-+a*+aaaaaaa--/-*aaaaaaa-[19]   = 380.8585
```

Note that chromosome 0 is the clone of the best individual of the previous generation. In this generation, a new individual was created, chromosome 7, considerably better than the best individual of the initial population. The descendants of the individuals of this generation are shown below:

```
Generation N: 2
012345678901201234567890120123456789012
*aa-+-aaaaaaa+a/-+/aaaaaaa*+--++aaaaaaa-[ 0]  = 961.8512
*/**+aaaaaaaa*/-+**aaaaaaa***/-*aaaaaaa-[ 1]  = 446.2061
+-*-**aaaaaaa*+a//-aaaaaaa-/+*+*aaaaaaa-[ 2]  = 323.1036
+--+//aaaaaaa+*+/*-aaaaaaa/*-*-+aaaaaaa-[ 3]  = 551.2066
*aa-+-aaaaaaa+a/++/aaaaaaa***+-*aaaaaaa-[ 4]  = 567.9289
++a/*aaaaaaaa*/-+-*aaaaaaa*+--++aaaaaaa-[ 5]  = 0
+-*-**aaaaaaa+*+/*aaaaaaaa*/*a**aaaaaaa-[ 6]  = 386.6484
++a/*aaaaaaaa+-+/*-aaaaaaa+aa++aaaaaaaa-[ 7]  = 466.1533
+-*-a*aaaaaaa*/-+**aaaaaaa*a*a**aaaaaaa-[ 8]  = 194.0452
/***/+aaaaaaa*+/+-aaaaaaaa-a--*aaaaaaaa-[ 9]  = 541.4829
+-*-+*aaaaaaa+-+a*-aaaaaaa***/-*aaaaaaa-[10]  = 346.2235
--*+*-aaaaaaa*aa-+-aaaaaaaaa/-+/aaaaaaa-[11]  = 467.0862
*/-+**aaaaaaa+-*-*+aaaaaaa*/*a**aaaaaaa-[12]  = 672.877
*aa+*/aaaaaaa+a/-+/aaaaaaa*+--++aaaaaaa-[13]  = 961.8512
```

```
*+++/+aaaaaaa*++/+-aaaaaaa-a/-*aaaaaaaa-[14]  = 395.858
/***-/aaaaaaa/--///aaaaaaa+-+a-aaaaaaaa-[15]  = 467.0862
*aa-+-aaaaaaa+a/++/aaaaaaa***+-*aaaaaaa-[16]  = 567.9289
+-+aaaaaaaaaa*+++-/aaaaaaa/--///aaaaaaa-[17]  = 661.5933
+/-*//aaaaaaa*/a+**aaaaaaa*+--++aaaaaaa-[18]  = 903.8886
*/**+aaaaaaaa+*+/*aaaaaaaa+/aa/aaaaaaaa-[19]  = 423.885
```

Note that none of the descendants surpassed the best individual of the previous generation. And finally, in the next generation, an individual with maximum fitness was created:

```
Generation N: 3
012345678901201234567890120123456789012
*aa+*/aaaaaaa+a/-+/aaaaaaa*+--++aaaaaaa-[ 0] = 961.8512
*aa-+-aaaaaaa+a/-+/aaaaaaa/--///aaaaaaa-[ 1] = 560.9289
*aa-+-aaaaaaa-++/+-aaaaaaa-a/-*aaaaaaaa-[ 2] = 558.2066
*+++/+aaaaaaa**a/-+aaaaaaa++--++aaaaaaa-[ 3] = 569.0469
/+++/+aaaaaaa*++/+-aaaaaaa-a/-*aaaaaaaa-[ 4] = 699.5153
+-+aa/aaaaaaa++++-/aaaaaaa***+-*aaaaaaa-[ 5] = 466.1533
*aa-+-aaaaaaaaa--**aaaaaaa*+--++aaaaaaa-[ 6] = 957.9443
--++*-aaaaaaa*a+/*-aaaaaaa+aa++aaaaaaaa-[ 7] = 337.7807
*aaa*/aaaaaaa+a+-+/aaaaaaa*+-/++aaaaaaa-[ 8] = 953.9443
/***/-aaaaaaa*+/+-aaaaaaaa-a--*aaaaaaa-[ 9] = 0
*aa-+-aaaaaaa+a/-+/aaaaaaa*/--++aaaaaaa-[10] = 560.9289
*aa-+-aaaaaaa+a/++/aaaaaaa/--///aaaaaaa-[11] = 567.9289
+-+a-aaaaaaaa/***-/aaaaaaa*+--++aaaaaaa-[12] = 676.0663
+/**//aaaaaaa*/a+**aaaaaaa*+--++aaaaaaa-[13] = 1000
*/-+**aaaaaaa+-*-*+aaaaaaa*/*a**aaaaaaa-[14] = 672.877
/***/+aaaaaaa/+*+/+aaaaaaa-a*/--aaaaaaa-[15] = 498.3734
+/-*//aaaaaaa*/a+-*aaaaaaa*+--++aaaaaaa-[16] = 0
--*+--aaaaaaa*/a-+-aaaaaaa/a/-+/aaaaaaa-[17] = 506.1233
++a/*aaaaaaaa+-a-+-aaaaaaa-a*-+/aaaaaaa-[18] = 815.7772
*+a//-aaaaaaa+a/-+/aaaaaaa-/+*+*aaaaaaa-[19] = 412.5237
```

Note that this chromosome is a descendant, via mutation, of chromosome 18 of the previous generation: their chromosomes differ only in one position (the '-' at position 2 of gene 1 was replaced by '*'). The expression of this chromosome shows that it codes for a perfect solution, the test function (4.1).

5. Summary

Gene expression programming is the most recent development on artificial evolutionary systems and one that brings about a considerable increase in performance due to the crossing of the phenotype threshold. For the first time in artificial evolution, with GEP, the phenotype threshold is fully crossed, allowing the unconstrained exploration of the search space. In GEP, the implementation of high-performing search operators such as point mutation, transposition and recombination, is a child's play as any modification made in the genome results always in valid phenotypes or programs. The structural and functional organization of GEP chromosomes and the new language (Karva language) especially developed to read and express the information encoded in the chromosomes, were thoroughly presented, allowing the easy understanding and implementation of the algorithm.

Furthermore, the workings of the algorithm were analyzed step-by-step with a simple problem of symbolic regression, where entire populations were thoroughly analyzed so that the simple yet wondrous ways of evolution could be completely understood.

References

1. Cramer, N. L., A Representation for the Adaptive Generation of Simple Sequential Programs. In J. J. Grefenstette, ed., *Proceedings of the First International Conference on Genetic Algorithms and Their Applications*, Erlbaum, 1985.

2. Dawkins, R., *River out of Eden*, Weidenfeld & Nicolson, 1995.

3. Ferreira, C., 2001. Gene Expression Programming: A New Adaptive Algorithm for Solving Problems. *Complex Systems*, forthcoming.

4. Goldberg, D. E., *Genetic Algorithms in Search, Optimization, and Machine Learning*, Addison-Wesley, 1989.

5. Holland, J. H., *Adaptation in Natural and Artificial Systems: An Introductory Analysis with Applications to Biology, Control, and Artificial Intelligence*, University of Michigan Press, 1975 (second edition: MIT Press, 1992).

6. Koza, J. R., *Genetic Programming: On the Programming of Computers by Means of Natural Selection*, Cambridge, MA: MIT Press, 1992.

7. Mitchell, M., *An Introduction to Genetic Algorithms*, MIT Press, 1996.

Studies of the XCSI Classifier System on a Data Mining Problem

Chunsheng Fu [1], Stewart W. Wilson [2], Lawrence Davis [1]

[1]NuTech Solutions, Inc. 28 Green, Newbury, MA 01951, USA
[2]Prediction Dynamics. 30 Lang, Concord, MA 01742, USA

Abstract. Several machine learning techniques have been applied to the Wisconsin Breast Cancer (WBC) database, a publicly available data mining benchmark problem. Among these, a version of the classifier system XCSI achieved performance comparable to the best published results (Wilson 2000). This paper describes some modifications to the XCSI algorithm and to the parameter settings of XCSI that improved its performance noticeably on that problem. The modifications are robust in the presence of noise and appear to reduce the algorithm's tendency to overtrain. This paper describes some of the experiments carried out in the course of the work, and characterizes those changes that seem to have made the greatest impact on the change in performance of the classifier system.

1 Introduction

XCS is a recently developed classifier system (Wilson 1995). Like its predecessors, XCS was created to classify records consisting of binary attribute values. The definition of fitness used in XCS is based on the accuracy of a classifier's payoff prediction, instead of the prediction itself. XCS includes a number of changes over prior classifier systems, and these changes appear to significantly improve its generalization ability and prediction accuracy. XCSI is Wilson's extension of XCS to carry out classification for problems with inputs that are vectors of integers rather than bits (Wilson 2000).

The principal difference between XCS and XCSI lies in their encoding and matching strategies. XCS matches an n-bit input with classifiers containing n fields that consist of 0, 1, or "#" (the "don't care" symbol). XCSI matches an n-integer input with classifiers consisting of 2n fields organized into pairs. To determine whether an XCSI classifier matches, each pair of classifier fields is applied to the corresponding input value. If the input value is not less than the first member of the pair and not greater than the second member of the pair, the input value is matched. In other words, the two classifier members associated with an input establish the minimum and maximum values of that input if the classifier is to match that input. As an example, the integer signal "2 4" would be matched by the classifier 1 5 2 6, since 2 lies in the interval between 1 and 5, and 4 lies in the interval between 2 and 6. This signal would not be matched by the classifier 2 5 1 3, since 4 does not lie in the interval between 1 and 3.

The Wisconsin Breast Cancer database, donated by Prof. Olvi Mangasarian, is a database of real-world data collected by Dr. William H. Wolberg to serve as a test case for classification data mining systems (Blake 1998). There are 699 records in the database, and each contains values for 9 attributes. The attribute values are integers, and each ranges between 1 and 10. The attributes have to do with properties of tissue samples, such as: clump thickness, uniformity of cell size, etc. Each record is classified as either benign or malignant. The task of a data mining system on this database is to use the attributes of records whose classification is known ("training records") to learn to predict whether an unseen case (a "test re-cord") is benign or malignant. In other words, the task is to discover patterns and regularities in the data that allow reliable prediction of an unseen record's classification. The measure of performance of a system on this task is the system's accuracy at predicting records that it has not seen during training. It should be noted that a small number (16) of the records in the WBC database have some missing attributes. The current version of XCSI followed the procedure in Wilson's version by regarding a missing attribute as matched by any classifier.

Wilson tested XCSI on this problem (Wilson 2000) using a standard statistic methodology called stratified tenfold cross-validation (STCV) (Witten and Frank 2000). When using STCV, the data records are partitioned randomly into ten sets of roughly equal size and equal distribution of classifications in each set. The classification system then does ten different, independent training and test runs. In each run, one of the data sets is withheld as the test set and the other nine data sets are combined and used as the training set. The performance of the system on a single training run is the percentage of test records that it classifies correctly. The performance of the system on one run of tenfold stratification is the average of its performance on the ten training runs. The result of this process is that in executing STCV, every record in the database is used as a test record once, and every record is used as a training record nine times. For the experiments reported here, each partitioned set has roughly 70 records and the ratio of benign/malignant for a set is as close as possible to the average for the whole data set.

Wilson's implementation of XCSI was tested on one run of STCV, in which it achieved a prediction accuracy of ~95.5%. This level of performance was comparable to the performance of the best-published system on this database. The XCSI using the modifications reported in this paper achieved a prediction accuracy of ~96.5%. This paper will compare and contrast the modifications and parameter settings of this version of XCSI with Wilson's version on the WBC database.

Statisticians recommend carrying out ten runs of STCV in order to increase the reliability of the results. Wilson noted in his paper that he would have done this had he had the time. The current implementation of the modified version of XCSI was tested with ten runs of STCV, achieving results that are summarized in the Table 1.

Table 1. 10 runs of STCV (rounded to 0.1)

Run Number of STCV	Performance(%)
1	96.4
2	96.0
3	97.0
4	96.1
5	96.9
6	96.9
7	96.7
8	97.0
9	96.3
10	96.1

Note that there is significant variation in the average of the system performance across the different STCV runs. In this discussion, the mean value of 10 runs of the STCV is regarded as the most important metric of performance for comparison. Note also that in these runs (and many others carried out in other experiments) the minimum value found is higher than the value achieved by Wilson's version on this task.

Experiments with a number of changes to XCSI were carried out in an attempt to improve the performance of XCSI even more. The starting point for these experiments was a version of XCS implemented to solve the multiplexer problem (Wilson 1998). Modifications that improved performance on the multiplexer problem were imported to XCSI and tested on the WBC database. Most of these modifications improved performance on the WBC database as well. Parameter values that were tuned in XCS on the 11-multiplexer (a comparable-sized problem) were adapted to the WBC, and generally produced good results in that domain as well.

The remainder of this paper describes the algorithmic and parameter differences between the version of XCSI that produced this performance and Wilson's prior version. The paper also describes those changes that seemed to make the greatest contributions to the improvement in prediction performance, and describes some apparently promising modifications that did not produce improvements on the WBC database.

2 Algorithmic modifications

The implementation of the algorithm described here is based on the description of XCS in Butz and Wilson (2001). A great deal of experimentation produced some modifications with better efficiency or better learning performance. Some of these modifications primarily impact the efficiency of XCS rather than its performance, and will not be discussed here.

The version of XCSI described here is based on Wilson's description (Wilson 2000). A number of experiments were carried out to determine the impact of modifications to that algorithm, and to determine the robustness of the modified algorithm in the face of noise. The results of ten runs of STCV for these modifications are shown in Table 2.

Table 2. Performance of some algorithm modifications.

Modification	Max	Min	Mean
GA applied to single fields	96.9	95.3	96.1
GA applied to pairs of related fields	96.9	95.7	96.1
Give new chromosomes initial parameter values	97.0	95.4	96.4
Give new chromosomes average parameter values	96.7	95.0	96.2
Shuffle records before presenting them	97.1	95.2	96.2
Pick records randomly	96.7	95.6	96.2
5% persistent noise	90.8	89.6	90.0
5% transitory noise	96.9	95.6	96.1
Reorder after mutation	96.9	95.7	96.3
None-reorder after mutation	97.0	95.4	96.3

(These comparisons were carried out early, and are not based on the best XCSI parameters that were ultimately found. Therefore, the mean values are not as high as ~96.5%. These results are presented for the comparison in performance that they embody, rather than the actual final value)

2.1 GA applied to pairs of related fields

Experiments were done with crossover and/or mutation on pairs of attribute matching values. For example, crossover was not allowed to split the minimum and maximum values for an attribute on a classifier. Improved performance with this technique was found on other problems, but was not noticeable here, as is shown in Table 2.

2.2 Shuffle records

Wilson's implementation of XCSI presented training records to the system in the same order in each pass through the training set. The version described here was modified to shuffle the order of the training records on each pass. No significant change in performance was noticed for this problem using this technique. This is probably understandable. Since the training set as a whole has about 600 records, using 40,000 training steps essentially gives each record many opportunities to be studied. However, the shuffle strategy is less likely in general to produce mistraining as a result of sequences of records that are unrepresentative of the whole data set. The shuffle strategy did at least as well as the random selection strategy in our experiments. Based on these and other experiments on other domains, it is probably better to modify the order of records from a database rather than presenting them to the learning system in the same order on each pass.

2.3 Redefine the concept of "most general" classifier

For the subsumption procedure, the algorithm described here employs a modified computation of generality in the determination of the "most general" classifier. Instead of summing the allowable ranges of each attribute to compute generality, this implementation multiplies the ranges, which is likely to give a more accurate measure when one field has a very restrictive range. However, on this database this modification did not change the performance of the algorithm significantly.

2.4 Using exploitation in the training phase

One parameter is used in the system described here that is not common in the literature. This parameter describes the relative probability that the system will use an exploration rather than an exploitation action selection regime on a given training step to form the action set. This parameter was dubbed "randomAction-ChoiceProb", and was not given a Greek name. The standard classifier system practice has been to train the system only on exploration steps, and track its performance without doing any system training on exploitation steps—those in which the action with strongest fitness is chosen. Work with XCS on multiplexer experiments suggested that training the system with pure exploitation or with a high portion of exploitation steps (>90%) degrades the performance, but using about equal amounts of exploration and exploitation can improve performance. On the WBC database, performance is slightly improved with an equal mix of choosing actions randomly and choosing the action the system believes is best. It is likely that this parameter would have a much greater effect for problems with more than two possible actions.

2.5 Overtraining

The system terminated each training run after 40,000 training steps, where a training step is a presentation of a data record to the system followed by system updates. 40,000 is the termination point most often used by Wilson in his experiments in (Wilson 2000). Experiments showed that termination at 40,000 time steps is also good for the XCSI described here. However, Wilson terminated his runs manually when his system had achieved 100% performance for some time on the training records. It is possible that the requirement of 100% performance causes overtraining. Wilson has recently run his version of XCSI to investigate this point. Wilson's version of XCSI used on the WBC database sometimes fails to match one or more records in the test when it is terminated with a 100% performance criterion on the training phase. In other experiments of the modified XCSI carried out to 200,000 training steps, the system performance was reduced by only one or two tenths of a percent in prediction accuracy using STCV. The current system appeared to be highly resistant to the overtraining phenomenon. Experiments were also conducted on the parameter of the population size (3200/6400). No significant difference was observed under 40,000 training steps. The reason for the reduced tendency to overtrain is not clear.

2.6 Performance under noise

Some experiments were carried out introducing noise into the WBC database to determine the sensitivity of the current algorithm to noise. These experiments involved the introduction of two types of noise.

The first set of experiments involved a noise level of 5% that was permanently induced – persistent noise. That is, before beginning any run of STCV, the outcome of 5% of the training records was reversed and those changes were held constant throughout the training run and in the test of the trained system. The performance of the run was the performance of the trained classifiers on the test data, which included 5% of records that were artificially misclassified. The performance of the system, as shown in Table 2, is about 6.5% lower than its performance without this type of noise, as would be expected, if the original records were correctly classified. The degradation in performance is more than the amount of noise, but not significantly so.

The second experiment involved a transitory noise level of 5% random noise. That is, in each presentation of a training record, there was a 5% chance that the environment would give incorrect reinforcement. In these experiments, however, the test records did not have any noise introduced. The version of XCSI described here was relatively immune to this type of noise, as shown in Table 2. The performance of the system with 5% noise is about .5% lower than without the noise. Evidently, the algorithm modifications and parameter changes described here do not degrade the performance of XCSI significantly when transitory noise is added during the training process.

2.7 Reorder after mutation

Mutation may cause the minimum value for a parameter to be greater than the maximum value. In such a case, a classifier would match no records. The current version of XCSI switches the two values of the pair when this occurs. No significant change was observed for the current settings when this was done, as opposed to leaving the values in place, with no possibility of matching messages as shown in Table 2. However, it may be good to use this technique in general. Otherwise, the system will be devoting resources and population space to classifiers that can take no part in the learning process.

2.8 Experience of GA offspring

This version of XCSI does not use average population values as the initial values for parameters such as prediction, error, fitness, experience, last time in GA, and action set size for new classifiers created by the GA. Instead, these values are each set to initial values. With this modification, those new classifiers are considered to have no knowledge about the problem. This technique caused significant improvement during prior experiments on multiplexer problems. Various experiments were carried out on this factor for the WBC database. The results are as shown in Table 3. Using the scheme of resetting to the average values of population under Wilson's parameter settings using our implementation significantly decreases the performance. However, the scheme under the current version's settings only slightly decreases the performance.

Table 3. Performance under different GA offspring experience.

Modification	Performance1	Performance2
Set GA offspring to initial parameter values	96.4	94.6
Set GA offspring to average of population	96.2	86.0

(Performance1 is the mean prediction accuracy obtained under current setting. Performance2 is obtained under Wilson's setting.)

3 Parameter Modifications

The parameter settings used in the version of XCSI reported here were derived on experiments on multiplexer problems of size 11 and 20, and were then carried over to the WBC domain. Some of Wilson's recommended parameter settings for those problems are used here, and some are modified. Table 4. shows the comparison of parameter settings for the two systems.

Table 4. Parameter setting comparison.

Problem	Wilson's settings	Our settings
Population size	6400	3200
Learning rate	0.20	0.25
Alpha	0.1	0.1
Epsilon zero(e0)	1.0	1.0
Nu	5	5
GA threshold	48	48
Crossover probability	0.8	0.8
Mutation probability	0.04	0.04
Deletion threshold experience	50	50
Deletion threshold fitness	0.1	0.1
Subsumption threshold experience	1000	100
Random action choice probability	1.0*	0.5
Minimum action set actions	2	2
Do GA subsumption	Yes	Yes
Do ation set subsumption	Yes	Yes
Payoff	1000/0	100/-100
Mutation range(m0)	4	6
Covering range(r0)	2	2

(*Wilson's XCSI specification did not include a parameter controlling the ratio of training with exploitation versus exploration. Instead, all training steps used exploration. That regime is equivalent to a random action choice probability of 1.0 here.)

662

3.1 Error threshold (e0)

e0 is the error threshold: if the classifier's error is less than e0, the classifier is deemed as "accurate". Figure 1 shows the mean performance for ten runs of STCV using values of e0 ranging from .01 to 2.0. The system described here used 1.0 as the value of e0 except when epsilon-zero was the subject of experimentation. Under the current payoff setting range (100/-100), performance is not significantly affected when e0 is less than 1% of the payoff range.

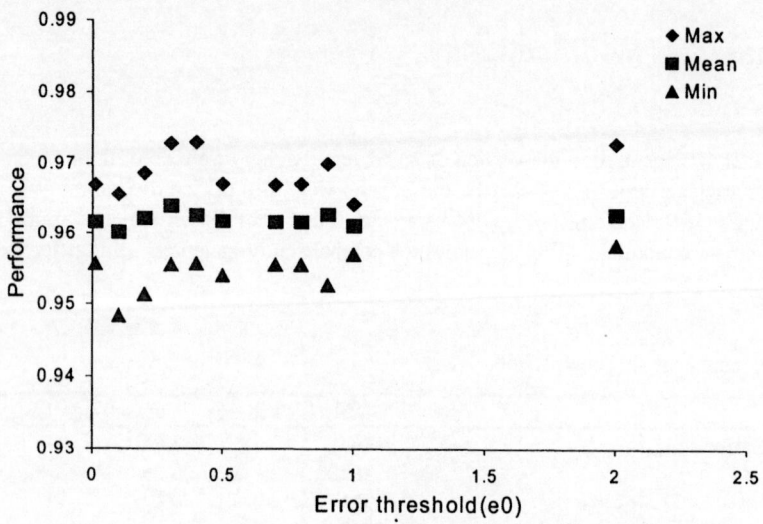

Figure 1. Performance under different error thresholds (e0)

3.2 Learning rate

The learning rate is .25 instead of .20. This change was a bit better on the multiplexer problems, and it is a bit better here, too, as shown in Figure 2. There is a peak in performance at 0.25.

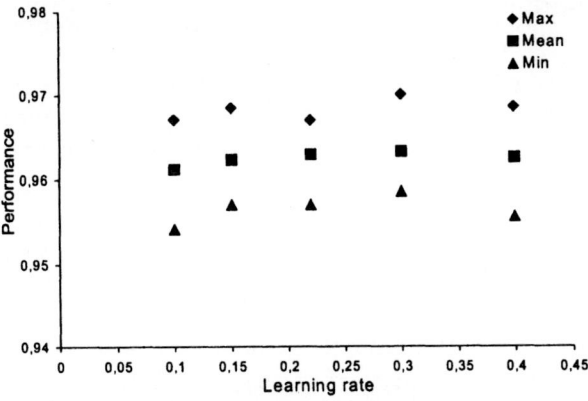

Figure 2. Performance under different learning rates

3.3 Minimum action set size

Table 5. Comparison of the minimum action set size parameter.

Parameter Name	Value	Max	Mean	Min
Minimum action set size	1	97.0	96.4	95.9
Minimum action set size	2	97.1	96.4	95.9

Minimum action set size can be either 1 or 2, since the performance of this system was not significantly different for either choice as shown in Table 5.

3.4 Payoff and e0 combination

Some experiments used symmetric value pairs for payoff and set e0 to be 0.5% of the range. Performance was severely degraded when both payoff and e0 are small as shown in Table 6. A clear peak at the best parameter setting was observed.

Table 6. Comparison of payoff and e0 combination.

Payoff (N/-N)	e0	Performance(%)
1	0.01	85.4
10	0.1	66.5
50	0.5	96.1
70	0.7	95.8
100	1	96.5
500	5	96.2
700	7	96.1
1000	10	96.2

4 Conclusions

The comparison of parameter values used by the version of XCSI described here and those used in Wilson's implementation is contained in Table 4. The results of the experiments showed a clear difference in performance, but it is hard to say just why the improvement occurred. On the basis of a number of comparative experiments, it appears that the following factors contribute most significantly to the difference in performance between the Wilson's version of XCSI and the version described here:

- GA offspring have zero experience
- The good combination of Payoff and e0
- Learning rate
- Termination criterion during the training phase

Wilson's application of XCSI to the WBC database showed that integer-based classifier systems could be competitive in the real-world data-mining arena. This paper's modifications and extensive tests further confirm that XCSI does extremely well on the WBC database. The rules incorporated in the best classifiers often are very meaningful to physicians, and the predictive accuracy of the system was better than any other reported approach. This paper has described some modifications to that algorithm and its parameter settings, resulting in improvements of performance of XCSI on the WBC database. It is worth noting that most of those modifications also improve the performance of XCS on the multiplexer problem, a binary problem that is most quickly solved by classifier systems when the systems evolve compact solution representations. This fact implies that the modifications described here have more general application than to the WBC database problem, or to integer-based data mining problems.

References

1. Blake, C. and C. Merz (1998). UCI repository of machine learning databases. http://www.ics.uci.edu/~mlearn/MLRepository.html
2. Butz, M. V. and Wilson, S. W. (2001). An Algorithmic Description of XCS. In Lanzi, P. L. , Stolzmann, W., and S. W. Wilson (Eds.) *Proceedings of the International Workshop on Learning Classifier Systems (IWLCS-2000)*. Springer-Verlag. Or see http://prediction-dynamics.com
3. Wilson, S. W. (1995). Classifier Fitness Based on Accuracy. *Evolutionary Computation*, 3(2), 149-175
4. Wilson, S. W. (1998). Generalization in the XCS Classifier System. *GECCO: Proceedings of the Third Annual Conference*, J. Koza et al. (eds.) San Francisco, CA: Morgan Kaufmann
5. Wilson, S. W. (2000). Mining Oblique Data with XCS. In: Technical Report No. 2000028, Illinois Genetic Algorithms Laboratory, University of Illinois at Urbana-Champaign.
6. Witten, I.H. and E. Frank (2000). *Data Mining: Practical Machine Learning Tools and Techniques with Java Implementations*. San Francisco, CA: Morgan Kaufmannn.

Artificial chemistry - a new metaphor for evolutionary algorithms

Vladimír Kvasnička

Department of Mathematics, Slovak Technical University, 812 37 Bratislava, Slovakia, email: kvasnic@cvt.stuba.sk

Abstract. An *artificial chemistry* is a branch of contemporary computer science that uses a chemical metaphor as a new highly parallel approach to computations. It can be defined by (1) a set of objects (a chemostat composed of molecules) and (2) a set of transformation rules (chemical reactions), which specify how the objects are transformed into other objects. The purpose of the present short communication is to use the well-known Eigen's chemical system of replicators as a new (chemical) metaphor. It is then applied in design of the so-called *replicator algorithm*. This algorithm has some common features with genetic algorithms, but as an advantage of the presented replicator algorithm we consider an existence of relatively simple proof that the algorithm offers globally optimal solutions.

1 Introduction

Recently, an idea of using chemical systems for computation became very popular in computer science as an interesting alternative possibility how to realize massive parallel computations. It was initiated by Adleman's seminal paper [1] published in 1994 on a DNA computation (he solved experimentally a simple traveling salesman problem composed of seven vertices). Immediately after Adleman's paper, this chemical computational metaphor was generalized to universal molecular computers based on graphs [18,19] and molecular Turing machines [13,3,21].

The chemical metaphor for computation has been discussed at various places in the literature. Haken [12] proposed to use chemical reactions for information processing. In 1988 the Γ-language was introduced by Benatre et al. [4]; their primary motivation was the inadequacy of the imperative programming for massively parallel computers. Later, the idea was further developed by Berry and Boudol [5]; they created the so-called *chemical abstract machine*. Another extension of chemical computation metaphor was done by Fontana et al. [11]. They studied a computation system based on the so-called λ-calculus. In their series of highly abstract papers the chemostat elements are interpreted as functions which react with each other. Through intermediate states, λ-expressions are evaluated with the reaction partner as an argument and the result of this computation is considered as an analog of a reaction product. Dynamical aspects of systems based on the chemical computation metaphor have been considered by a series of papers of Banzhaf and Dittrich [2,7]. These authors studied interesting chemical systems of autoreplicators and hypercycles (introduced by Eigen and Schuster [8,9] as proper intermediates for an understanding of processes running on a border of biotic and abiotic systems). Banzhaf and Dittrich have studied a dynamics of these systems in such a way that molecules were represented by binary strings. The main outcome of their results is the fact, that the phenomenological results of Eigen and Schuster's differential equations may be simply simulated by the chemostat filled by simply structured objects - molecules. Similar results have been obtained almost simultaneously and independently by many other authors as a part of artificial life studies [7]. Here should be mentioned a seminal book of Voigt [23], who already at the end of eighties formulated many basic principles of evolutionary calculations based on replicator networks.

The purpose of this communication is to demonstrate that the well-known Eigen's replicators [8,14] may be used as a new metaphor for a design of a new stochastic optimization algorithm called the replicator algorithm. Its main advantage consists in a theoretical area. In particular, it offers relatively simple and straightforward arguments why the globally optimal solution can be achieved. The same property was observed for the simulated annealing [6,16,20,22] optimization method, where a general convergence property to globally optimal solution can be done by making use simple arguments from statistical thermodynamics.

2   Chemical metaphor - chemostat

Let us consider a *chemostat* (chemical reactor) composed of formal objects called the molecules. There is postulated that the chemostat is not spatially structured (in chemistry it is said that the reactor is well stirred, see Fig. 1). Molecules are represented by formal structured objects (e.g. token strings, rooted trees, λ-expressions, etc.). An interaction between molecules is potentially able to transform information, which is stored in the composition of the molecules. Therefore a chemical reaction (it causes changes in the internal structure of reaction molecules) can be considered as an act of information processing. The capacity of the information processing depends on the complexity of molecules and chemical reactions between them.

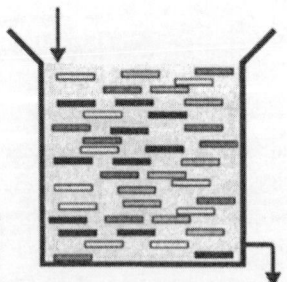

Fig. 1. The chemostat contains a homogeneous "solution" (mixture) of molecules, a few molecules are fully randomly selected (with a probability determined by their concentrations), then these molecules undergo a chemical reaction that consists in a transformation of some of them. The produced molecules are returned to the chemostat in such a way that they may eliminate some other molecules.

3. Eigen's replicators

Manfred Eigen published in the beginning of seventies a seminal paper entitled *"Self organization of matter and the evolution of biological macro molecules"* [8,14], where he postulated a hypotetical chemical system composed of the so-called replicators. This system mimics Darwinian evolution even on an abiotic level. Eigen and Schuster [9] discussed the proposed model as a potentially possible abiotic mechanism of a driving force for an increase of complexity on a border of abiotic and biotic systems.

Let us consider molecules (called the *replicators*) $X_1, X_2, ..., X_n$ that are capable of the following chemical reactions:

$$X_i \xrightarrow{k_i} X_i + X_i \quad (i = 1, 2, ..., n) \tag{1a}$$

$$X_i \xrightarrow{\phi} \varnothing \quad (i = 1, 2, ..., n) \tag{1b}$$

The first reaction (1a) means that a molecule X_i is replicated onto itself with a rate constant k_i, whereas the second reaction (1b) means that X_i becomes extinct with a rate parameter ϕ (this parameter is called the *"dilution flux"* and will be specified further). Applying the mass-action law of chemical kinetics, we get the following system of differential equations

$$\dot{x}_i = x_i (k_i - \phi) \quad (i = 1, 2, ..., n) \tag{2a}$$

The dilution flux ϕ is a free parameter and it will be determined in such a way, that the following condition is satisfied: A sum of time derivatives of concentrations $\dot{x}'s$ is vanishing, $\sum \dot{x}_i = 0$, we get

$$\dot{x}_i = x_i \left(k_i - \sum_{j=1}^{n} k_j x_j \right) \quad (i = 1, 2, ..., n) \tag{2b}$$

where the condition $\sum x_i = 1$ was used without a loss of generality of our considerations. Its analytical solution looks as follows

$$x_i(t) = \frac{x_i(0) e^{k_i t}}{\sum_{j=1}^{n} x_j(0) e^{k_j t}} \tag{3}$$

This solution satisfies the following asymptotic property, where only one type of molecules (with the maximal rate constant k_{max}) is surviving while other ones become extinct

$$\lim_{t \to \infty} x_i(t) = \begin{cases} 1 & (\text{for } k_i = k_{max} = max\{k_1, ..., k_n\}) \\ 0 & (\text{otherwise}) \end{cases} \tag{4}$$

Loosely speaking, it means, that each type of molecules may be considered as a type of species with a fitness specified by the rate constant k. In a chemostat "survive" only those molecules - species that are best fitted, i.e. that have the highest rate constant k_{max}, and all other molecules with smaller rate constants become extinct, see Fig. 2. The condition of invariability of the sum of concentrations (i.e. $\sum x_i = 1$) introduces a "selection pressure" to replicated molecules, there will survive only those molecules that are best fitted with the maximal rate constant.

668

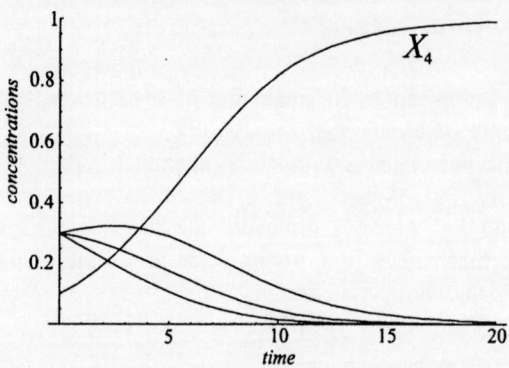

Fig. 2. A plot of concentrations of four component system with rate constants $k_1=1$, $k_2=2$, $k_3=3$, and $k_4=4$. We see that only molecules X_4 survive at the end.

The proposed model may be simply generalized in such a way that mutations are introduced in process of replications, the system (2) is modified as follows

$$\dot{x}_i = \sum_{j=1}^{n} k_{ij} x_j - x_i \phi \quad (i=1,2,...,n) \tag{5}$$

where k_{ij} is a rate constant assigned to a modified reaction (1a)

$$X_i \xrightarrow{k_{ij}} X_i + X_j \quad (i,j=1,2,...,n) \tag{6}$$

There is postulated that a rate constant matrix $K = (k_{ij})$ has dominant diagonal elements, i.e. nondiagonal elements are much smaller than diagonal ones. This requirement directly follows from an assumption that replications (6) are very rare, that is the product X_j is considered as a weak mutation of X_i, $X_j = O_{mut}(X_i)$. The dilution flux ϕ from (5) is determined by the condition that a sum of time derivatives of concentrations is vanishing, $\sum \dot{x}_i = 0$, we get

$$\phi = \sum_{i,j=1}^{n} k_{ij} x_j \tag{7}$$

Analytical solution of (6) with dilution flux specified by (7) is [14]

$$x_i(t) = \frac{\sum_{j=1}^{n} q_{ij} x_i(0) e^{\lambda_i t}}{\sum_{m,j=1}^{n} q_{mj} x_j(0) e^{\lambda_j t}} \tag{8}$$

where $Q = (q_{ij})$ is a nonsingular matrix that diagonalizes the rate matrix K, $Q^{-1} K Q = \Lambda = diag(\lambda_1, \lambda_2, ..., \lambda_n)$.

Since we have postulated that nondiagonal elements of K are much smaller than its diagonal elements, its eigenvalues λ's are very close to diagonal elements, $\lambda_i \square k_{ii}$, and the transformation matrix Q is tightly related to a unit matrix, $q_{ij} \square \delta_{ij}$ (a Kronecker's delta symbol). It means that an introduction of weak mutations does not change dramatically the general properties of the above simple replicator system without mutations. In particular, the final (for $t \to \infty$) chemostat will be composed almost entirely of molecules with the greatest rate constant k_{max}.

These molecules are weakly accompanied by other replicators with rate constants k's slightly smaller than k_{max}.

4 A binary function optimizer

Let us consider a binary function

$$f : \{0,1\}^n \to R \qquad (9)$$

This function $f(x)$ maps binary strings $x = (x_1, x_2, ..., x_n) \in \{0,1\}^n$ of the length n onto real numbers. We look for an optimal solution

$$x_{opt} = arg \min_{x \in \{0,1\}^n} f(x) \qquad (10)$$

Since the cardinality of the set $\{0,1\}^n$ of solutions is equal to 2^n, a CPU time necessary for solution of the above optimization problem grows exponentially.

$$t_{CPU} \approx 2^n \qquad (11)$$

It means that the solution of the binary optimization problem (10) belongs to a class of hard numerical NP-complete problems. This is the main reason why the optimization problems (10) are solved by the so-called evolutionary algorithms [10], that represent very efficient numerical techniques how to solve binary optimization problems. The purpose of this subsection is to demonstrate that a metaphor of replicator provides an efficient stochastic optimization algorithm.

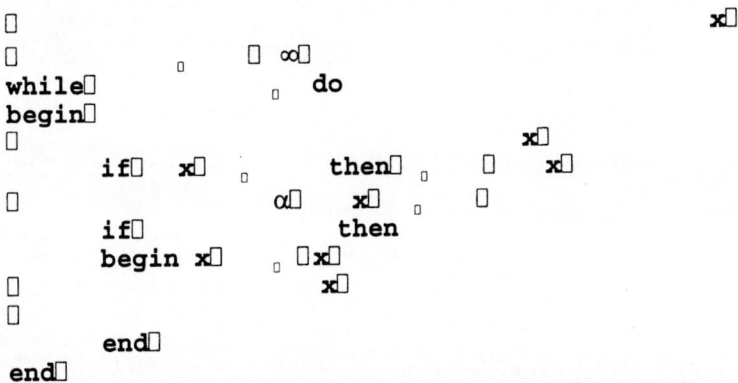

Algorithm 1. A pseudo Pascal implementation of chemostat optimization. The algorithm is initialized by randomly generated chemostat molecules, i.e. binary strings of the fixed length. A randomly selected molecule x is transformed with a probability *prob* by a mutation onto a new molecule x'. This new mutated molecule substitutes a randomly selected molecule.

Let a chemostat be composed of molecules that are realized by binary strings $x = (x_1, x_2, ..., x_n) \in \{0,1\}^n$. A monomolecular reaction is considered (cf. eq. (6))

$$x \xrightarrow{prob} x + x' \qquad (12)$$

670

where the formed molecule x' substitutes a randomly selected molecule from the chemostat. A term *prob* assigned to the chemical reaction is interpreted as a probability (rate constant) of a performance of reaction (12). This probability is determined by function values of x and a temporarily best solution $x_{temp,opt}$

$$prob = exp\left(-\alpha\left[f(x) - f\left(x_{temp,opt}\right)\right]\right) \qquad (13)$$

where α is a positive constant. If the functional value of x is much greater than the value for the temporally optimal solution, $f(x)? f\left(x_{temp,opt}\right)$, then the entity *prob* is a small number. On the other hand, if functional values $f(x)$ and $f\left(x_{temp,opt}\right)$ are close, then *prob* is closely related (from below) to one. In evolutionary algorithms a selection pressure in population of solutions (chromosomes) is created by a reproduction process based on chromosome fitness.

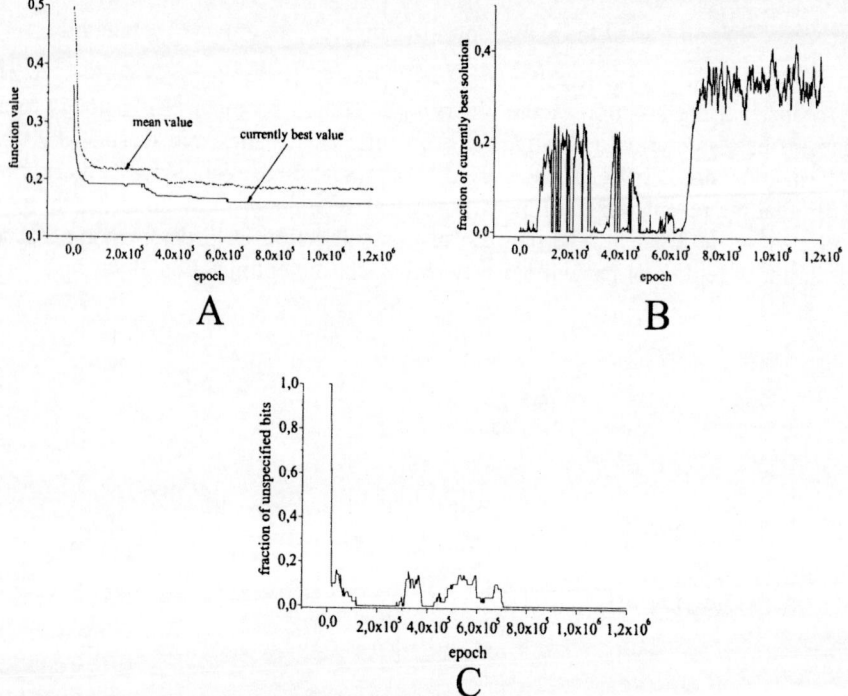

Fig. 3. Plots of results obtained by Algorithm 1 for parameters: Chemostat size 1000; length of binary strings 50, mutation probability P_{mut}=0.01. The optimized function f is specified as Kauffman's [18] function with parameters N=50 and K=5. The probability *prob* is specified by (13), where the constant α=5. Diagram A represents plots of the mean value of function values of all molecules in chemostat and a function value of the temporarily best solution. Diagram B shows a plot of the fraction of temporarily best solution, we see that when after 7×10^5 epochs the globally best solution is achieved (see also diagram A), roughly 30% of the chemostat consists of these solutions. Diagram C is a plot of the fraction of bits in molecules that are ambiguously specified. We see that when going from one temporarily optimal solution to another temporarily optimal solution there exist a big rebuilding of bits in molecules.

Chromosomes with a greater fitness have the greater chance to take part in a reproduction process (a measure of quality of chromosomes); on the other hand, chromosomes with a small fitness are rarely used in the reproduction process. This simple manifestation of Darwin's natural selection ensures a gradual evolution of the whole population. In the present approach of the replicator algorithm the principle of fitness selection of solutions is substituted by a probability of performance of reaction (12). A molecule incoming to the reaction is randomly selected from the chemostat. After an evaluation of a quality of the selected molecule with respect to the temporarily best solution is then stochastically decided whether the reaction is performed or not (see Algorithm 1).

Finally, we specify a product x' from the right-hand side of (12) as a mutation [10] of an incoming molecule x

$$x' = O_{mut}(x) \tag{14}$$

where O_{mut} is a stochastic mutation operator that changes single bits with probability P_{mut}. A pseudo Pascal code of a replicator algorithm is specified by Algorithm 1. The obtained results that are achieved for a highly-multimodal binary Kauffman's function [15] are outlined in Fig. 3. We see that the present chemostat approach is able to achieve correct global minimum for Kauffman's function specified by $N=50$ and $K=5$ (according to our numerical experiences, this function represents an extremely difficult optimization problem) .

5   Discussion

In this communication it was demonstrated how a chemical metaphor (based on Eigen's replicators) may be applied in design of the so-called replicator algorithm. This algorithm resembles in many features standard evolutionary algorithms, but the most important feature of the present metaphor is its introduction of unexpected new possibilities how to prove a convergence of the proposed algorithm to a globally optimal solution. In particular, since the proposed algorithm is based on a chemical metaphor, an application of a standard "chemical kinetics" offers simple and straightforward arguments why an asymptotic chemostat state should be mainly composed of globally optimal solutions. A very similar situation exists in the well-known simulated annealing method, that represents a very efficient stochastic optimization method based on a metaphor of physical process of annealing [6,16,20,22]. This fact has a very important impact on a theory of simulated annealing that may be simply elaborated based on arguments adopted from statistical thermodynamics.

Other important aspect of artificial chemistry may be "in silico" performance of hypothetical chemical processes that could not be realized by the standard "in vitro" chemistry. Recently, many chemical processes running on the border of abiotic and biotic systems are simulated by "in silico" techniques. This approach represents a substantial part of the current artificial life area of computer science. In particular, Eigen and Schuster's [8,9] autoreplicators and hypercycles are easily modeled by chemostats. This possibility was extensively discussed in our recent paper [17].

Acknowledgement. The work was supported by the grants 1/7336/20 and 1/8107/01 of the Scientific Grant Agency of the Slovak republic.

References

1. Adleman L.M. (1994) Molecular Computation of Solutions of Combinatorial Problems. *Science* **266**, 1021
2. Banzhaf W., Dittrich P., Eller B. (1999) Topological Interactions in a Binary String System. *Physica D* **125**, 85
3. Beaver D. (1995) Universal Molecular Computer, unpublished manuscript, available at ftp://math.chtf.stuba.sk/pub/vlado/dimacs95.ps
4. Benatre J.-P., Le Metayer D. (1990) The Gamma Model and its Discipline of Programming. *Sci. Comp. Progr.* **15**, 55.
5. Berry G., Boudol G. (1992) The Chemical Abstract Machine. *Theoret. Comp. Sci.* **96**, 217
6. Černý J. (1985) Thermodynamical Approach to the Traveling Salesman Problem: An Efficient Simulation Algorithm. *J. Opt. Theory Appl.* **45**, 41
7. Dittrich P. (1999) Artificial Chemistries (tutorial material). A tutorial held at *ECAL'99*, 13-17 September 1999, Lausanne, Switzerland, available at ftp://math.chtf.stuba.sk/pub/vlado/Artificial_Chemistry/Dittrich_tutorialAChemECAL99.pdf .
8. Eigen M. (1971) Self organization of matter and the evolution of biological macro molecules. *Naturwissenshaften* **58**, 465
9. Eigen M., Schuster P. (1977) The Hypercycles: A Principle of Natural Evolution. *Naturwissenschanften* **64**, 541; **65**, 7; **65**, 341
10. Fogel D. B. (1995) *Evolutionary Computation*. IEEE Press, New York
11. Fontana W. (1991) Algorithmic Chemistry. In: Langton C.G. (ed.) *Artificial Life II* Addison Wesley, Reading, MA, p. 159
12. Haken H. (1979) Pattern Formation and Pattern Recognition - An Attempt to a Synthesis. In: Haken H. (ed.) *Pattern Formation by Dynamical Systems and Pattern Recognition.* Springer Verlag, Heidelberg
13. Hjelmfelt A., Weinberg E.D., Ross J. (1991) Chemical Implementation of Neural Networks and Turing Machines. *Proc. Natl. Acad. Sci. USA* **88**, 10983
14. Jones B.L., Enns R.H., Rangnekar S.S. (1976) On the theory of selection of coupled macromolecular systems. *Bulletin of Mathematical Biology* **38**, 15
15. Kauffman S.A. (1993) *The Origins of Order: Self-Organization and Selection in Evolution.* Oxford University Press, New York
16. Kirkpatrick S., Gelatt Jr. C.D., Vecchi M.P. (1983) Optimization by Simulated Annealing. *Science* **220**, 671
17. Kvasnička V., Pospíchal J. (2001) Autoreplicators and Hypercycles in Typogenetics, *Journal of Molecular Structure (Theochem)* **547**, 119
18. Lipton R.J. (1995) Speeding up Computation via Molecular Biology, unpublished manuscript, available at ftp://math.chtf.stuba.sk/pub/vlado/DNA_computing/bio.ps
19. Mayoh B. (1995) Biological Computation is Universal. Unpublished manuscript, available at ftp://math.chtf.stuba.sk/pub/vlado/DNA_computing/biocomputationNat.ps
20. Otten R.H.J.M., van Ginneken L.P.P.P. (1989) *Annealing Algorithm.* Kluwer, Boston
21. Rothemund P.W.K. (1995) A DNA and Restriction Enzyme Implementation of Turing Machines, unpublished manuscript, available at ftp://math.chtf.stuba.sk/pub/vlado/Rothemund_dimacs.ps

22. van Laarhoven P.J.M., and Aarts E.H.L. (1987) *Simulated Annealing. Theory and Applications*. Reidel, Dordrecht
23. Voigt H.-M. (1989) *Evolution and Optimization: An Introduction to Solving Complex Problems by Replicator Networks*. Akademie-Verlag, Berlin

Immune System Simulation through a Complex Adaptive System Model

António Grilo, Artur Caetano, Agostinho Rosa

LaSEEB-ISR, Instituto Superior Técnico
Av. Rovisco Pais 1 - Torre Norte 6.21 - 1049-001 Lisboa - Portugal
amg@amadeus.inesc.pt, ampc@amadeus.inesc.pt, acrosa@laseeb.ist.utl.pt

Abstract. Evolutionary algorithms and cellular automata are two computational approaches to model complex adaptive systems. Here is described an immune system simulator that uses a cellular automaton to model the physical environment along with an evolutionary genetic algorithm to attain adaptation and selection. Agent genetic coding comprised within the genotype is a set of rules which expresses behavior. Moreover, an agent includes a collection of operators which use the genetic code in order to interact with other agents and physical sites. We also depict the system's methodology as well as some of the obtained results.

Keywords: Artificial Immune Systems, Complex Adaptive Systems, Evolutionary Genetic Algorithms, Cellular Automata, Artificial Life.

1 Introduction

Immune systems, ecological systems as well as many others, are difficult to control or describe using traditional computational methods. Two main difficulties are ensued when modeling such a system. The first problem arises from nonlinear interactions among system components. The second is issued when system's agents can evolve, or change their specification, over time. Systems with these properties are sometimes called Complex Adaptive Systems (CAS), and comprise the following [34]:

1. A collection of primitive components which will stand for the artificial entities, called 'agents.'
2. Interactions among agents and between agents and their environment.
3. Unanticipated global properties often result from the interactions.
4. Agents adapt their behavior to other agents and environmental constraints.
5. System behavior evolves over time, as a consequence of the previous properties.

Creating a model of a CAS is complex for several reasons. First, both nonlinearities and the changing behavior of the system restrain predictive mathematical models. Second, to model detailed simulations, computational problems exist, since expressing every detail is virtually impossible. For example, the vertebrate immune system can express, at a given time, over 10^7 receptors. Modeling such level of detail is computationally overwhelming. Therefore, in every large complicated system, precise computational modeling is virtually impossible, not only due to the former issue, but also because nonlinear system are highly dependent on small inaccuracies. To solve this problem, removing all the possible detail from the model, retaining only the essential interactions is a possibility. The goal is then to develop models whose behavior is sound with respect to the details of the interactions, and which produces the desired behavior classes. The major drawback with this approach is that such models will very seldom make precise quantitative predictions. Then, what are the expectations of a model that does not correspond directly to any real system? Patterns of behavior can be studied, such as how agents interact and cooperate under given circumstances, which can be difficult to obtain in real systems. On the other hand, it is much easier to run what-if experiments than to conduct real system experiments. With a well designed model, theoretical reasoning can be built about evolution dynamics, such as agent dependencies and interactions.

In silico simulations are becoming powerful research tools through the definition of biological system models, namely immune system models, since a better understanding of important immunological phenomena is required to withhold the rising threat provoked by the offspring of new viruses and the contagious property of others, such as the HIV-I [17, 19].

This paper presents an Artificial Immune System (AIS) constructed using an hybrid approach that supports the evolving of an heterogeneous population of agents over an artificial environment. A genetic approach is used to model the agents while the underlying layer is supported by an object oriented cellular automaton. Genotypes are formed by tagged rules which express an agent's behavior, upon the interpretation of an operator. The immune responses comprised in the development of Acquired Immuno-Deficiency Syndrome (AIDS), provoked by HIV-I, are simulated using this AIS.

The structure of this paper is the following: in section 2 is presented some other work related with this system, namely some work on artificial life and on artificial immune systems. Section 3 discusses the methodology of the proposed system. Section 4 starts with an overview of the human immune system and the HIV-I virus. Follows the model used in the artificial immune system simulation. This section ends with the results obtained with this simulator. Finally, section 5 outlines some conclusions and future work.

2. Related Work

2.1 Cellular Automata

Although cellular automata (CA) have been used to model ecological systems, they offer a number of limitations when used *per se*.

The first comes from the fact that each CA site is static in space and can only change its state according to a set of local rules, not being able to stand for an autonomous agent. Another problem is related to spatial and temporal scaling. While CA models usually assume each place to be sized for one individual, this assumption proves inappropriate when modeling several ecological environments. Nevertheless, cellular automata are adequate to model some classes of environments. In [27] is presented a CA model for the study of competition between grass species. [26] addresses the effects of fire and dispersal on spatial patterns in forests. Various aspects related to the use of

CA in the study of emergent behavior and Artificial Life in general, are addressed in [28].

When used along with other computational approaches, the cellular automata model proves best. An example is the Object-Oriented Cellular Automaton (OOCA) [23]. The OOCA uses a conventional CA as its lowest layer, where a set of agents, or devices, may interact with its sites, modifying or using their current state to decide what action to perform. Moreover, communication through message-passing mechanisms, resulting from object independence, allows actions to be directly performed over other devices. Thus, an OOCA model is easily extendible through the definition of new devices that possess different action rules. The modeling and simulation of biological systems which are prone to frequent changes, is one of the applications where the OOCA may be used to model the artificial environment. The Cellular Device Machine (CDM), presented in [5], is a system that comprises this approach. It includes a development environment that uses SLANG, a dedicated object-based programming language, to express space and time relationships in large complex adaptive systems.

There are many examples of the use of the CA model that include genetic processing. In fact, all the systems described in the next subsection fall into this category.

2.2 Artificial Life

Genetic Algorithms (GA) are computational models of evolution which play a central role in many artificial life models. Fitness functions can be implicit, expressing the performance of an agent bounded by its environment, in the same way as natural selection. Moreover, genetic mechanisms can be easily modeled by way of genetic operators, and population dynamics can be simulated without complex mathematical models, as the set of simple local rules within a GA is usually sufficient. Consequently, it's not surprising that most evolving Artificial Life applications use the GA as a basis [33]. Several examples are described next.

A model for an ecological system simulation is proposed in Echo [30], [31], [32], [34]. In this system, a population of agents inhabits an heterogeneous lattice, where each site can be shared. The lattice serves also as a resources repository of different types. Agents can interact through mating, trading and fighting. A set of rules define the interaction among agents. Likewise, a set of attributes, corresponding to its external "appearance", is encoded in the string that forms its genotype. The fitness function is endogenous, and reproduction is made by cloning, with low mutation probability. Genetic material is exchanged between agents by mating.

In [2] is addressed the simulation of bean and weevil population, emphasizing the role of contest and scramble competitions on its evolution. The system uses a diploid GA, coding all life parameters in a bit string.

The evolution of a size-structured predator-prey community is studied in [24]. Organisms are grouped accordingly to their size: those of the smallest class are autotrophs; all the others are heterotrophs which can kill organisms of smaller classes. The attributes of an organism are coded on a bit string, and their surrounding environment consists of a lattice through which the organisms move in random directions depending on the organism size.

Game theory is used to model evolution and interaction between species in food webs that result from explicit resource flow [25]. The fitness of a species - which constitutes its rate of reproduction - is proportional to the amount of accumulated energy. The latter is distributed among the organisms taking into account their score in the game. The main part of a genome is the strategy gene which defines an organism's behavior. Genomes are subject to three kinds of mutation: single bit mutation, gene duplication and split mutation. While gene duplication increases agent memory, split mutation performs the opposite by dividing the genome into two parts and randomly choosing one of them to be kept.

Several other Artificial Life articles focus on other important issues, such as the impact of environmental topologies in artificial ecological systems. Topology impact in various kinds of communities formed by two interactive species is studied in [1]. Among the conclusions, are the role of an homogenous topology: fully connected graphs which promote rapid growth and higher population levels than local and random-graph topologies. It is suggested that while local topologies, such as the nearest neighbor, offer special interaction restrictions, homogeneous ones offer no barrier for the individuals inhabiting the same place.

A survey on the role of Genetic Algorithms in Artificial Life can be found in [33].

2.3 Artificial Immune Systems

Some important work is being developed at the University of California at San Diego and at Santa-Fe Institute on Artificial Immune System models, including HIV-I simulation.

The previously referred CDM is the platform of research at the University of California at San Diego. In [4], is presented a Cellular-Device Machine simulation of the HIV infection in an artificial immune system, where the state machines which depict model behavior are described. A detailed analysis of the results is presented as concentration progresses for both immune system cell agents as well as for soluble substances. When compared to the data obtained *in vivo*, the results were surprisingly accurate. Observed responses include: systemic patterns, like the predominance of the macrophage reservoir; increased IL-2 responsiveness on CD4 T cells; higher productions of both IL-1 and γ-IFN when matched with IL-2 and IL-5.

The hu-SCID (Severe Condition Immuno Deficiency) mouse lymph node simulation is publicly available at [23], and also uses the CDM. Its set of devices comprises T lymphocytes, macrophages, and stoma cells of the lymphatic tissue. The SCID mouse presents initial immuno-deficiency and xenografts of human cells are used to complement the immune system. It is observed the lymph node's response to virus infection, and results have already been obtained for HIV-I. Some of them helped to conclude about the role of macrophage-tropic HIV-I in AIDS development [7].

Another Artificial Immune System is being developed at the Santa-Fe Institute. It uses a set of differential equations whose parameters are adapted by a neural network. Like the human nervous system, the immune system performs pattern recognition tasks, where antigen presentation to the neural net induces system learning. Some issues of this work are presented in [29].

3. Methodology

Our system addresses two different questions: how to model artificial independent individuals; and how to set up an artificial environment, able to provide feedback and support interaction.

Individuals are modeled by way of self-ruling devices, called agents, which can interact with the system as well as with other agents. Comprised within each, lies every mechanism required to express its internal behavior and to respond to system feedback. Since each agent is an independent and closed device, heterogeneous agent populations are directly considered. Communication can be either the result of direct agent interaction, or indirect, using the underlying environment as platform. Each entity has a particular set of rules which determines its external interaction mechanisms, or behavior. These rules are encoded within each gene of the agent's chromosome. Hence, a gene is a partial state controller, where if-then or other type of rules may specify the behavior. Each gene can keep a different format rule, along with other data, such as the conditions under which the rule is to be applied. Therefore, the agent's genotype is the fusion of all the current genes within the chromosome.

Behavior relies in a set of operators which decode specific genetic information and act accordingly. This operators use no explicit fitness function, instead they use environment and genetic information to guide the agent. Thus, global behavior is obtained through each local operator's evaluation of the genetic rules comprised within each agent. Moreover, each agent can present the environment and other agents with a set of external features, which may be also encoded in the genetic code. On the other hand, the number of genes and operators is not fixed, allowing dynamics during the agent's evolution. With this approach, each agent comes as an autonomous device, where its behavior is given through a function of the current set of rules, environment feedback and external information presented in other agents. Furthermore, genetic rules are mapped into a variable length chromosome, which can hold an arbitrary number of autonomous genes at each instant. This proves to be an attractive feature, especially in the modeling of immune systems where genetic code modification exists.

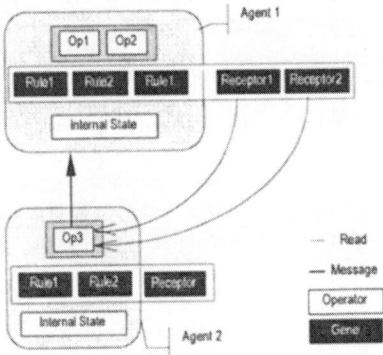

Fig. 1. Agent internal structure and inter-agent relationships. Particular genes, included in the agent, may be visible to the surrounding environment, acting as receptors. Operators can interfere with other agents using message-passing mechanisms, while receptors may serve as input devices.

The agent population is distributed over a set of sites comprised in the lattice, which, along with the corresponding spatial relationships, can be expressed through an object-oriented cellular automaton. This allows a detachment between each physical location and the lattice that comprises them. Moreover, agents and environment will also become separated. Since each lattice site can be different from all others, it enables the construction of heterogeneous site worlds.

A further extension to the usual OOCA container location allows each site to be active rather than to simply hold data. This is accomplished by assigning a local processing function to each site, which is used to decide what local action to take upon its current information, providing a measure of locality within each site. Many agents can cohabit at the same site, and, since each location has a degree of activity, it can interact with the devices it currently comprises. This allows a twofold mechanism between devices and the artificial environment, providing the means to model environmental hazards or other features.

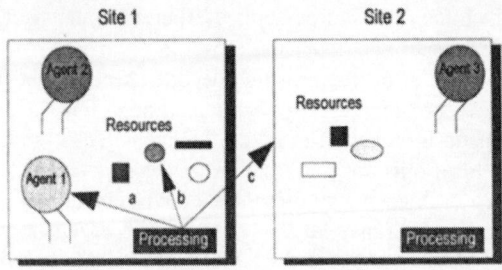

Fig. 2. Each site is self-contained and can hold several agents simultaneously. The measure of locality is provided through a local processor which can interact with the agents comprised within (a); interact with the local resources (b); and interact with other independent sites (c).

The outcome of binding both mechanisms above described resulted in an hybrid architecture which has, on the one hand, the advantages of outlining each entity through a genetic approach, and on the other, the usefulness of having an autonomous layer controlling and providing mechanisms to simulate artificial world interactions. The genetic approach allows natural selection to be achieved through an endogenous fitness mechanism, emerging from the actions and interactions with the system as a whole.

Artificial life simulations can be accomplished with this methodology through the definition of rules which depict behavior. Likewise, the creation of an heterogeneous artificial world is straightforward, for each site's local 'behavior' can be expressed through the outlining of a local response function.

The cellular automaton layer uses a synchronous update policy to express the simultaneous transitions that would exist in a biological environment if observed as a discrete system.

Every cycle the population is prompted to interact with the system. A notification is then sent to every agent, which has the external result of mapping the genotype into the phenotype through its set of operators. First, recall that every entity is self-contained and each chromosome may vary in the number and type of genes. Therefore, behavior results as function of the device's operators over the required genetic code included in the entity's chromosome. With this approach, every entity can hold several operators along with variable-length chromosomes, as formerly stated.

The sequence of events in each simulation cycle consists of the following:

1. Every agent located at each site is prompted to evaluate the current environment along with the visible features of other agents. This will update the agent's set of triggering conditions.

2. Every operator executes the rules contained in the corresponding genes, using the current conditions' state, set in the previous step. In this way, agent-agent and agent-environment interactions take place and external features are presented.

3. The environment interacts with the agents upon site local data and specific rules.

4. Environment-environment interaction. Site resources are modified upon local rules.

5. 'Dead' agents are removed from the lattice.

As an example, suppose the simulation of a system where two different agent classes were comprised (Fig. 3). Each entity moves conformably with a specific pattern. Crossover can be performed by entities of different sexes, and is followed by a gene mutation in the offspring. The artificial world is composed of two types of locations: L1, where substances spread to same class sites where its amount is scarcer; in L2 locations substances do not spread but dissolve locally at a certain rate.

To simulate this artificial world one needs to define two different features: first, define the behavior of each one of the agents. This is accomplished through the definition of the genetic rules and operators which hold its behavior. An agent is then able to autonomously perform the task subdued to its set of rules; second, define the underlying environment. This is done by creating the local rules for each independent site.

682

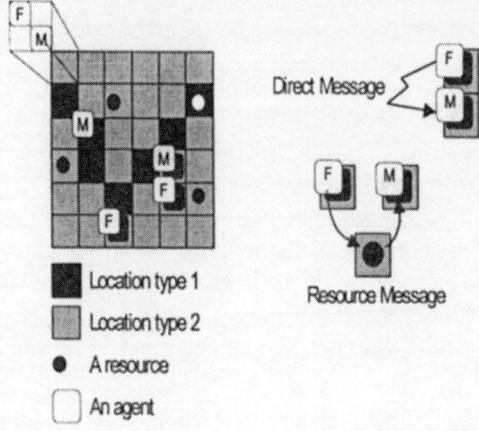

Fig. 3. An example of an artificial world. Each entity's chromosome holds the current behavior rules. Interaction among agents can take place directly using message-passing mechanisms, though the cellular automata's locations enable indirect interaction through resources. Any resource or agent within a site is bound to its local behavior rules, enabling resource local handling and environment-agent communication.

4. The Artificial Immune System

4.1 Human Immunology

The human body is repeatedly attacked by several infectious intruders present in the surrounding environment. Although these intruders can cause disease and eventually lead to death, just a few engender injury to the human body, since the immune system is able to deal with numerous microorganisms. The human immune system consists of a series of layers, from the most external one, the skin, to the internal immune system. The latter comprises a large group of cells whose primary function is to eliminate infectious agents and to minimize the damage they might cause.

An immune system response involves, in the first place, recognizing a pathogen and secondly triggering a reaction in order to eliminate it. These reactions or immune responses fall into two categories: innate (or non-adaptive) and adaptive responses. The innate responses can successfully destroy many pathogens on first encounter using substances present in the blood stream such as complement. However, this is not enough to provide protection against all infections, as the rapid evolving of the pathogenic agents provides mechanisms able to evade the innate immune defense. To counteract, most vertebrates including man, are able to respond to any foreign substance even if it has never been faced before. The adaptive immune system involves two main features: the specificity, leading to a highly precise response for a particular pathogen; and the memory, causing a response improvement when dealing with the same infectious agent once again.

After a pathogenic agent successfully breaks through the external defense layers, peptides called antigens make possible its recognition and later eradication. Each immune system cell is able to recognize a single specific antigen, and only the system as a whole can detect thousands of it. Yet, upon antigen recognition, a rapid proliferation of these cells is induced to render an adequate immune response, the so called clonal selection.

Immune responses are produced by leukocytes, of which there are several types. One group is the phagocyte cells, such as the macrophages. These cells bind to the foreign substances, internalize and then destroy them. Since they use non-specific antigen recognition they all produce innate responses and act as the first line of defense against infectious agents. Another group of leukocyte cells are the lymphocytes. Broadly, they fall into two categories: the T cells and the B cells.

The B cells, after recognizing a particular antigen, rapidly divide and differentiate into plasma cells, which can produce large amount of specific antibody. This substance binds to the target antigen that initially made the B cells active. Other B cells remain in a so called memory state for they retain immunological memory for a particular antigen. Memory cells provide the means for a lasting immunity to a pathogen, and the basis for vaccine development. Hence, the aim of vaccination is to modify a specific pathogen in such a way that they become innocuous without losing their antigen presentation. This is possible because antibodies and T cells recognize antigen and not the pathogenic organism as a whole. In other words, vaccination is a method that increases the immunological response against a pathogenic substance, by taking full advantage of the specificity and memory of the adaptive immune system.

The second lymphocyte group, the T cells, are involved in a larger set of activities. The T cells bearing the CD4 marker, also named T helpers, further influence the response of T and B cells and induce the immune system's cytotoxic function. The CD8 T lymphocytes, also called T killer lymphocytes, eliminate cells that display antigen on their surface. This is accomplished by releasing cytotoxic substances that will rupture the cell's cytoplasm leading to its destruction.

When a pathogenic substance is internalized in a cell, it releases non-self peptides which are presented through specific proteins called the major histocompatibility complex (MHC). These molecules are divided into two classes, named MHC-I and MHC-II, being the former found in almost all types of the body cells. However, the latter appears only in cells related with the immune response. It is the presence of foreign peptides in the MHC that tells the immune system whether a cell is infected, since in a healthy cell all the peptides come from self proteins. The foreign peptides in this way presented can be then recognized by T cells, namely CD8 cells.

The two classes of MHC molecules present peptides that arise from different places within the cell. Class I molecules bind and present proteins resulting from the continuous peptide processing and renewal that takes place inside a cell. These proteins are then carried to the cell's surface, where they can be later recognized by CD8 cells. If the peptide presented is non-self, say from a virus infection, the CD8 cell releases cytotoxic elements that will destroy the host cell. This response is the only effective way to prevent the creation of more viruses by the infected cells and hence avoid the infection's outspread.

On the other hand, MHC class II have the ability to seize any peptides they find inside the cell and then deliver it to the cell's surface, unlike class I molecules, which are limited to the non-nuclear compartment and can only bind to a specific peptide after a molecular reorganization. The peptide presented can only be recognized by T cells which have the CD4 marker. Therefore, the identification of an infected cell through the MHC-II complex does not lead to the cell annihilation, rather, CD4 cells activate those which have displayed the foreign peptide. For example, a CD4 cell can stimulate a macrophage to destroy the pathogenic elements contained in its structure. Also, the helper T cells after identifying a non-self peptide on the cell's surface, produce cytokines, namely interleukins, involved in cell differentiation and division.

A large diversity of molecules is present at the onset and throughout the development of immune responses, including complement, a key substance in the non-adaptive response, and other soluble mediators of immunity such as the cytokines and the antibody.

The cytokines hold a wide variety of molecules which are bound in cell signaling and triggering during immune responses. Although there is high diversity in this group of molecules, they fall into a number of categories.

Interferons (IFN) are proteins that are characterized as strong immune regulators and growth factors. They fall into three classes: IFN_α, the largest variant, produced by leukocytes; IFN_β made by fibroblast in response to viruses or nucleic acids and IFN_γ. IFN is produced by activated T cells as the outcome of immune activation. Therefore, this substance leads to an increase of the antigen presenting cell (APC) function, and will further activate other T cells and macrophages as well. IFN_γ is responsible for regulating the APC function of many cell types, and an excessive production of this substance is a factor of the auto-immunity induction [35].

Interleukins are in majority molecules made by leukocytes which act on leukocytes, mainly T and B cells, represented by the abbreviations IL-1 through IL-11.

IL-1 or catabolin, is made by many cells, including B cells, but in its majority is produced by macrophages. It stimulates T and B cells providing a means to perform immuno-regulation, and induces inflammatory responses and fever. Virtually every cell in the human body can respond to this molecule.

Another interleukin, IL-2, previously known as T cell growth factor, is produced by T cells, mostly by CD4 cells. Its range of response is limited to T cells where it acts as a powerful growth factor and activator (e.g. enabling the T cells to release IFN_γ). It also acts on B cells, inducing growth and differentiation, and further activates macrophages.

IL-6, also known as B cell differentiation factor, is produced by T and B cells, macrophages and other cells. It induces B cell differentiation, which will cause anti-body-forming cells (AFCs) to be produced.

Colony-stimulating factors (CSF) are involved in the division and differentiation of stem cells. The balance of these cells determines the proportions of the different cell types to be produced. Tumor necrosis factors (TNF) and transforming growth factors (TGF) are particularly important in mediating inflammation and cytotoxic reactions.

The interaction between T cells and the antigen-presenting cells (APCs) is the most important issue in immunological response. Only if CD4 cells are in sufficient number and are successfully triggered, the activation and subsequent response of B cells will follow. Otherwise no immunological response will take place.

Both T and B cells are activated upon a successful bind to an antigen. B cells can bind to free antigen but generally need T cell's help to become activated, whereas T cells can only bind to antigen presented in MHC molecules. After the interaction with the cell's specific antigen, a number of internal biochemical reactions will modify the cell's DNA. At the same time, the cell develops the ability to respond to specific cytokines, such as IL-2, by producing receptors on its surface. The response to those cytokines will cause cell proliferation and maturation.

Clonal selection is an aftermath of the offspring of specific cells, able to recognize a specific antigen. In this procedure, called primary response, some cells develop into effectors or activated cells while some others become memory cells. The second type of cells, the memory cells, will home to certain areas of the lymphoid tissues where they remain ready to respond to the same antigen if it comes across again – the secondary response. Therefore, the secondary response against a specific type of pathogen will prove to be more effective as it develops an immune response more rapidly.

4.2 The HIV Virus

Viruses are the smallest known organisms, and yet they constitute one of the greatest threats to human health. However, they hold a minimal design: a protein capsule, holding RNA or DNA, where the genetic code for the virus survival and offspring remains. Thus, the success of a virus depends on infecting and then modifying a cell's genetic code in order to produce more viruses.

Certain virus, such as the HIV-I, can have a peculiar property: it can take both active and latent forms. During the active phase, the virus interferes with the cell's normal metabolism, causing the symptoms associated with the disease. When in the latent phase, the virus remains in quiescent state in the infected host cell, although the host is a symptom-free carrier of the disease. The latter state can endure for several years.

The Human Immunodeficiency Virus (HIV-I), also known as Human T-Lymphotropic Virus-III (HTLV-III), usually enters the serum or blood stream within a foreign macrophage or helper T cell. These foreign cells are correctly recognized as non-self particles and are encapsulated by the body macrophages. The virus remains unaffected inside the carrier cell until it pierces the host cell's membrane and waits for the antigen to be presented on the macrophage surface. When the host's helper cell binds to the antigen, the virus will infect it. This cell-type specific attack is unique among retrovirus.

The HIV-I virus may also stay inside the original macrophage, where it may reproduce and bud into vacuoles which are kept within the macrophage's cell membrane. In this manner, the virus will not be detected or recognized by the body defense mechanisms as it is still encapsulated within a body cell. This allows amplification of the virus without the body becoming aware of its presence.

686

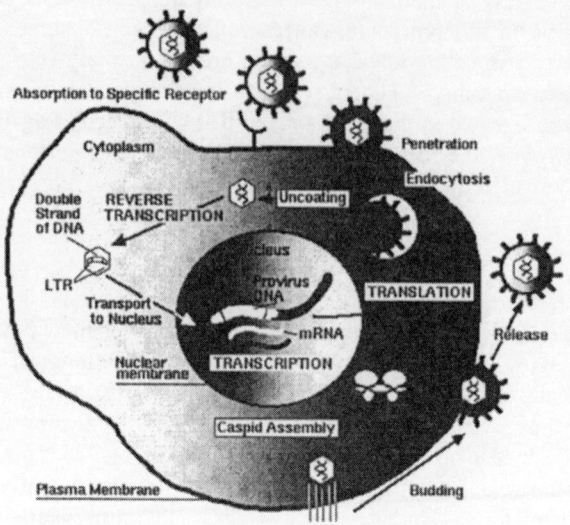

Fig. 4. HIV-I retrovirus replication cycle

Usually, IL-1 is released by a macrophage that has seized a non-self particle. An infected macrophage demonstrates a reduction in chemotaxis and releases a substance that disables the activation message from getting to T cells, preventing the immune response cascade. The infected macrophage or helper cell will eventually home to a lymph node. Here the virus will proliferate, transfer and infect other macrophages and helper cells, and eventually collapse the lymph node, decreasing the total circulating lymphocytes.

The normal infection cycle involves the cycling through the macrophage, positioning itself on the macrophage surface and infecting the helper cell. The last is entered through endocytosis and injected with the HIV provirus. The viral RNA is then converted by reverse transcription into DNA which then inserts itself into the original DNA of the host cell. Here it may remain latent for several years, or become active.

The HIV attack is specific to the CD4 and related inducer cells. These cells normally compound 60-80% of the circulating T cells. After a successful infection, this number can be reduced to such a degree that it is impossible to detect their presence. As the active virus multiplies, it pierces the membrane of the CD4 cells, killing them. Infected cells also secrete a toxin which is fatal to other non-infected CD4 cells, but which does not kill those already infected.

The HIV virus not only reduces the cell population, but also modifies their function. Each CD4 cell usually produces about 1.000 clones when properly stimulated. An infected cell can produce as about 10. This loss of reproductive ability eventually collapses the helper T cell population except for the few HIV infected cells, following the disruption of the immune system.

For several reasons the HIV virus is especially complex for the body to attack. First, the virus undergoes rapid enveloping protein mutations. This means that the antibodies produced to restrain the recognized antigen will try to bind to a virus that no longer exists. For the same reason, multiple strains of the virus will coexist within the body, each carrying a different protein coat, turning the virus immune to the original antibody. Secondly, the HIV virus will incorporate into its protein envelope, part of the host cell membrane as it ruptures and kills its host. This membrane makes it nearly impossible for the body defenses to recognize it as non-self. On the other hand, the virus blocks the binding of molecules to MHC-II in helper T cells. The MHC-II will not bind, immuno-reactions will decline and the level of IL-2 will fall, decreasing the global immune response level. This means that the helper T cells are not going to be activated against an antigen, which in turn stops the T cell immune system coordinator response. No helper, killer or B cell multiplication and no antibody production. The damage done to the immune system as a whole is permanent and irreversible. The effect of all this devastation removes all immune system protection from the body. Hence, any pathogenic invasions normally present within the body are no longer held in check. The body would be wide open to any external attack.

It should be pointed out that in all cases the immune system will mount an immune system response. It usually even goes so far as to produce effective antibodies against the virus. Unfortunately, when the response comes it is already too late.

4.3 AIS Simulation Model

Artificial immune system simulations aim two broad areas: hypothesis generation and experiment prototyping. Since every complex system comprises a large parameter space and a variable set of emergent behaviors, a computer simulation provides guidance in order to identify the system's dynamics from the basic immunologic data. Modeling hypothesis in disease processes and therapeutic intervention is a natural outcome. On the other hand, laboratory experiments are not able to uphold the parameter settings necessary to fully resolve the problem considered. Therefore, in parallel with the *in vivo* investigation, *in silico* experiments can be used to bound the parameters that will most likely yield interesting laboratory resultsband to classify the global behaviors found in the whole parameter domain.

To simulate the immune system, specific agents and an adequate underlying environment were first identified and later modeled. Several behavior rules were then tested in order to obtain a set of results which related to those from a real system.

The artificial immune system agents here considered may be separated into two main classes. The first one contains the immune system cells, namely four types of leukocytes: B lymphocytes, CD4 T lymphocytes, CD8 T lymphocytes and macrophages. The other class is a set of pathogenic agents, specifically the HIV-I and two other theoretical viruses, RB and V*, whose function is to present the system with a series of situations, allowing the study of different immune system responses. The RB virus simply releases specific antigen, leaving the immune system cells unscathed, while V* infects TH cells and remains latent for a random period. When activated it kills its host cell.

A subset of immune system soluble mediators is included in this model, comprehending Interleukin-1, Interleukin-2, Interleukin-6, γ-Interferon and cytotoxic agents produced by T killer cells. Antibody and antigen is produced or secreted accordingly to the pathogen's type.

688

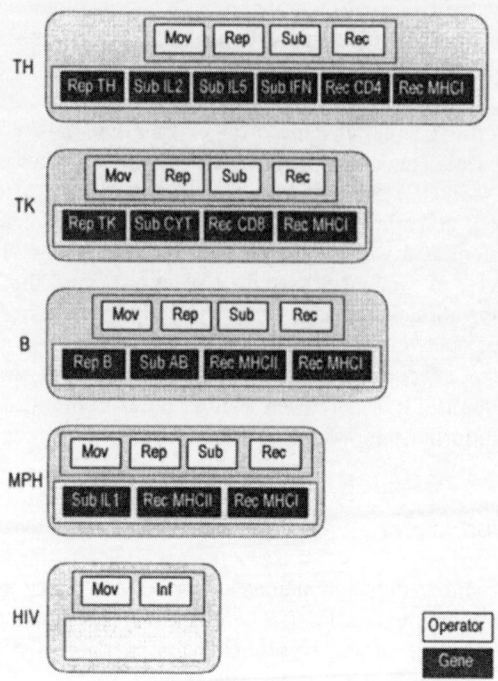

Fig. 5. Initial genotypes and operator sets of the AIS agents

Agent operators are guided by their host's genotype. To model the artificial immune system, several types of genes were considered, each one including a specific part of genetic code:

1. Reproduction or division genes. Contain the condition code, triggering probability and data about the new cell.
2. Substance secretion genes. Keep the condition code along with the release probability for a specific soluble mediator.
3. Receptor presentation genes. Contain the condition code and presentation factor of a specific receptor.

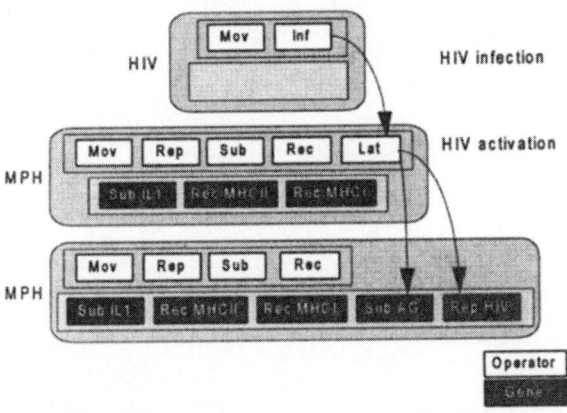

Fig. 6. HIV infection mechanism. Virus infection is accomplished through the injection in the host cell of a pathogenic operator, which remains in latent state. Following activation, the operator modifies the genetic data of the cell using the virus' essential behavior rules. The cell's phenotype, which is the interpretation of the genetic data using the current operator set, will then include the virus' behavior. Throughout this process, the cell remains unaware of the pathogenic agent.

The agent's operators verify the state of the condition's codes in the genes they use to perform their specific action. Healthy cells of the immune system start with a set of four operators:

1. Reproduction operator. This operator creates new agents using the information within the Reproduction genes.
2. Substance secretion operator. This operator releases the substances expressed in the Substance genes.
3. Receptor operator. Presents proteins expressed in the Receptor genes to the environment
4. Movement operator. Models cell movement, but as every agent rely on fluid movements to perform this action, this operator needs no director genetic code.

Each pathogenic agent holds a specific Infection operator, which may append an extra operator to the cell when a virus agent infects a cell. In the specific case of the HIV-I, this operator, the Latent operator, remains idle until virus activation conditions apply. This model considers that HIV-I is activated upon an immune response of the host cell. Once activated, this operator modifies the host cell's genotype. This is attained by adding two extra rules: HIV-I antigen secretion and HIV-I virion reproduction. The cell behavior will then comprise the virus' rules, every time the genotype is interpreted. Additionally, the MHC-I receptor condition is set, once the foreign genetic material is detected. Initial operators and genotypes for every agent are shown in Fig. 5. In Figure 8 is the resultant behavior for each modeled agent. Finally, the model for HIV-I infection mechanisms is depicted in Fig. 6.

The artificial environment is modeled using a toroidal manifold lattice, since the environments here considered are closed. The soluble mediators are spread using a concentration ratio basis, flowing to where they are less clustered. Substance dissolution results from decreasing each one a fixed value each simulation cycle.

The computational application developed to support the artificial immune system simulator comprehends a set of features:

- User-defined graphics can be constructed from all the substances and agents comprised in the simulation. Scale adjustment is possible.

- Graphical data file output complies with CSV (Comma Separated Values) format, which can be used in common spreadsheets.

- Simulation parameters can be adjusted, enabling dynamic settings. This includes environment settings and agent parameters.

- System configuration and parameter settings may be stored and later retrieved.

- A view of the environment is available where the agents comprised in the system can be observed. Moreover, a graphical notation is used in order to follow the agents' status, enabling the trace of agent's distribution patterns.

- Substances and agents in any state can be added to the system through a syringe tool.

- Events related with user interference are traced into a log, which can be stored, and later be used to identify specific system responses in previously obtained graphics.

- Simulation temporal control, step and breakpoint conditions can be controlled by the user.

5. Results

Facing the simulator with a set of typical situations, a strong resemblance was found between the obtained results and those from a real immune systems, validated through several sources [3, 4, 7, 23, 35].

The system was firstly submitted to a series of tests in order to verify the role of the modeled soluble mediators in the regulation of immune activities. The first test aimed Interleukin-1 and Interleukin-5, and the obtained results are in Fig. 7. In point A, T helper cells were introduced in an otherwise empty system. The system was challenged with antigen upon the addition of RB virus, in instant B. No immune response was observed, for T helper cells are not able to recognize antigen in free form, only MHC-II/antigen pairs, thus requiring the help of antigen presenting cells. APC cells, namely B cells, were introduced in instant C. T helper response was still unobserved, since IL-1 is required to attain a successful activation, and is produced by macrophages, which were not considered in this system. On the other hand, B cells only react in presence of IL-5, which in turn is secreted by activated T CD4 cells. To overcome this deadlock, IL-1 was injected in instant D, which led to the chain of the immune response. As an aftermath, the pathogenic agents were successfully eliminated.

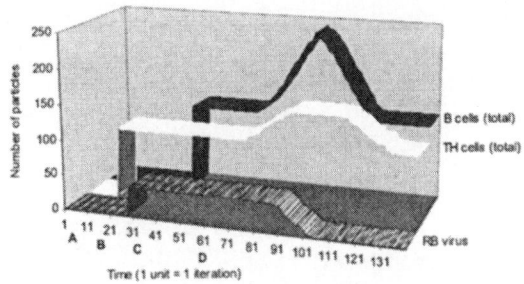

Fig. 7. IL-1and IL-5 as regulators of the immune system. IL-1 was presented in instant D, resulting in IL-5 production. A typical immune response followed. The importance of cell cooperation and immune mediators was asserted.

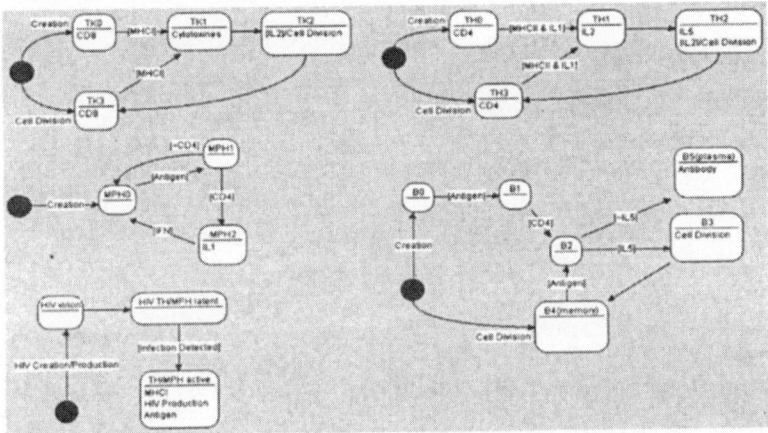

Fig. 8. AIS rule induced state transition diagrams, expressing device behavior. In $a, b, c,$ and d is shown behavior for T Killer, T helper cells, Macrophage and B cells. In e is shown HIV activity in the immune system.

The second experiment evaluated the role of IL-2, and was parted in two. In the first case-study (Fig. 9), the system comprehended no cells at the onset. In A, a number of HIV infected macrophages were added. The addition of T killer cells followed in B, which led to a progressive destruction of the infected macrophage population. The second part of the test is depicted in Fig. 10, where the former process was followed, apart of Interleukin-2 being injected in instant C. IL-2 acted as a growth factor, increasing the number of T CD8 cells. The outcome: a more rapid destruction of infected macrophages.

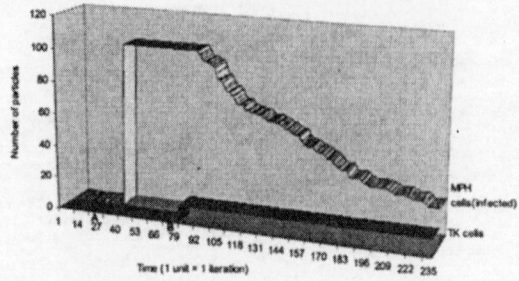

Fig. 9. T killer cell response without Interleukin-2. The immune system cells were able to recognize and destroy the HIV infected macrophages.

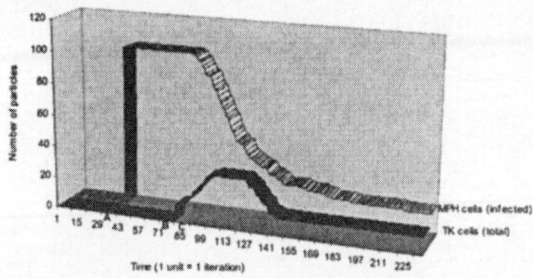

Fig. 10. The importance of Interleukin-2 as a T cell growth factor is here emphasized. In *C* IL-2 was added to the system which promoted T cell division. The lysis of infected macrophages developed more rapidly, when compared against Figure 9.

Other simulation presented the immune system the same antigen twice in order to compare both immune responses. A stronger immune response was observer during the second antigen challenge (Fig. 11). At the outset, immune system cells were the only artificial immune system agents. Following the release of RB virions, an immune response was developed, increasing the number of T helper cells and B cells with specific RB-antigen receptors. T CD8 cell number remained constant, since this virus does not infect cells, therefore no MHC-II presentation was attained. When the system was again challenged with the same number of pathogenic RB virions, an immune response developed more rapidly and stronger than the first. This proved that the model possesses one of the key features of a real immune system: specific antigen memory.

HIV-I and AIDS dynamics also proved to match real quantitative results (Fig. 12, Fig. 13 and Fig. 14). At the beginning of the simulation, every immune system agent was healthy and no pathogenic agents were within the environment. HIV-I infected macrophages were introduced in the system, and, since the virus was still in latent state, no immune response was observed until the system was injected with specific RB antigen.

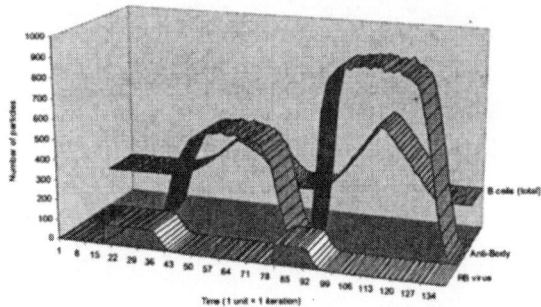

Fig. 11. Secondary immune response. When the immune system is challenged a second time with the same pathogenic agent (around cycle 80), a typical secondary response follows: a stronger cell and antibody development. As can be observed, the second infection was easier to overcome.

Macrophage IL-1 production lead to the activation and increase of T helper cells. On the other hand, B cells were triggered by IL-5 which was secreted by the activated T helpers, resulting in specific antibody production. This secondary infection activated the latent HIV-I. Moreover, T helper activation led to IL-2 production, along with MHC-I antigen presentation by HIV infected T helpers and macrophages. Meanwhile, IFNγ kept stimulating macrophages to secrete IL-1, closing the immune response cycle.

IL-2 produced by T helpers caused T killer cells to divide and kill the infected cells. HIV production was mainly macrophage-tropic, accentuating even more T CD4 cell depletion. In a short time interval, the immune system was unable to respond to further infections.

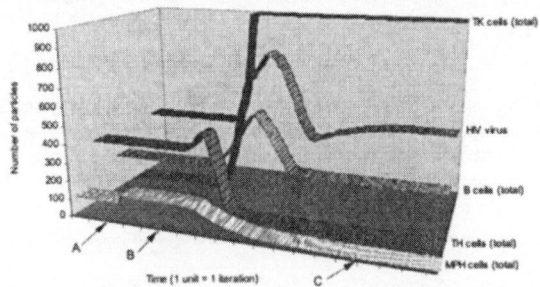

Fig. 12. HIV-I infection, T CD4 and macrophage depletion. Firstly, the healthy system received a number of HIV infected macrophages (A). Later, the introduction of RB antigen caused an immune reaction, increasing the number of B and T helper cells. However, the outcome of the immune response was the activation of the HIV virus, which led to a systematic production of HIV virions on the part of the infected cells, spreading the infection. As more and more T helpers and macrophages became infected, T killer cells took their toll from them, leading to helper and macrophage reduction, resulting in AIDS. In C the system was again presented with RB virus antigen. This time it yielded no response.

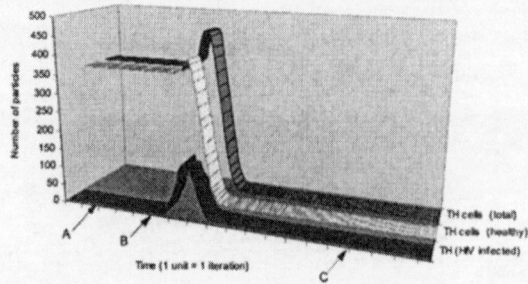

Fig. 13. T helper depletion is the main cause of AIDS during an HIV infection. The initial augment of T helpers can be clearly seen in moment B, as can be also seen the effect of HIV activation and production by the increase of infected helpers. A rapid decrease follows, eliminating the capacity of the immune system to respond to further infections.

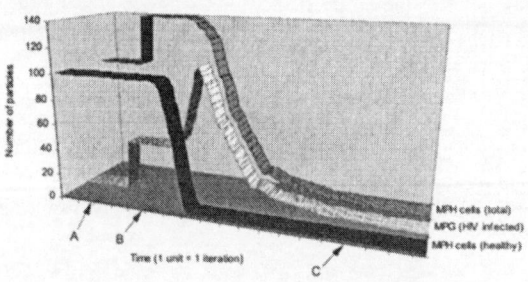

Fig. 14. At the onset, no macrophages were infected with HIV-I. The introduction of HIV infected macrophages saw no hostile reaction, as the HIV virus subsisted in latent state inside the infected cells, and thus undetected. The introduction of RB antigen gave rise to the infection to become productive, and to spread among the healthy macrophages. What followed was the action of T killers in order to destroy the infection. As can be seen, macrophage depletion is somewhat lighter than helper depletion, causing HIV production to be kept even after T helper destruction.

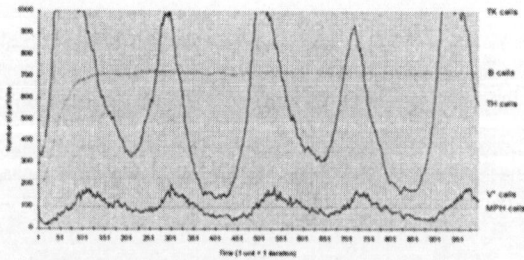

Fig. 15. V* infection. The response pattern resembles that of a predator-prey community

The AIS was also examined under V* virus infection. In this model, the V* virus infects TH cells and stays within, in latent state, for an undetermined interval. Upon activation, the V* operators change the host's genotype in order to promote its reproduction. The host cell is killed after creating a new virus agent.

The obtained response is shown in Fig. 15: the resultant T helper immune response and V* production are almost periodic and their pattern resembles the one found in predator-prey communities.

After V* infection and later activation, its production is enabled and the infection spreads. This comes as consequence of TH cell response, since an increase in helper cell population will expand the V* number, resulting on a major infection in the TH compartment. On the other hand, V* production also leads to a decrease in TH population, which in turn causes a progressive diminishing in the V* population. However, the V* virus will continue to activate the immune system cells, including the T helper cells, leading to another cycle.

The system is also used in a more serious simulation of the immune system [37], where the behavior and type of immune responses currently believed to be the result of cross regulation of $CD4^+$ T lymphocyte populations are investigated. Many debates have arisen concerning the way the immune system is able to provide an immune response and tolerate self simultaneously. Classical theories try to explain these phenomena through the specificity of T cell receptors. Nevertheless, observations showed us that the specificity of the immune system cells can be quite degenerated, providing a different scope on the understanding of the immune system's balance.

We proposed and implemented in this simulation system a computational model for the dynamics of Th_1 and Th_2 $CD4^+$ T lymphocyte sub-populations. The aim is the study of diversity and multiple responses. Using this model we are able to identify some experimental observations which are poorly understood. Some of the results showed us that the immune system's balance can be related to a measure of locality, helping to explain the paradigm of concomitant responses and tolerance. The model enabled both TCR and MHC/peptide diversity considering different matching coefficients with the same antigen, which makes it suitable for simulations concerning *in vitro* and *in vivo* test protocols. Moreover, the ontogeny and the dynamics of Th lymphocytes appear to be one of the major regulatory mechanisms that upholds the tolerance and rejection of the immune system. We note that the establishment and maintenance of local equilibrium states, a process in which cytokines appear to have a decisive role, may help to understand one of the many important phenomena in the immune system. The proposed model gives interesting clues on the learning process the immune system is submitted during its ontogeny and on how a population that presents a very degenerated specificity can recognize both self and allogeneic antigen. These results somewhat contradict the more orthodox theories which put exclusive specificity as the main factor for immune regulation and immunity.

6. Conclusions and Future Directions

Summarizing, the system presents an hybrid architecture that allows the modeling of systems involving multiple heterogeneous agents, physically distributed on a lattice made up of dissimilar elements. Agents are autonomous devices, containing a chromosome whose length may dynamically change, and a set of operators that express behavior based on a set of rules encapsulated throughout the gene structure. A collection of operators interpret the genetic rules and may also use external information, whether seized from external information presented by the environment or by other agents.

Interaction is not limited to the usual agent-agent and agent-environment, since each environment site is an active entity. Thereupon, a site can also interact with the agents comprised within it. In addition, environment-environment interaction is made possible since the site lattice is an active network, where each location has a local processing engine. Thus, the system architecture consists of three main layers, with the physical lattice forming the lowest level. At the second level, the network of active sites keeps local data and uses the processing engine either to interact with the data or to modify it. Finally, at the highest level, is the multiple agent population.

Artificial immune system simulators aim the domain of hypothesis generation and experiment prototyping. This class of systems can help to design rational therapeutic intervention as well as understanding the process of disease. Moreover, the system's large parameter set can be constructed upon what-if hypothesis, otherwise difficult to attain in laboratory. The resulting data, obtained from *in silico* simulations, can support clinical trials and diagnosis and further bound *in vivo* laboratory tests to a set of experiments which will probably lead to attractive outcomes.

The system's complex parameters may be optimized through a Evolutionary Genetic Algorithm, whose fitness function equates both the current output and the data from real immune systems. The resulting data the parameter chromosome holds, can be used in a twofold manner. On the one hand, it comprehends the settings for a suitable model, according to the given data. On the other, the resulting parameters can be used to formulate theoretical assumptions on how the settings induce the system's behavior.

Besides the HIV-I simulation, other viruses, can be simulated with respect to the class of macroscopic responses here modeled. To attain this task, one first requires to identify and outline the pathogenic agent's behavior, and then construct a set of rules which model the virus.

The results showed how the complex and emergent behavior can be the result of the interactions among immune system agents. It was asserted that the model possesses antigen memory, one of the main features of a real immune system. The system was also submitted to HIV-I infection which developed into AIDS, with similar patterns to those observed in real conditions. Likewise, it was tested the effect of the soluble immune mediators in several events.

Simulation of other artificial systems, with heterogeneous physical environments and populations, are possible through the object-oriented support layer. This is accomplished by the redefinition of the specific genes and operators. Moreover, since twofold agent-environment interaction is possible, it can be used to simulate physical phenomena effects. Such flexibility is already demonstrated in the simulation with success of the immune response of the CD4+ T Lymphocyte Sub-Populations as described in [37].

The underlying physical medium can also be modeled as a complex adaptive system, enabling the simulation of detailed environment features, such as lymph nodes.

A possible extension to the current system, would aim one current drawback: the runtime redefinition of new models for both agents and environment. To offset this, a special-purpose language script containing agent rules and site local processing, among other settings, could be used, using a similar approach to that proposed in [5].

7. Acknowledgments

We would like to give thanks to Dr. Hans Sieburg from University of California at San Diego also gave us some support and current development on artificial immune systems. Dr .Raquel Gouveia and Dr. Diogo Santos, from Faculdade de Medicina de Lisboa, gave us essential help the immunology research.

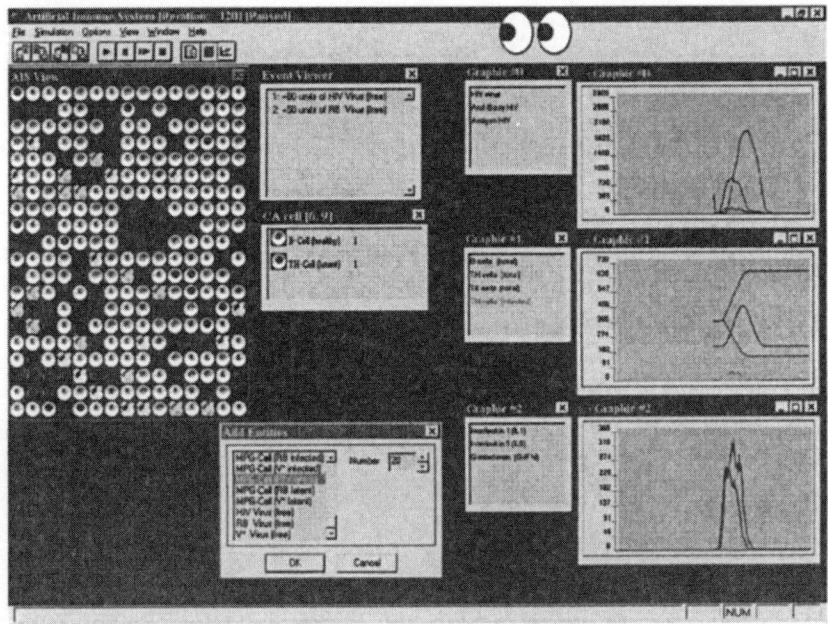

Fig. 16. Screen dump from an AIS simulation. Here we can see the environment lattice and the agents within it, where some of their partial state is shown. Some quantitative graphics are also available along with other information.

References

1. J. Kephart. *"How Topology Affect Population Dynamics"*. In Artificial Life III, XX USA, 1993.
2. Y. Toquenaga, M. Ichinose, T. Hoshino, K. Fujii. "Contest and Scramble Competitions in an Artificial World: Genetic Analysis with Genetic Algorithms". In Artificial Life III, XX, USA., 1993.
3. I. Roitt, J. Brostoff, D. Male. "Immunology". 3rd edition, Mosby-Year Book Europe Limited, 1993.
4. H. Sieburg, J. McCutchan, O. Clay, L. Caballero and J. Ostlund. "Simulation of HIV Infection in Artificial Immune Systems". In Physica D 45, 1990.
5. H. Sieburg, O. Clay. "The Cellular Device machine Development System for Modeling Biology on the Computer". In Complex Systems 5, 1991.
6. H. Sieburg and C. Baray. "Testing HIV Molecular Biology in in Silico Physiologies". In Proceedings of the First International Conference on Intelligent Systems in Molecular Biology, Washington D.C, July 1993.
7. D. Mosier and H. Sieburg. "Macrophage-tropic HIV - critical for AIDS pathogenesis?". In Immunology Today 15, pp. 332-339, 1994.
8. G. Nossal. "Life, Death and the Immune System". In Scientific American, vol. 263, number 3, pp. 20-31, September 1993.
9. I. Weissman and M. Cooper. "How the Immune System Develops". In Scientific American, vol. 263, number 3, pp. 32-39, September 1993.
10. C. Janway Jr.. "How the Immune Recognizes Invaders". In Scientific American, vol. 263, number 3, pp. 40-47, September 1993.
11. P. Marrack and J. Kappler. "How the Immune System Recognizes Invaders". In Scientific American, vol. 263, number 3, pp. 48-55, September 1993.
12. W. Paul. "Infectious Diseases and the Immune System". In Scientific American, vol. 263, number 3, pp. 56-65, September 1993.

13. W. Greene. "AIDS and the Immune System". In Scientific American, vol. 263, number 3, pp. 66-73, September 1993.

14. L. Steinman. "Autoimmune Disease". In Scientific American, vol. 263, number 3, pp. 74-83, September 1993.

15. L. Lichtenstein. "Alergy and the Immune System". In Scientific American, vol. 263, number 3, pp. 84-93, September 1993.

16. H. Wigzell. "The Immune System as a Therapeutic Agent". In Scientific American, vol. 263, number 3, pp. 94-101, September 1993.

17. A. Mitchison. "Will We Survive ?". In Scientific American, vol. 263, number 3, pp. 102-115, September 1993.

18. G. Cowley and S. Begley "More Than Magic: Living Longer with the HIV.". In Newsweek, pp. 43-44, February 12, 1996.

19. "Antiviral Agents Bulletin, September 1995". In http://www.covesoft.com/biotech/.

20. "The 7.001 Hypertext Book". In http://esg-www.mit.edu:8001/esgbio/7001main.html

21. Several Cellular Automata Articles. In http://alife.santafe.edu:80/alife/topics/cas/ca-faq/ca-faq.html

22. Several Genetic Algorithms Articles. In http://alife. santafe.edu:80/alife/topics/ga/.

23. Several Biological Informatics and Theoretical Medicine Articles at BITMed. In http://bitmed.ucsd.edu

24. A. Johnson. "Evolution of a Size-Structured, Predator-Prey Community". In Artificial Life III, XX, USA., 1993.

25. K. Lindgren, M. Nordahl. "Artificial Food Webs". In Artificial Life III, XX, USA., 1993.

26. D. Green. "Simulated effects of fire, dispersal and spatial patterns on competition within forest mosaics.". In Vegetation, vol.82, pp. 139-154, 1982.

27. J. Silvertown, S. Holtier, J. Johnson, P. Dale. "Cellular Automaton models of interspecific competition for space - the effect of patterns on processes ". In Journal of Ecology, vol. 80, pp. 527-534, 1992.

28. C. Langton. "Studying Artificial Life With Cellular Automata". In Physica D 22, pp.120-149, 1986.

29. Jeanne and Joseph Sullivan. In http://www.santafe.edu/projects/Immuno-logy. Theoretical Immunology program.

30. J. Holland. "Adaptation in Natural and Artificial Systems". MIT Press, Cambridge, MA, 1992.

31. J. Holland. "Echoing emergence: Objectives, rough definitions, and speculations for Echo-class models". Technical Report 93-04-023, Santa Fe Institute, 1993.

32. T. Jones, S. Forrest. "An introduction to sfi echo". Technical Report 93-12-074, Santa Fe Institute, Santa Fe, N.M., 1993.

33. M. Mitchell, S. Forrest. "Genetic Algorithms and Artificial Life". Santa Fe Institute Working Paper 93-11-072, 1994.

34. T. Jones, S. Forrest. "Modeling Complex Adaptive Systems with Echo". In Complex Systems: Mechanism of Adaptation, R.J.

35. Stonier and X.H. Yu (Eds), IOS Press, pp. 3-21, September 1994.

36. Chantry D., Feldmann M. "The role of cytokines in auto-immunity", Biotech 1991 ; I(4); 361-409.

37. Caetano A, Grilo A, Rosa A. "Modeling Thymic Selection and Concomitant Immune Responses on CD4$^+$ T Lymphocyte Sub-Populations", ALife VI, C Adami, R Belew, H Kitano, C Taylor (eds), MIT Press, pp 143-150,1998.

The Effectiveness of Mutations: Simulation of Natural Mechanisms

Ivanoe De Falco[1], Antonio Della Cioppa[2], and Ernesto Tarantino[1]

[1] ISPAIM, National Research Council of Italy
    Via P. Castellino 111, 80131 Naples, Italy
    email: {defalco.i,tarantino.e}@irsip.na.cnr.it
[2] Department of Computer Science and Electrical Engineering, Università di Salerno
    Via Ponte don Melillo 1, 84084 Fisciano, SA, Italy
    email: dean@unina.it

Abstract. Current versions of Genetic Algorithms make use of point mutation only, in spite of the existence in nature of many different forms of mutations. In this paper we present an Evolutionary Algorithm based on the simulation of natural mutation mechanisms acting at gene level and chromosome level. Experimental trials are effected on several test functions. The effectiveness of the approach proposed is demonstrated through a comparison with the results obtained by classical crossover–based Genetic Algorithms.

1 Introduction

During the last three decades much research on Evolutionary Algorithms (EAs) has been devoted to investigation of importance of the involved operators. In Genetic Algorithms (GAs) [1,2] crossover has been considered the leading operator, while mutation has always been confined to the role of a secondary operator. In Evolution Strategies (ESs) [3,4] and in Evolutionary Programming (EP) [5], instead, mutation has been seen as the main operator driving evolution, and particular attention has been given in designing suitable models of it. However, independently of the approach, any of the mutation operators taken into account often seems to represent either an oversimplified model of biological mutation or a completely new "artificial" operator, specifically tailored to deal with problem variables, without any correspondence with nature. In the last years the search for more realistic mutation operators has involved many researchers, like for instance Mitchell [6] and Banzhaf [7], who have both pointed out the importance of considering new evolution operators. As a result, several operators relying on natural mechanisms and some mutation–based models of evolution have been developed.

In [8,9] we introduced two brand–new mutation operators suitable for GAs, i.e. the *frame–shift* and the *translocation*, based on the homonymous biological events. In fact, the former tries to mimic as closely as possible biological base insertions and deletions, while in the latter chromosomal segments move from a location to another in two nonhomologous chromosomes. Starting from our previous research, in this paper we wish to make use of these operators and of classical point mutation and inversion to design an EA based on two sequential mutation steps which can simulate the *gene–level mutation* (replication phase) and the *chromosome–level mutation* (mutagens phase). We aim to compare this algorithm with a number of well–known EAs on a testbed, in order to investigate the effectiveness of a closer simulation of biological mutation in terms of solution quality.

The outline of the paper is as follows. Firstly, we give in Section 2 an overview of natural mutation. Section 3 reports on some papers attempting at introducing "natural" mutation operators in the EAs. In Section 4 we describe the operators and the EA we make use of. Section 5 shows the experimental findings obtained when applying this algorithm to test functions. The results are compared with those achieved by using the classical GA or some of its variants. Finally, our conclusions and foreseen future work are left to Section 6.

2 Mutations in biology

There exist in nature two broad classes of mutations, namely those which occur in gametes (*germ mutations*) and those which take place in body cells (*somatic mutations*). In the following we shall make reference to the latter only [10]. These can be divided into two classes, let us denote them by *gene–level mutations* and *DNA rearrangements* (*chromosome–level mutations*). They both result in DNA information being damaged.

2.1 Gene–level mutations

Gene–level mutations occur during DNA replication, and result in mistakes being inserted in DNA information. They are classified as substitution (point mutation), insertion or deletion mutations. Insertions or deletions can be small (one or a few nucleotides) or large (more than a few dozen nucleotides) and result in correct DNA sequence being shifted (frame–shift mutations).

Substitutions from one pyrimidine (A and G bases) to another or one purine (T and C bases) to another are called transitions. Mutations from a pyrimidine to a purine, or the reverse, are called transversions and are less frequent than transitions.

The sum of all genes in a human cell, the human genome, is estimated to be approximately 3 billion base pairs, and a single DNA chain might contain up to 250 million pairs of bases. In spite of the immense size of human DNA, an error occurs only about once in each 10–100 billion bases. This is because multiple systems for repairing damage to DNA have evolved, probably due to the importance of the DNA information content. Among them we can cite at least base excision repair, in which each of a variety of bases that could result from DNA damage can be removed by an enzyme specific for the damaged base. Cell division does not happen until damage is repaired.

2.2 DNA rearrangements

In addition to the changes that genomes undergo due to mutations arising from unrepaired damage to DNA and replication errors, genomes change by a variety of other mechanisms, loosely classified as rearrangements. These rearrangements imply DNA breakage taking place because of mutagens (i.e. anything that causes a mutation). Some well–known environmental examples are ultraviolet radiation, X-rays, tars from tobacco, asbestos. After breakage DNA segments glue together in a wrong way. DNA rearrangements are often called illegitimate recombination to distinguish them from general or homologous recombination in which exchanges occur between homologous DNA sequences. Two types of rearrangements can be distinguished:

- in site–specific recombination, specific target sequences on each of two DNA segments are sites of strand exchange, resulting in segment inversions, deletions and insertions.
- in translocation, a change in position of a chromosomal segment to another region of the same chromosome or to another chromosome takes place.

As regards the first type of DNA rearrangements, it is catalyzed by site–specific recombinases. These catalyze a reciprocal double–stranded DNA exchange between two DNA segments which can be on the same or on different strands. If on different strands an insertion results. If on the same strand, the outcome depends on alignment of the target sequences. In one alignment inversion occurs. In the other alignment, deletion of a circular DNA from a linear DNA results. Intrachromosomal translocations involve the movement of a chromosomal segment from one location in the chromosome to another. This is normally nonreciprocal, that is another segment does not exchange places with the first segment. Interchromosomal translocations involve the movement of a chromosomal segment(s) between chromosomes. Reciprocal translocations occur when chromosomal segments are exchanged between two nonhomologous chromosomes and is the most typical type of translocation. Nonreciprocal translocations are a one-way transfer of a chromosomal segment to another chromosome. In the following, we shall make reference only to interchromosomal tranlocations.

3   Review of research on biologically inspired mutation operators

The awareness of the importance of the mutation is growing within the EA community and interest is increasing in a closer consideration of its features. Researchers are striving to precisely measure and compare the effectiveness of crossover and mutation, and to find their strengths and their limitations as well [11–14]. Some researchers have defined new biologically–inspired mutation operators and have tested them on real problems. In the following, we shortly report on the aspect.

Furuhashi and others [15] proposed a new coding method based on biological DNA, called "DNA coding method". Their work heavily relied on the actual mechanisms taking place in nature at DNA level, so they used concepts like bases, codons and aminoacids. Thanks to this approach they were able to define crossover and mutation operators which worked very similarly to the biological ones. Furthermore, they defined operators corresponding to natural viruses and enzymes.

Nawa and others [16] developed a Pseudo–Bacterial Genetic Algorithm (PBGA), based on bacteria and on transduction. In this process bacteriophages can carry a copy of a gene from a host cell and insert it into a chromosome of an infected cell. The results showed that the strategy adopted by the PBGA of improving parts of chromosomes using the bacterial operation is beneficial.

Faulkner [17] defined several biologically–inspired operators aiming at effectively solving the Travelling Salesman Problem. Those operators were based on concepts like inversion, swap and mutation as they can be found in nature, so they changed the positions of groups of genetic material within the chromosome.

In [8] a frame–shift mutation operator trying to mimic as closely as possible biological insertions and deletions was introduced. This operator was studied with respect to sharp–peak functions with the aim to compare its behavior about error threshold [18] with that of point mutation.

In [9] a translocation operator based on the homonymous biological event was introduced and tested on a wide set of classical functions to evaluate its effectiveness in the solution of multivariable optimization problems. The achieved results assessed its competitivity and its robustness with respect to classical GAs.

Simoes and Costa [19] defined a transposition operator which was strongly inspired by biology. Their operator can work both as a sexual and as an asexual mechanism. The latter always achieves better results than standard crossover on the testbed they used.

4 Simulation of Natural Mutation

In this paper we wish to take into account both the gene–level mechanisms, i.e. point mutation and frame–shift, and the chromosome–level ones, i.e. translocation and inversion, and to use them in an integrated way, so as to obtain a "global" mutation operator able to simulate with a high degree of accuracy the natural mutation processes. It should be noted that we neglect the insertion and deletion at chromosome level in order to keep the genome length of an individual constant. In the following a string in the population represents for us a whole genome. Each of the n_v variables of the problem represents a chromosome, and each of the n_b bits needed to code that variable is a gene in the chromosome. This means that in our approach the notions of gene and locus coincide. The total string length is then $l = n_v \cdot n_b$.

4.1 The Gene–level Mutation Operator

As previously stated, gene–level mutations are characterized both by single base substitutions (point mutation), and by insertions and deletions involving one or more base pairs (frame–shift). In the following, we shall consider only the insertion or the deletion of one single base. Our basic frame–shift mechanism can operate in two different modes, the *delete–first* mode and the *insert–first* mode. Formally, the *delete–first* mode can be represented as:

$$\sigma = \sigma_1 \ldots \sigma_{p-1} \sigma_p \boxed{\sigma_{p+1} \ldots \sigma_{p+k}} \sigma_{p+k+1} \ldots \sigma_l \Longrightarrow$$

$$\sigma' = \sigma_1 \ldots \sigma_{p-1} \boxed{\sigma_{p+1} \ldots \sigma_{p+k}} \sigma'_{p+k} \sigma_{p+k+1} \ldots \sigma_l$$

while the *insert–first* mode works as it follows:

$$\sigma = \sigma_1 \ldots \sigma_{p-1} \boxed{\sigma_p \ldots \sigma_{p+k-1}} \sigma_{p+k} \ldots \sigma_l \Longrightarrow$$

$$\sigma' = \sigma_1 \ldots \sigma_{p-1} \sigma'_p \boxed{\sigma_p \ldots \sigma_{p+k-1}} \sigma_{p+k+1} \ldots \sigma_l$$

where σ and σ' denote the whole genome before and after the frame–shift respectively.

Let us suppose that a frame–shift mutation is to take place starting from a position p. An equiprobable coin is tossed to decide whether deletion or insertion must be performed first. In the former case the allele in position p is not copied in the offspring, then the allele in the locus $p+1$ of the parent genome is copied in the locus p of the offspring genome, and so on, for a number of times k randomly chosen, and lower than or equal to a maximum block size bf_{max}, set before execution. Now a random value is inserted in offspring position $p + k$, so as to preserve genome length l. If, instead, insertion must be performed first, the locus p is filled with a random value, then the allele in position p of the parent is copied in position $p + 1$ of the offspring and this is repeated for a random number of k loci ($k \leq bf_{max}$) until offspring position $p + k + 1$.

Starting from this basic operating mode, we have implemented the gene–level mutation operator. This operator integrates point mutation and frame–shift so as to make the error copy mechanisms as similar as possible to gene–level mutations. We start to copy parent alleles into the offspring from left; when copying each allele a random real number in the interval $]0,1[$ is chosen. If this number is higher than mutation probability μ, then the offspring locus is filled with the parent allele. Otherwise, a new random real number in the interval $]0,1[$ is chosen. If this new number is lower than bit–flip mutation probability μ_b the basic bit–flip mechanism occurs, if not the frame–shift mechanism takes place (with probability is $\mu_f = 1 - \mu_b$) on a block whose length is random, yet lower than or equal to bf_{max}. Then, the copy continues from the end of the shifted block until the right end of the string is reached.

In our operator the minimum block length may equal zero. This means that there is no block shifted, rather a bit–flip mutation occurs in the position p to be mutated. It should be noted that bit–flip means to toss an equiprobable coin, therefore the allele contained in p can either change or remain unchanged as well.

We must remark here that the deep difference between this operator and the real frame–shift consists in having set an upper limit to the block size for a shifted genotype: such a limit does not exist in nature.

4.2 The Chromosome–level Mutation Operator

The sequential action of both translocation and inversion simulates the genome rearrangements and represents our chromosome–level mutation operator. In the following, we describe the operators which perform this natural mechanism.

The Translocation Operator The translocation mutation operator moves chromosomal segments from a location to another in two chromosomes. It takes place both in reciprocal (exchange of segments) and nonreciprocal mode (a segment copy moves to a new location without exchange). The reciprocal mode can be represented as follows:

$$\sigma = \sigma_1 \ldots \boxed{\sigma_{p+1} \ldots \sigma_{p+l_{tr}}} \ldots \boxed{\sigma_{q+1} \ldots \sigma_{q+l_{tr}}} \ldots \sigma_l \Longrightarrow$$
$$\sigma' = \sigma_1 \ldots \boxed{\sigma_{q+1} \ldots \sigma_{q+l_{tr}}} \ldots \boxed{\sigma_{p+1} \ldots \sigma_{p+l_{tr}}} \ldots \sigma_l$$

while the nonreciprocal mode works as follows:

$$\sigma = \sigma_1 \ldots \boxed{\sigma_{p+1} \ldots \sigma_{p+l_{tr}}} \ldots \boxed{\sigma_{q+1} \ldots \sigma_{q+l_{tr}}} \ldots \sigma_l \Longrightarrow$$
$$\sigma' = \sigma_1 \ldots \boxed{\sigma_{p+1} \ldots \sigma_{p+l_{tr}}} \ldots \boxed{\sigma_{p+1} \ldots \sigma_{p+l_{tr}}} \ldots \sigma_l$$

where, as usual, σ and σ' denote the whole genome before and after the translocation respectively.

The chromosomes and the starting points p and q are randomly chosen, and so is the segment size l_{tr}. This latter must be lower than or equal to a maximum block size bt_{max} set *a priori*. It should be noted that it is not important whether or not the two chosen segments start from equally positioned genes. Instead, special care must be devoted not to trespass the bounds of the two chosen chromosomes, otherwise more than two chromosomes would be modified.

Our implementation of reciprocal translocation, based on swap between equally long substrings, is due to the need to keep all the chromosomes with constant length. Such a limitation does not exist in nature. The operator takes place after the replication of the whole genome with a probability μ_t for each individual in the population. The reciprocal and the nonreciprocal modes are mutually exclusive and are applied with the same probability.

The Inversion Operator Inversion occurs when a piece of a chromosome breaks off and reattaches itself in reverse order. It requires two breaks in the chromosome followed by a 180° rotation of the chromosomal segment and rejoining of the ends. Formally, it can be represented as follows:

$$\sigma = \sigma_1 \ldots \sigma_p \boxed{\sigma_{p+1} \ldots \sigma_{p+l_{in}}} \sigma_{p+l_{in}+1} \ldots \sigma_l \Longrightarrow$$
$$\sigma' = \sigma_1 \ldots \sigma_p \boxed{\sigma_{p+l_{in}} \ldots \sigma_{p+1}} \sigma_{p+l_{in}+1} \ldots \sigma_l$$

where σ and σ' denote the whole genome before and after the inversion respectively.

Also in this case, the segment size l_{in} is randomly chosen and must be lower than or equal to a maximum block size $bi_{\max}$ set *a priori*. Obviously, the maximum segment size is equal to the length of the chromosome. The inversion operator is applied with a probability μ_i and takes place, as the translocation, after the replication of the whole genome. The importance of these mutations in evolution is that they change the linkage relationships between genes.

4.3 The Mutation–based Algorithm

In this paper we make use of an EA which is a binary GA without crossover. Rather, we only apply mutation. Following nature, this happens in two separate and sequential phases. In the first step (corresponding to DNA replication), gene–level mutation is applied with a low probability. In the second (corresponding to DNA rearrangements) chromosome–level mutation takes place. The classical tournament selection is taken into account. The resulting "Mutation–based Genetic Algorithm" (MGA) is reported in the following:

Mutation–based Genetic Algorithm (MGA)
begin
  randomly initialize a population of n elements
  while (a termination criterion is not fulfilled) **do**
    evaluate the population by using fitness function
    for $i = 1$ to n **do**
      randomly choose one individual with tournament
      apply gene–level mutation to it
      apply chromosome–level mutation to it
    end for
  end while
end

---

5 Experimental Results

With the aim to test MGA effectiveness in solving multivariable optimization problems, we have taken into account a set of test functions. Furthermore, we have compared our results with those achieved by other versions of GAs, as long as these are available in literature. The gene–level mutation probability value chosen for all the experiments has been $\mu = 1/l$, with $\mu_b = 0.7\mu$ and $\mu_f = 0.3\mu$. As far as the chromosome–level mutation rates are concerned $\mu_t = \mu_i = 0.2$. As regards the block sizes of frame–shift, translocation and inversion, we have decided to set $bf_{\max} = bt_{\max} = bi_{\max} = n_b$ for all the functions faced. Gray coding has been used throughout the paper.

5.1 Whitley's F8F2 function

Whitley and others [20] stated that a good test suite set should contain problems that are resistant to hill–climbing, nonlinear, nonseparable, nonsymmetric and scalable. The classical F_1-F_8 testbed, instead, is not as difficult as previously supposed, since most of those functions do not have at least one among the above properties. They decided to build new functions by composing the already existing F_1-F_8 functions so as to take advantage of the good features possessed by each of them and to get rid, at the same time, of the absence of some of the above properties. As a result, they obtained a wide set of new, more challenging, test functions. We have taken into account Whitley and others' F_8F_2 scalable function (see [20]) which is a very challenging minimization problem with the global optimum equal to 0. We have decided to be consistent with their experiment environment, so comparison can be made with the results they reported, which are related to Elitist–SGA with Tournament (ESGAT), Genitor and two hillclimbers, CHC and RBC. As a consequence of the above, the total number of function evaluations has been set equal to $500,000$ and 12 bits have been used to represent each of the variables. For this set of experiments only we have used an elitist strategy, since all the available results in literature have been attained with elitist algorithms.

We have performed 30 runs for each of the sizes 10, 20 and 50 of the problem so as to test Elitist–MGA (EMGA) and its scalability. As regards population size, given that the mutation mechanism does not recombine genetic material among individuals, it is not granted that the best population size for a crossover–based GA is the same in a mutation–based GA. In fact, preliminary experiments have shown that the most suitable population size is 1,000, resulting in 500 generations allowed. Since in the above cited paper ESGAT has been run with a tournament size of 2 for a 200–sized population, we have decided to maintain the same 1% selective pressure, resulting in a tournament size of 10. Table 1 reports the related results which are as a function of the average final best value *aver* and of the related variance *var*. Also results for ESGAT, Genitor, CHC and RBC are shown (taken from [20]). Performance of EMGA is always superior to that of all of the

Table 1. The results on the F_8F_2 test function

	10		20		50	
	aver	*var*	*aver*	*var*	*aver*	*var*
ESGAT	4.07	2.74	47.99	32.61	527.10	176.98
Gen.	4.36	2.74	21.45	19.45	398.12	220.28
CHC	1.34	0.92	5.63	2.86	75.09	49.64
RBC	0.13	0.42	7.24	11.28	301.56	72.74
EMGA	1.71	1.12	16.99	13.06	347.49	149.86

evolutionary algorithms considered (i.e. ESGAT and Genitor) at all of the different problem sizes, and is inferior only to that shown by the hill–climbers. Furthermore, EMGA has shown sufficient scalability on this set of problems. An interesting feature is its variance, quite low with respect to those of ESGAT and of Genitor. This implies that EMGA is quite robust when compared to these algorithms.

5.2   The Royal Road

Furthermore, we have investigated the goodness of our scheme at solving the Royal Road function [11]. In it, we assign a target string ts. This latter can be seen as a list of schemata [2]. We denote with s_i the i–th schema for ts (i.e. the schema representing the i–th block making up the string), and with $o(s_i)$ the order of s_i, i.e. the number of defined bits in this schema. Then, given a binary string x, its fitness is defined as follows:

$$R_1(x) = \sum_{i=1}^{ns} \delta_i(x) o(s_i) \tag{1}$$

where $\delta_i(x) = 1$ if $x \in s_i$, 0 otherwise, and ns is the number of schemata forming ts. This function is well known for its difficulty and is a noticeable example of a frustrated problem. In fact, the fitness is incremented by $o(s_i)$ if and only if a block as a whole is completely found. Its global maximum value is equal to $n_v \cdot n_b$.

We have taken into account two different such functions, namely one with 8 blocks of 8 bits each and another with 10 blocks of 10 bits each. The algorithm has been run 30 times on each problem. A population size of 128 individuals and a maximum number of 2,000 generations have been used, resulting in at most 256,000 evaluations allowed. These choices have been made to be consistent with the experiments described in [11]. The tournament size chosen is 10% of the population. Also 30 runs with the same parameter setting and tournament selection have been made for SGA (SGAT) and crossover rate equal to 0.9. In this case both MGA and SGAT have been able to find the global best solution in all of the 30 runs, for the two different–sized problems.

In this last set of experiments MGA shows better behavior than SGAT while converging towards the global best. In fact, MGA initially decreases slightly slower than SGAT, but diminishes more quickly in the final part of evolution, resulting in the global best value achieved in a by far lower number of generations on average (about 99 for MGA and 262 for SGAT, so that the former needs on average about one third of the evaluations required by the latter). This can be seen in Fig. 1 (top), where we report the average best value over the 30 runs at each generation for both algorithms when solving the $8 \cdot 8$ problem. Shown results are distances from the optimum, rather than the raw values of $R_1(x)$.

Similar results are shown in Fig. 1 (bottom) for the $10 \cdot 10$ problem. Especially in this case the MGA convergence to the global best takes place in a much lower number of generations than SGAT can do (about 250 for MGA vs. 1520 for SGAT), resulting in a speed–up factor of about 6. If we take into account the $8 \cdot 8$ function, we can compare our results with those shown in [11]. They reported that all out of

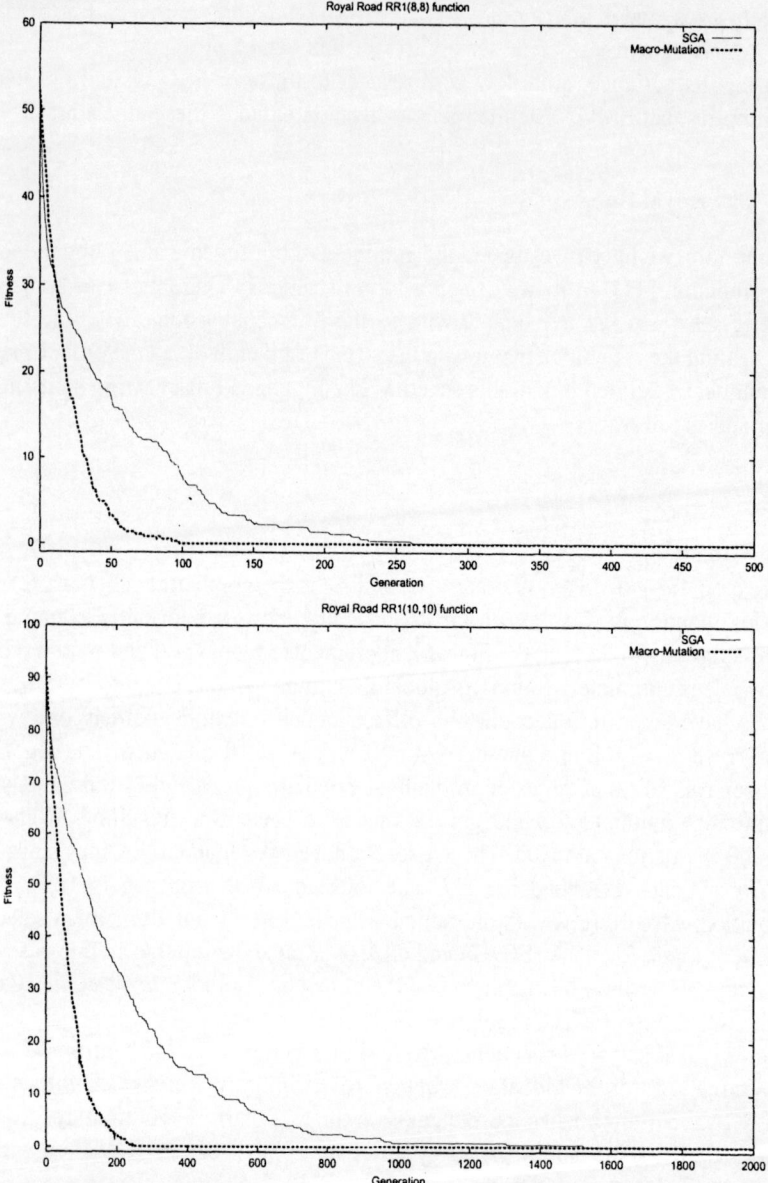

Fig. 1. The averaged best value as a function of the generation number for $8 \cdot 8$ Royal Road (top) and for $10 \cdot 10$ Royal Road (bottom)

200 runs of an SGA with Roulette Wheel (SGARW) converged to the global best, though in a quite high number of function evaluations. Furthermore, they noticed that most of hill–climbers they tested, i.e. Steepest Ascent Hill–Climber (SAHC) and Next Ascent Hill–Climber (NAHC), did not succeed in finding the optimum, while Random Mutation Hill–Climber (RMHC) showed excellent behavior. Those

Table 2. The average number of function evaluations needed for convergence to the best for the 8 · 8 Royal Road test function

	SGARW	SAHC, NAHC	RMHC	MGA	SGAT
η	61334	> 256000	6179	12672	33536

results are summarized in Table 2 together with those achieved by us, in terms of average number η of function evaluations needed to converge to 0. As it can be seen MGA and SGAT are far more efficient than SGARW. This gives hint that the selection mechanism is very important in facing this function. MGA and SGAT are worse than RMHC, but much better than all of the other hill–climbers. MGA shows much better performance than SGAT, suggesting that the operators chosen might be important for efficiently facing this problem.

If we took into account the 10 · 10 Royal Road the superiority of MGA with respect to SGAT would be much clearer, as Fig. 1 (bottom) shows.

6 Conclusions and Future Work

In this paper we have defined a Mutation–based Genetic Algorithm (MGA) in which "natural" operators simulate both gene–level and chromosome–level mutations. This algorithm has been tested on a set of functions with the aim to evaluate its effectiveness in the solution of multivariable optimization problems. The achieved results assess MGA competitivity with respect to classical GAs, and its robustness as well. The main difference is that in our case there is no crossing–over between individuals belonging to the same population. It is obvious that the MGA performance can be very dependent on the problem faced. The preliminary results encourage to further investigate forms of natural mutation other than the classically used point mutation.

¿From a theoretical point of view we wish to understand whether the good performance shown by MGA depends on the integration and positive interaction of all used mutation operators, or on just one or some of them for all the problems (the other ones being about useless), or on one or some in turn for any specific problem. Therefore, we plan to face the same problems presented here by using one or two mutation operators at a time.

¿From a practical point of view, since the test functions aim to evaluate algorithms behavior when they face artificial landscapes simulating those of real–world problems, the effectiveness and the robustness shown by our algorithm urge us to make use of it when dealing with real problems.

References

1. Holland J. H. (1975) Adaptation in Natural and Artificial Systems, Univ. of Michigan Press, Ann Arbor

2. Goldberg D. E. (1989) Genetic Algorithms in Search, Optimization, and Machine Learning, Addison–Wesley, Reading, MA
3. Rechenberg I. (1973) Evolutionsstrategie: Optimierung Technischer Systeme nach Prinzipien der Biologischen Evolution, Frommann–Holzboog, Stuttgart
4. Schwefel H. P. (1977) Numerical Optimization of Computer Models, Wiley and Sons, New York
5. Fogel L. J. (1966) Artificial Intelligence through Simulated Evolution, Wiley and Sons, New York
6. Mitchell M., Forrest S. (1994) Genetic Algorithms and Artificial Life. Artificial Life 1(3):267–289
7. Banzhaf W., Nordin P., Keller R. E., Francone F. D. (1998) Genetic Programming – An Introduction – On the Automatic Evolution of Computer Programs and its Applications, Morgan Kaufmann
8. De Falco I., Della Cioppa A., Iacuelli A., Iazzetta A., Tarantino E. (1999) Towards a Simulation of Natural Mutation. In: Banzhaf W. et al. (Eds) Genetic and Evolutionary Computation Conference, Vol. 1, Orlando (FL), July 13–17, Morgan Kaufmann, San Francisco, 156–163
9. De Falco I., Della Cioppa A., Iazzetta A., Tarantino E. (2000) Towards a Closer Simulation of Natural Mutation: the Translocation Operator. In: Martikainen J., Tanskanen J. (Eds.) Fifth IEEE World Conference on Soft Computing Methods in Industrial Applications, Finland, September 4-18, IEEE Finland Section, 198–206
10. Melcher U. (1999) Molecular Genetics: Overview. In: R. Mc Clenaghan (Ed.) Encyclopedia of Genetics, Salem Press, Pasadena (CA), 406–412
11. Mitchell M., Holland J. H., Forrest S. (1992) When Will a Genetic Algorithm Outperform Hill Climbing?, In: Advances in Neural Information Processing Systems 6, Morgan Kaufmann
12. Culberson J. (1993) Crossover versus Mutation: fueling the Debate: TGA versus GIGA. In: Fifth International Conference on Genetic Algorithms, Morgan Kaufmann, 632–639
13. Horn J., Goldberg D. E., Deb K. (1993) Long Path Problems. In: Parallel Problem Solving from Nature – PPSN III, Springer–Verlag, 149–158
14. Spears W. M. (1993) Crossover or Mutation?. In: Foundations of Genetic Algorithms, Vol. 2, Morgan Kaufmann, 221–237
15. Furuhashi T., Uchikawa Y., Yoshikawa T. (1996) Emergence of Effective Fuzzy Rules for Controlling Mobile Robots using DNA Coding Method. In: IEEE International Conference on Evolutionary Computation, Nagoya, Japan, May 20-22, IEEE Press, Piscataway (NJ), 581–586
16. Nawa N. E., Furuhashi T., Hashiyama T., Uchikawa Y. (1999) A Study on the Discovery of Relevant Fuzzy Rules Using Pseudo-Bacterial Genetic Algorithm. In: IEEE Transactions on Industrial Electronics
17. Faulkner G. (1995) Genetic Operators using Viral Models. In: IEEE Internation Conference on Evolutionary Computation, Vol. 2, IEEE Press, 652–656
18. Nowak M. Schuster P. (1989) Error Thresholds of Replication in Finite Populations: Mutation Frequencies and the Onset of Muller's Ratchet. Journal of Theoretical Biology 137:375–395
19. Simoes A. .B., Costa E. (2000) Using Genetic Algorithms with Asexual Transposition. In: Whitley D. et al. (Eds.) Genetic and Evolutionary Computation Conference, Las Vegas (NV), July 10–12, Morgan Kaufmann, 323–330
20. Whitley L. D., Beveridge R., Graves C., Mathias K. (1995) Test Driving Three 1995 Genetic Algorithms: New Test Functions and Geometric Matching. Journal of Heuristics 1:77–104

Root-Finding of Monotone Nonlinear Functions with Fuzzy Iterative Methods

Peter Planinsic, Marjan Golob

University of Maribor, Faculty of Electrical Engineering and Computer Science, Smetanova 17, 2000 Maribor

Abstract. There are numerous of applications where numerical determination of the roots of an nonlinear function is required. In this paper the generalization of numerical iterative methods for function root-finding is presented through the application of fuzzy set theory, and its application to image coding. Two novel fuzzy iterative methods are introduced and compared to the classical bisection and Newton's iterative method. These iterative methods are alternatively considered as discrete control algorithms.

1   Introduction

The problem, studied here, is to find the roots of an equation in the form of $f(x) = 0$. In many practical problems it turns out that it is impossible to find an exact solution ξ, so a numerical method is employed. For example, such a problem appears, when a nonlinear relationship $f(x)$ is unknown in advance and must be obtained point-wise by measurement and/or calculations from real process. We will mainly limit our discussion to an interval in which the function has only one root and is monotone decreasing or increasing. Novel fuzzy iterative algorithms, applied for image quality control will be described and experimentally compared to the usually used bisection and Newton's method [1].

The paper is organized as follows. In section 2 classical iterative methods are briefly reviewed. In sections 3 and 4 the introduced novel fuzzy iterative methods are described. In section 5 the application of these methods in image coding, and in section 6, the experimental results are presented. The conclusions and suggestions for future work then follow.

2   Review of some classical iterative methods

The classical methods are closely connected to each other [1]. We have reviewed some of them to show their connection to our introduced fuzzy iterative methods.

712

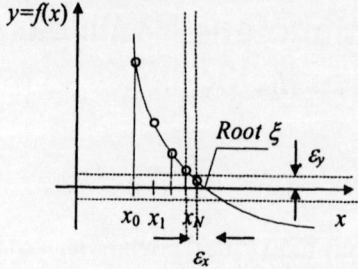

Fig. 1. Iterative root-finding

Figure 1 shows monotone nonlinear function $f(x)$. The problem is, to find the crossing point between $f(x)$ and the horizontal line, i.e. the root of this function. Iterative searching method starts in the initial point on the curve, which is then iteratively approaching the root. The numerical solution is defined by the point on the curve $f(x)$, which lies in the prescribed tolerance: $x_N - x_{N-1} < \varepsilon_x$ (or $y_N < \varepsilon_y$). It is desired that the solution is obtained in small number of iterations N, i.e. the algorithm converges fast. Using fixed point iterations a new estimate of the root is obtained in the iterative form

$$x_{n+1} = g(x_n) \tag{1}$$

where n is the iteration index. To get the iterative algorithm, we rewrite the equation $f(x) = 0$ in the form

$$x = g(x). \tag{2}$$

This method is also called the general iterative method. When choosing different $g(x)$ special methods are obtained. Some of the possibilities are:

$$g(x) = x - f(x), \quad g(x) = x - C \cdot f(x) \quad \text{or} \quad g(x) = x - h(x) \cdot f(x) \tag{3}$$

The main drawback of this method is convergence rate dependence on the slope of the curve $g'(x) = dg(x)/dx$ near the root. It can also diverge.

2.1 Bisection method

The bisection method brackets a root at an interval $[a_n, b_n]$. At each iteration, the current interval is bisected and the subinterval containing the root is kept. New endpoints of interval end estimate of the root are

$$b_{n+1} = (a_n + b_n)/2 \quad \text{and} \quad a_{n+1} = a_n. \tag{4}$$

If the root lies on the opposite site, a_n is calculating using first equation and b_n using second equation. This method is very stable (also in the case of using finite arithmetic), but its convergence rate is linear. This algorithm is often used in image coding for quantization steps adjusting to achieve the desired bit rate or distortion.

2.2 Newton's method

This method is also called the tangent method. The new estimate of the root is found at the intersection between the x-axis of the line tangent to $f(x)$ at the current state estimate of the root:

$$x_{n+1} = x_n - f(x_n)/f'(x_n). \qquad (5)$$

This method can be viewed as a general iterative method in which the generating function $g(x)$ contains the term $h(x)=1/f'(x)$. Its drawback is the demand to know exactly the function derivation $f'(x)$.

Secant method is a very economical method and can be seen as Newton's method in which the derivation $f'(x_n)$ is replaced by its approximation

$$x_{n+1} = x_n - f(x_n) \cdot \frac{x_n - x_{n-1}}{f(x_n) - f(x_{n-1})}. \qquad (6)$$

This is very useful in applications, in which the function's derivation $f'(x)$ is unknown in analytic form, as in our image coding application. Later in this paper it is referred as "Newton's" method.

3 Proposed Fuzzy iterative algorithm 1 (Fuzzy iterative)

The root finding methods can be generalized in the sense that the crossing point of function $y_1=f(x)$ with the desired horizontal line $y_2=y_{set}$ has to be found, where y_{set} can be viewed as the set point, which is searched in the closed loop iteration. The root of $f(x)$ is obtained by setting y_{set} to zero. In this way the iterative algorithm can be considered alternatively as a control algorithm, in our case fuzzy control algorithm. The algorithm starts in the initial point $(x_0, f(x_0))$ on the curve $f(x)$. In each iteration the error is considered:

$$e_n = f(x_n) - y_{set}. \qquad (7)$$

The point on curve $f(x)$ is forced to the set point by the iterative equation

$$x_{n+1} = x_n + dx_n. \qquad (8)$$

where increment dx_n is the output of fuzzy system. Using this generalization the root of error function $e(x)=f(x)-y_{set}$ is in fact searched for, i.e. the root of function $f(x)$, shifted for constant y_{set}. The iteration process stops when the error is smaller then the prescribed control tolerance ε_y. The proposed fuzzy iterative algorithm is optimized for the family of monotone convex functions with different slopes, as typically appear in image compression systems. Operational curves PSNR(Q) which practically cover the region of interest, obtained by our image coding system, are shown in Figure 2. The particular function PSNR(Q) corresponds to the particular function $y=f(x)$. Here PSNR (Peak Signal to Noise Ratio) is an image quality which depends on the quantization factor Q.

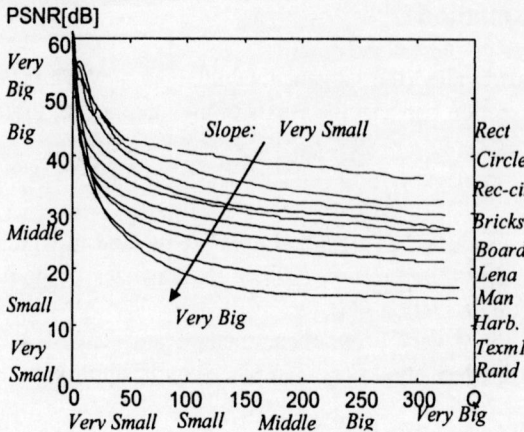

Fig. 2. PSNR(Q)-curves for some typical 512x512, 8-bit, gray scale images

The idea of the proposed Fuzzy iterative method 1 is that the increment dx_n depends simply on the control error (distance between current function value $f(x_n)$ and set point value) and is obtained by the fuzzy rule system. If the error is large, the decrement is big and the point on the curve moves more to the set point value. To speed up the convergence, the slope of the curve $f(x)$ is additionally considered. For example, if the particular curve slope is smaller, at the same error, the increment dx is bigger. These principles are incorporated in the fuzzy rules, presented in linguistic form in Table 1.

Table 1. Fuzzy system 1 rules.

No.	Rules
1	If (e is NL) and ($slope$ is S) then (dx is PL)
2	If (e is NM) and ($slope$ is S) then (dx is PL)
3	If (e is NS) and ($slope$ is S) then (dx is PM)
4	If (e is ZE) and ($slope$ is S) then (dx is ZE_3)
5	If (e is PS) and ($slope$ is S) then (dx is NM)
6	If (e is PM) and ($slope$ is S) then (dx is NL)
7	If (e is PL) and ($slope$ is S) then (dx is NL)
8	If (e is NL) and ($slope$ is M) then (dx is PL)
9	If (e is NM) and ($slope$ is M) then (dx is PL)
10	If (e is NS) and ($slope$ is M) then (dx is PM)
11	If (e is ZE) and ($slope$ is M) then (dx is ZE_2)
12	If (e is PS) and ($slope$ is M) then (dx is NM)
13	If (e is PM) and ($slope$ is M) then (dx is NL)
14	If (e is PL) and ($slope$ is M) then (dx is NL)
15	If (e is NL) and ($slope$ is L) then (dx is PM)
16	If (e is NM) and ($slope$ is L) then (dx is PS) ·
17	If (e is NS) and ($slope$ is L) then (dx is PS)
18	If (e is ZE) and ($slope$ is L) then (dx is ZE_1)
19	If (e is PS) and ($slope$ is L) then (dx is NS)
20	If (e is PM) and ($slope$ is L) then (dx is NS)
21	If (e is PL) and ($slope$ is L) then (dx is NM)

The abbreviations correspond to two input sets and one output set of membership functions. For the input variable e: NL means negative large, NM is negative middle, NS is negative small, ZE is zero, PS is positive small, PM is positive middle and *PL* is positive large. Input variable *slope* has the membership functions: *S*-small, *M*- middle, *L*- large. Output variable *dx* has the following membership functions: *PL*- positive large, *PM*- positive middle, *ZE_1* is middle membership function (near zero) , *ZE_2* is narrower middle membership function and *ZE_3* is the narrowest middle membership function. The activation of appropriate middle membership function (*ZE_1*, *ZE_2* or *ZE_3*) is obtained automatically by rules. The input and output membership functions are shown in Fig. 3. For practical realization the fuzzy controller input and output variables are normalized and limited in the range: $e \in [e_{min}, e_{max}] = [-1, 1]$; $dx \in [dx_{min}, dx_{max}] = [-1, 1]$; $K_e = 1/25$; $K_{dx} = 500$; $slope \in [slope_{min}, slope_{max}] = [0, 1]$, where K_e and K_{dx} are normalization constants.

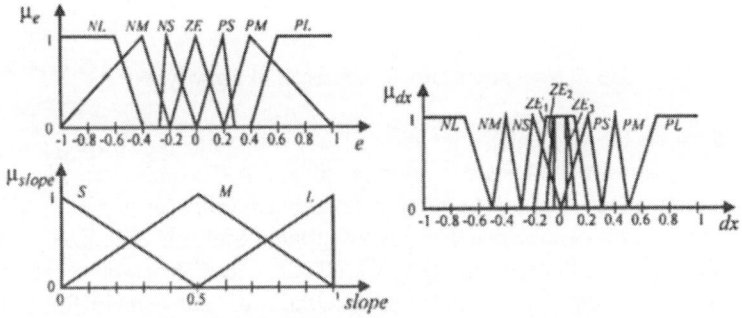

Fig.3. The membership functions of fuzzy system 1

The principle scheme of the proposed Fuzzy iterative algorithm 1 is shown in Fig.4. The center of gravity (COG) defuzzyfication method and the single tone fuzzyfication method were used. The MAX-MIN (Mamdani-type) inference mechanism [2] was implemented.

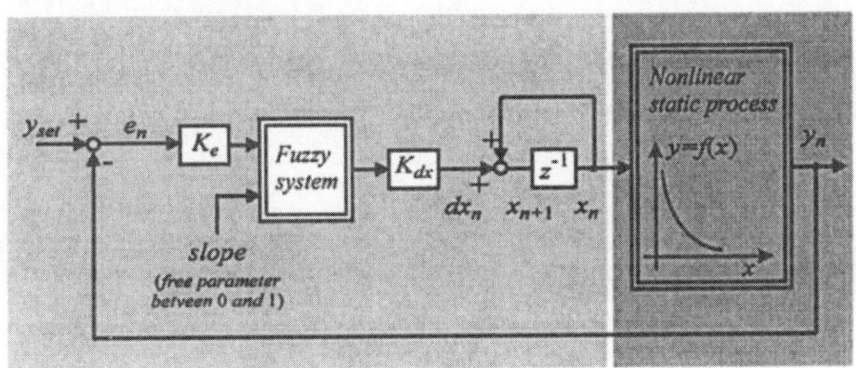

Fig. 4. Fuzzy iterative algorithm 1

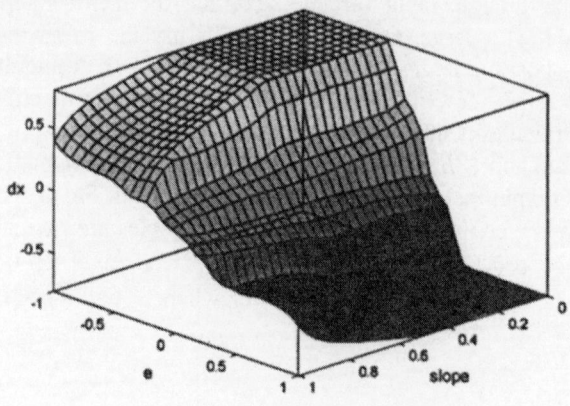

Fig. 5. Nonlinear static characteristics of fuzzy system 1

Figure 5 shows the static characteristic of the fuzzy system 1. The slope of actual function is unknown in advance in practical situations. The parameter *slope* can be estimated in the first iteration (assuming monotone convex function) or calculated from other process parameters. The proposed algorithm was optimized using the equal small initial value $x(0)$, i.e. $Q(0)$ for all functions. The static characteristics were optimized using appropriate rules and width of membership functions. To further reduce the number of iterations, it can be useful, if the initial point can be predicted close to the desired one, from the process parameters. Approximate prediction can again be realized using fuzzy logic. Additional intelligence can be included using on-line and a prior process information.

To see the fuzzy iterative algorithm (8) as a general iterative algorithm (1) for searching the root of $f(x)$ (in our general case of $e(x)$) we can rewrite it in the form:

$$x_{n+1} = g(x_n); \quad g(x_n) = x_n + dx_n = x_n + f_{nonl}(e_n, slope) \tag{9}$$

where $f_{nonl}(e, slope)$ is a two-dimensional static characteristic of the fuzzy rule system. The fuzzy iterative algorithm can be further expressed as:

$$x_{n+1} = x_n + k_{nonl}(e_n, slope) \cdot e_n \tag{10}$$

where k_{nonl} is from the input e dependent gain factor and the *slope* is the parameter of a particular image's curve $f(x)$. Introducing the expression (7) for error e in (10), we obtain the algorithm:

$$x_{n+1} = x_n + k_{nonl}(e_n, slope) \cdot (f(x_n) - y_{set}) \tag{11}$$

If we set $y_{set}=0$, then the root of $f(x)$ is found. When the static characteristic of the fuzzy system is linear, i.e. gain factor is constant and independent of e, we obtain a simple iterative algorithm:

$$x_{n+1} = x_n + k(slope) \cdot f(x_n); \quad k = -h = -C. \tag{12}$$

The nonlinear static characteristics can therefore be user friendly designed using fuzzy rules. The fuzzy iterative algorithm 1can be modified so, that the instantaneous slope $f'(x)$ of the curve in the instantaneous iteration point is considered. It can be estimated with the function derivation approximation:

$$slope = \frac{\Delta e_n}{\Delta x_n} = \frac{e_n - e_{n-1}}{x_n - x_{n-1}} = \frac{(f(x_n) - y_{set}) - (f(x_{n-1}) - y_{set})}{x_n - x_{n-1}} \tag{13}$$
$$\approx f'(x_n)$$

Therefore, the fuzzy iterative algorithm can be described by expression:

$$x_{n+1} = x_n + k_{nonl}(e_n, f'(x_n)) \tag{14}$$

where k_{nonl} is the nonlinear static function of two input variables. In this way similar behaviour to that with Newton's (secant) method can be achieved.

4 Fuzzy iterative algorithm 2 (Fuzzy Newton's)

The proposed fuzzy iterative algorithm 2 (Fuzzy Newton's method) can be seen as a variant of the adaptive Newton's method, where the additional multiplicative adaptive factor f_{nonl} is introduced and generated by the fuzzy rule system. The block scheme of the algorithm is shown in Fig. 6.

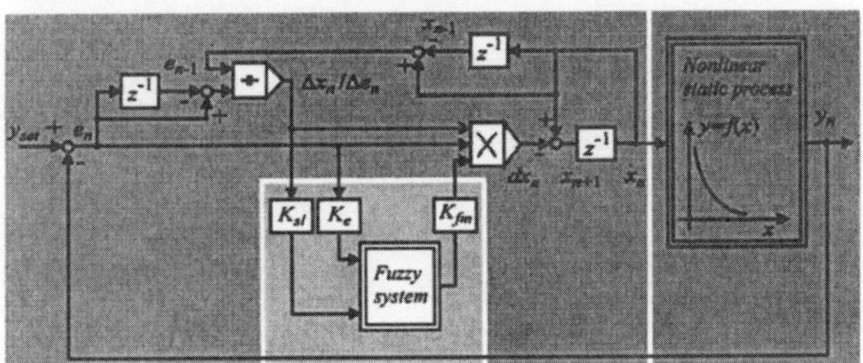

Fig. 6. Fuzzy iterative algorithm 2

The algorithm can be expressed as follows:

$$x_{n+1} = x_n + dx = x_n + e(x_n) \cdot \frac{1}{slope(x_n)} \cdot f_{nonl}(e(x_n), slope) \tag{15}$$

The multiplicative correction factor depends on the instantaneous slope of the curve at the instantaneous iteration point (approximation of $slope(x_n) = \Delta e_n / \Delta x_n$) and the control error. Near the set point the consideration of slope is important to avoid undesired oscillations (the correction factor is smaller than 1). If the error is greater, then the obtained increment dx is greater (using a correction factor greater than 1). See also Fig. 8. This principle is implemented in the fuzzy system for obtaining correction factor using simple linguistic description, given in Table 2.

Table 2. Rules of Fuzzy system 2

No.	Rules
1	If ($|e|$ is *S*) and (*slope* is *S*) then (*dx* is *M*)
2	If ($|e|$ is *S*) and (*slope* is *M*) then (*dx* is *M*)
3	If ($|e|$ is *S*) and (*slope* is *L*) then (*dx* is *S*)
4	If ($|e|$ is *M*) and (*slope* is *S*) then (*dx* is *L*)
5	If ($|e|$ is *M*) and (*slope* is *M*) then (*dx* is *L*)
6	If ($|e|$ is *M*) and (*slope* is *L*) then (*dx* is *M*)
7	If ($|e|$ is *L*) and (*slope* is *S*) then (*dx* is *L*)
8	If ($|e|$ is *L*) and (*slope* is *M*) then (*dx* is *L*)
9	If ($|e|$ is *L*) and (*slope* is *L*) then (*dx* is *L*)

All fuzzy variables have three equal width triangular membership functions, shown in Fig.7: *S*- small, *M*- middle and *L*-large. Fuzzy variables are normalized with normalization constants ($K_e = 1/25$, $K_{sl} = 1/50$ and $K_{fm} = 1$). Absolute values of *slope* and *e* are considered. The static characteristic of fuzzy system is in Fig. 8

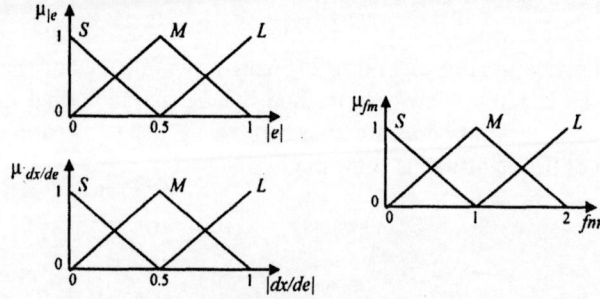

Fig. 7. The membership functions of fuzzy system 2

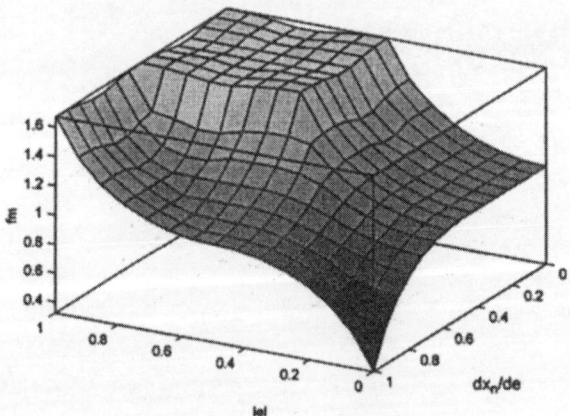

Fig. 8. The static characteristic of fuzzy system 2

5   Implementation in image coding

We implemented fuzzy iterative algorithms for control of the distortion or compression ratio in subband image coding. A simple version was implemented in [5], [6]. The scheme using Fuzzy iterative algorithm 1 is shown in Fig. 9. The static characteristics PSNR(Q) of some nonlinear coding processes for different 512x512, 8-bit, gray scale images are in Fig. 2. The problem in image coding is to ensure the desired image quality of the arbitrarily coded image. It is unknown in advance, which image will appear at the input of coding system. Different image quality measures can be used. In the presented application the standard image quality measure PSNR (Peak Signal to Noise Ratio) is used. The compression ratio (bit rate) can be also controlled.

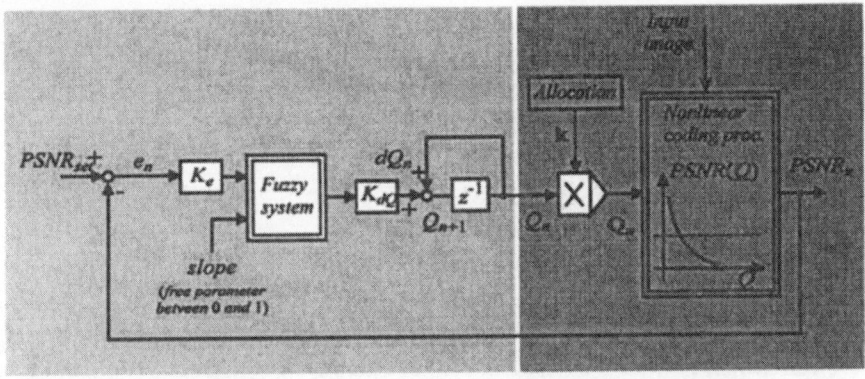

Fig. 9. Implementation of Fuzzy iterative algorithm 1 for coded image quality control

This scheme is practically the same as the scheme in Figure 4. The nonlinear static process is now the nonlinear relationship PSNR(Q) between image quality (measured in PSNR) and reference quantization step Q (quality factor). In this scheme, therefore, y corresponds to PSNR, and x to Q. As reference quantization step Q simply Q_0 was taken. Fuzzy rules in Table 1 are used. The membership functions of the fuzzy system also remain the same as depicted in Figure 3. To obtain vector of subband quantization steps $\mathbf{Q}=[Q_0,Q_1,...Q_{J-1}]$, the reference quantization step Q is multiplied with the allocation vector $\mathbf{k}=[k_0, k_1,...k_{J-1}]$. Factors k_i represent subbands scaling factors (weights). J is the number of subbands. We use a simple solution, equal quantization step for all subbands ($\mathbf{k}=\mathbf{1}$).

In the coding process the input original image (image matrix dimension of 512x512 pixels with 8-bit coefficients) is first transformed into different frequency subbands. The subband coefficients are then scalar quantized. Quantizing the subbands causes the distortion and so the lossy compression of images is obtained. The distortion (quality) of compressed images can be controlled by adjusting the quantization steps. This is done in this application using the fuzzy iterative algorithm (the greater the quantization Q the smaller is quality PSNR). The quantized coefficients are then loss-less entropy encoded. Using an orthogonal subband transform the distortion can be measured in the transform domain. The relationship between Mean Square Error (MSE) and quality (PSNR) for 8-bit images is $MSE = 255^2 \cdot 10^{0.1 \cdot PSNR}$.

An equal small initial quantization step $Q(0)$ was used for all images. To achieve faster but still stable convergence, the slope of the actual nonlinear function is considered. Parameter *slope* can be estimated in the first iteration or estimated using parameter c_{DR} in equations:

$$\mathrm{PSNR}(R) = 10 \cdot \log_{10} \frac{255^2}{\mathrm{MSE}(R)} \tag{16}$$

$$\mathrm{MSE}(R) \approx \prod_{i=0}^{J-1} \left(\frac{h \cdot \sigma_i^2}{k_i^2} \right)^{\alpha_i} \cdot \sum_{i=0}^{J-1} \left(\alpha_i \cdot k_i^2 \right) \cdot 2^{-2 \cdot R} = c_{DR} \cdot 2^{-2 \cdot R} \tag{17}$$

which describes the PSNR(R)-curves, where R is bit rate and h_i parameter, which depends on the statistics of the subband with variance σ_i (most subbands of images have Laplacian distribution: $h_i = 1.2$), and α_i are normalized areas of subbands. Compression ratio CR for 8-bit images is $8/R$. Shapes (slopes) of PSNR(CR)-curves correlate well to shapes of PSNR(Q)-curves, therefore, the parameter c_{DR} is well correlated to the slope of (PSNR-Q-curve).

6   Experimental results

When the optimization of the fuzzy algorithms was finished, a comparison of the convergence of the algorithms (Fuzzy 1, Fuzzy2, Bisection and Newton's) was made by compressing three typical images on three desired values, considering by this selection the typical situations in the whole region of interest. The obtained numbers of iterations are summarized in Table 3. The moving of iteration points on PSNR(Q) curve for image "Lena" is shown in Figure 10. Fig.11 shows original image and to the desired quality PSNR_set=28 dB compressed image.

Table 3. The obtained numbers of iterations N, control torelance 0.1dB

	Circle			Lena			Rand		
	PSNR_set_point			PSNR_set_point			PSNR_set_point		
	25 dB	28 dB	31 dB	25 dB	28 dB	31 dB	25 dB	28 dB	31 dB
Bisection	N=6	N=7	N=7	N=7	N=6	N=8	N=10	N=10	N=8
Fuzzy	N=6	N=5	N=3	N=4	N=2	N=5	N=4	N=2	N=3
Newton	N=6	N=6	N=6	N=6	N=6	N=6	N=5	N=4	N=4
Fuzzy Newton	N=5	N=4	N=5	N=6	N=5	N=5	N=3	N=3	N=3

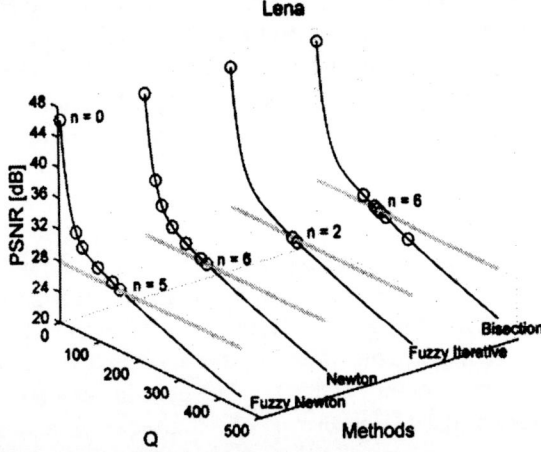

Fig. 10. The moving of iteration points on PSNR(Q)-curves for image "Lena" at different iterative algorithms; $Q(0)$=5, PSNR_set=28 dB, tolerance 0.1 dB. At bisection algorithm the maximum upper value for Q is set to 500, at Fuzzy iterative algorithm the parameter *slope* is 0.5.

| Image "Lena" (original, 512x512, 8-bit) | Image "Lena", compressed: PSNR=28 dB, Q=134. |

Fig.11. Test images, compressed to the same desired PSNR_set=28 dB.

Both proposed fuzzy iterative algorithms converged stable and outperforms the bisection and Newton's method in the obtained number of iterations (see Figure 10 and Table 3). Bisection algorithm needs in average approximately 7.6 iterations, Newton's method needs about 5.4, Fuzzy iterative algorithms 1 (Fuzzy iterative) approximately 3.7 and Fuzzy iterative algorithm 2 (Fuzzy Newton's) about 4.3 iterations at the same control tolerance (0.1 dB).

6 Conclusions

The possibility was presented of designing an application adapted fuzzy iterative algorithm for the root-finding of a nonlinear function (process). A connection with classical iterative algorithms was made and showed that fuzzy iterative algorithms can be viewed as enhanced classical algorithms. It was also shown that fuzzy iterative algorithms can be alternatively considered as fuzzy close loop control algorithms. A great advantage of decision for fuzzy logic is, that a iterative algorithm can be user friendly adapted (optimized)) to the particular application (for example minimizing the number of iterations) by simple using fuzzy rules. There are also a large number of available user friendly development tools. The example for optimizing the fuzzy algorithms for image quality control was reported and experimentally verified that fuzzy iterative algorithm outperforms the classical bisection algorithm and Newton's algorithm in the number of iterations, assuring stable convergence for different images (i. e. corresponding slopes of nonlinear functions). Comparisons between different algorithms must also take into account additional considerations such as robustness [2], [3], [4] rate of convergence, the difficulty of verifying assumptions made by an algorithm and the total computational efficiency. The computational complexity of evaluating the output of a fuzzy controller (fuzzyfication, inference machine and defuzzyfication) is approximately of the same order as other computations in classical controller. The so called look-up-table fuzzy controller can be implemented for real-time applications [8].

In the future the investigation of systematic theoretical studies and experiments can be made, extensions to more complex nonlinear functions, systems of nonlinear functions, as well to gradient search- and optimization-algorithms [7].

References

1. Conte SD, De Boor C (1972) Elementary numerical analysis: an algorithmic approach. McGraw-Hill, New York.
2. Driankov D, Hellendoorn H, Reinfrank M (1996) An introduction to fuzzy control, 163-196, Springer, Berlin.
3. Jenkins DF, Passino KM (1999) An introduction to nonlinear analysis of fuzzy control systems. Journal of Intelligent & Fuzzy systems, 7, no 1, pp 75-103.
4. Koprinkova P, Membership functions shape and its influence on the stability of fuzzy control systems. Cybernetics and Systems 31, pp 353-371.
5. Planinšič P (1998) Fuzzy quantization step controller for wavelet-subband image coding. J. Electrical Engineering 49, no 7-8, pp 179-185, Bratislava, Slovakia.
6. Planinšič P, Gergič B, et al. (1999) Efficient algorithms for control and optimisation of distortion or rate at wavelet-subband image coding. In: Proceedings of Picture coding symposium'99, pp 407-411, Portland, Oregon.
7. Salomon R, (1998) Evolutionary algorithms and gradient search: similarities and differences. IEEE Trans. on Evolutionary Computations 2, no 2, pp 45-55.
8. Surman H, Ungering AP, et al. (1994) What kind of hardware is necessary for a rule based system. In: Proceedings of the 3'rd IEEE int. Conference on Fuzzy Systems, pp 274-278. Orlando, FL.

On Partitioned Fitness-Distributions of Genetic Operators for Predicting GA Performance

Rafael Nogueras and Carlos Cotta

Dpto. de Lenguajes y Ciencias de la Computación, Univ. de Málaga
Campus de Teatinos (3.2.49), 29071 - Málaga - Spain
ccottap@lcc.uma.es
http://www.lcc.uma.es/~ccottap

Abstract. A statistical approach aimed at predicting the performance of a GA is presented. This approach is based on trying to mimic the fitness distribution of genetic operators. By modeling such fitness distributions, the effects of genetic operators can be simulated within the framework of virtual genetic algorithms (VGAs). An improved statistical model is provided for this purpose. A very important issue is the partitioning of the dataset used in the statistical analysis, so as to grasp better the structure of the real distribution.

Keywords: genetic algorithms, performance prediction, fitness distributions, statistical techniques, dataset partitioning

1 Introduction

Genetic algorithms (GAs) [8] are stochastic heuristic methods for search and optimization grounded upon the principles of natural evolution, namely adaption and survival of the fittest. A GA works by iteratively generating tentative solutions for the problem at hand; the GA maintains a population –pool– of candidate solutions, and repeatedly performs a basic cycle comprising selection, reproduction, and replacement stages. GAs constitute a valuable tool for tackling hard optimization problems whose resolution via classical exact or approximate techniques is unaffordable.

The application of a GA to a specific problem requires the instantiation of the generic template previously sketched. Such instantiation involves determining several algorithmic and parametric aspects. Regarding the former, the user/designer has to choose an appropriate representation of the problem, as well as adequate procedures for initialization and reproduction among others. As to the latter, several parameters must be given a value, i.e., the size of the population, the rate of mutation, etc.

There exist some tools that can be used for this purpose of GA design (e.g., fitness-landscape analysis [9,10], forma analysis [12,13], etc.), although these are mainly focused on algorithmic aspects (representation, recombination mechanism, ...). Furthermore (and despite their controversial implications), some theoretical results –e.g., the No Free Lunch Theorem [16]– clearly indicate the futility of trying to determine "universal" settings for the GA. Thus, the parameterization of the algorithm cannot be dismissed with a simple choice of "standard" values. In this sense, a very common approach is the use of a costly trial-and-error procedure, i.e., the user decides an initial parameterization, and embarks in a fine-tuning process by doing experimental tests.

In some situations, this trial-and-error scenario can be very inappropriate. Consider that GA could be being applied to a problem requiring complex mathematical calculations, simulations of a physical system, etc. Thus, the availability of models for the quantitative and/or qualitative prediction of GA performance is an important issue. In this sense, Grefenstette proposed an interesting approach, the so-called virtual genetic algorithms (VGAs) [7]. Essentially, VGAs constitute a statistical approach based on the use of fitness distributions. A possible extension of this model is presented in this work.

The remainder of the paper is organized as follows: first, the fundamentals of VGAs and some enhancements of the model are presented in Section 2. Then, the partitioning of the dataset used for the statistical analysis is tackled in Section 3. Subsequently, experimental results are presented in Section 4. Finally, Section 5 summarizes our results, and outlines future research.

2 The Underlying Model

The VGA model resembles a standard GA, but exhibits an important difference: individuals do not carry genetic information; they are used as place-holders for fitness information. This fitness information is transmitted from parents to offspring, using a statistical model to simulate the effects of genetic operators. To be precise, this statistical model is utilized to predict the fitness of a new individual in terms of the fitness of its parent (or the parents' mean fitness in the case of recombination). Thus, the reproduction (recombination plus mutation) and evaluation phases of a standard GA are joined in one single VGA phase, as show in Fig. 1.

The statistical model proposed by Grefenstette (denoted by B-VGA henceforth) is based on assuming that the fitness distribution of an operator can be modeled as a Gaussian distribution. Under this assumption, a linear regression approach is used to determine the parameters of this distribution. More precisely, the process (to be repeated for each operator) is as follows:

1. Generate a dataset of N pairs (f_p, f_d), where f_p is the fitness of a random solution (or the mean fitness of two randomly constructed solutions in the case of recombination), and f_d is the fitness of a descendant obtained via mutation (or recombination) of the previous (pair of) solution(s).
2. Perform a least squares fit of the linear model $f_d = a + b f_p$ to the N pairs (f_p, f_d).
3. Sort the dataset according to f_p, and group the pairs in bins of size T.
4. For each bin, calculate the standard deviation σ_f of the f_d values, and the mean $\bar{f}_p$ of the f_p values.

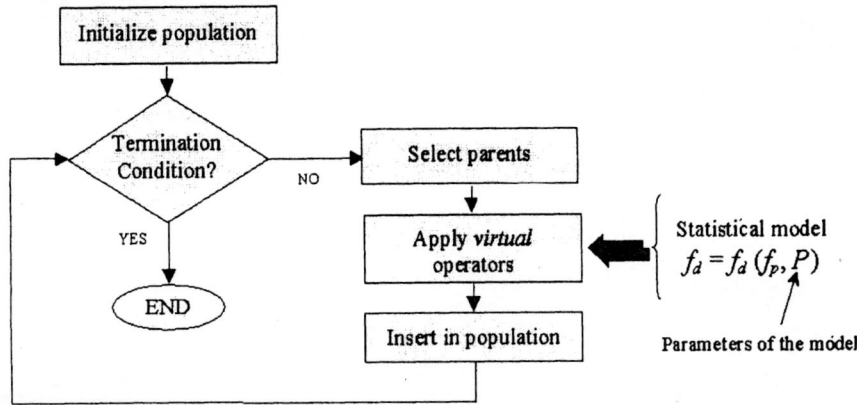

Fig. 1. Sketch of the VGA. A statistical model is used to transmit fitness from parents to offspring.

5. Perform a least squares fit of the N/T pairs $(\bar{f}_p, \sigma_f)$ to the linear model $\sigma_f = \sigma_a + \sigma_b \bar{f}_p$.
6. Output the model for the operator ω: $M(\omega) = \langle a, b, \sigma_a, \sigma_b \rangle$.

Once the parameters of the model are extracted, the fitness of the descendant can be estimated as a function of the parent's fitness (or the parents' mean fitness) as $f_d = N(a + b f_p, \sigma_a + \sigma_b f_p)$. This estimation is done in cascade for each genetic operator, e.g., assuming the typical situation of using recombination (ω_X) and mutation (ω_m), the parents mean fitness is used as an input to the $M(\omega_X)$ model, and the output of this model is fed to $M(\omega_m)$.

While this model can provide good results (in terms of prediction accuracy) in some situations, some previous experiments [11] have shown that this is not the situation in general. This has given rise to several modification in the basic model, intended to grasp two major difficulties identified: the statistical anisotropy of the distribution, and its asymmetry (see Fig. 2).

With respect to the first issue, a more careful generation of the dataset is needed, since it should comprise a representative sample of the fitness distribution of a particular operator. This can be achieved via the use of a pilot run of the GA [1], hence obtaining a sample of the fitness distribution of the operator across a larger region of the search space. This is similar to the mechanism proposed in [6], although with an important difference: rather that considering the rate of application of an operator as an internal parameter of the latter, it is consider an external parameter, thus being the dataset referred to *minimal* [13] applications of the operator.

As for dealing with the distribution asymmetry, the dataset is partitioned into two subsets L and U right after having calculated the linear fit for the Gaussian distribution mean. Each of these subsets respectively comprises the points of the dataset that

---

[1] Notice that the cost of this pilot run is roughly the same of performing a random sampling, since fitness calculation is assumed to represent the lion's share of the computational cost of the algorithm.

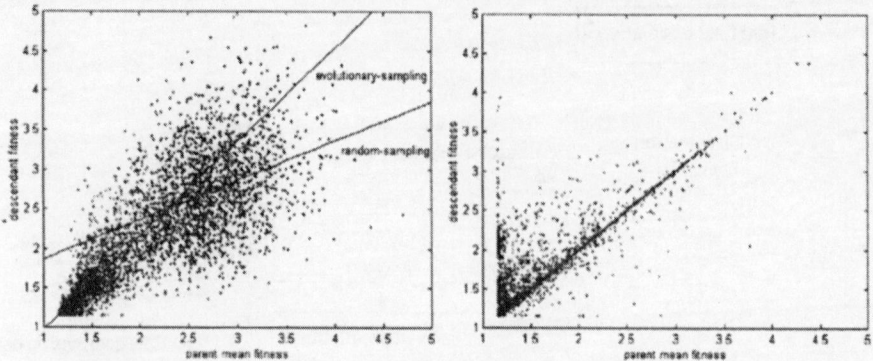

Fig. 2. (Left) Comparison of a random sample (dark dots) and the sample performed by a GA (light dots). (Right) Fitness distribution for a bit-flip mutation operator. In both cases, the data correspond to a brachystochrone design problem [14].

are below or above the central estimation. This is done in order to calculate different deviation parameters for each of these sets. Furthermore, deviations are measured with respect to predicted fitnesses rather than with respect to the descendant mean fitness. As a final addition, upper and lower limits of the fitness distribution are also calculated so as to truncate the Gaussian distribution, avoiding extreme values and thus making the VGA more robust. The complete pseudocode of the process is shown below:

COMPUTE-MODEL (*input:* dataset S)

1. Perform a least squares fit of data in S to the linear model $f_d = a + b f_p$.
2. Partition S into two subsets $S^U = \{(f_p, f_d) \in S \mid f_d > a + b f_p\}$
   and $S^L = \{(f_p, f_d) \in S \mid f_d \le a + b f_p\}$. Let $p = |S^U|/|S|$.
3. For each subset do
   (a) Group the pairs in bins of size T.
   (b) Perform a least squares fit to $\sigma_f^{U|L} = \sigma_a^{U|L} + \sigma_b^{U|L} \bar{f}_p$, where $\bar{f}_p$ is the mean of f_p in each bin, and $\sigma_f^{U|L}$ is the deviation of f_d with respect to $a + b f_p$.
   (c) For S^U calculate the maximum $f_p^{\max} = \max(f_p)$ in each bin (idem with the minimum $f_p^{\min} = \min(f_p)$ for S^L).
   (d) Perform a least squares fit to $f_p^{\max} = \lambda_a^U + \lambda_b^U \bar{f}_p$ (respectively to $f_p^{\min} = \lambda_a^L + \lambda_b^L \bar{f}_p$).
4. Output the model $M(\omega) = \langle a, b, \sigma_a^U, \sigma_b^U, \sigma_a^L, \sigma_b^L, \lambda_a^U, \lambda_b^U, \lambda_a^L, \lambda_b^L, p \rangle$.

This improved VGA model will be denoted as L-VGA$_1$, where the 'L' refers to the linear nature of the intervals (the meaning of the subscript will be evident in next section). As it can be seen in Fig. 3, a remarkable improvement with respect to the B-VGA is achieved by using the L-VGA$_1$ model, resulting in more accurate predictions of the GA performance.

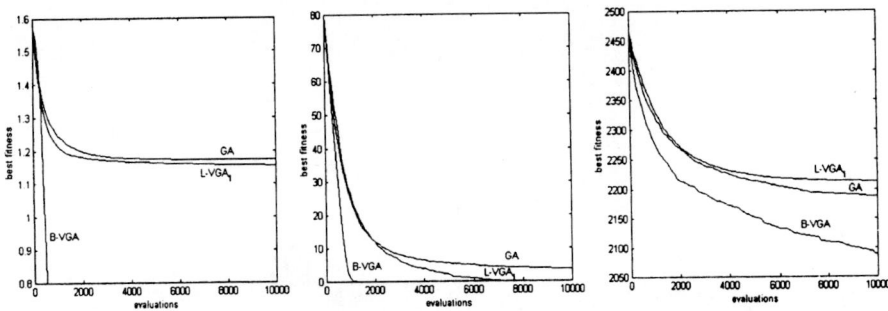

Fig. 3. Comparison of GA vs. B-VGA/L-VGA₁ in the brachystochrone design problem (left), in the Rastrigin function (middle), and in a flowshop scheduling problem (right).

There are further ways to enhance the performance of this modified VGA. To be precise, notice that a linear regression approach might be a not very appropriate approach for globally modeling the distribution of fitness. This is explored in the following section.

3   Partitioning the Dataset

As mentioned in the previous section, having an underlying linear model is somewhat inflexible. This is related to the different behavior of the GA in different regions of fitness space. Intuitively, the chances for obtaining a descendant better than the parent (or better than the parent's mean) are higher when the latter has a lower quality, i.e., there is more room for improvement than for degradation. This can be the situation during some initial stages of the GA run. The scenario is different when the population evolves since individuals tend to be better. In this case, it is more likely to obtain below average (i.e., worse) solutions when doing mutation for instance.

Several possibilities exist to improve the model. On one hand, it would be possible to use a higher-degree polynomial model. This would obviously redound in a better fit to empirical data, but it does not offer a conclusive improvement for a low-degree polynomial, and it is prone to over-fitting for a high-degree polynomial. Thus, a different alternative has been considered: partition of the dataset into several prediction intervals, and extract a model for each of these intervals. This way, the VGA would use statistical parameters adjusted to the region in fitness space in which it is located in a particular situation (e.g., see Fig. 4).

According to the above considerations, we have considered a mechanism for constructing the prediction intervals based on linear partitions along the parent-fitness axis, i.e., defining different models for different ranges of values for f_p. The pseudocode of this approach is shown below:

1. Sort the dataset according to f_p, and partition it into M subsets. Each of these subsets S_i is characterized by the interval $[L_i, R_i)$, where $L_i = R_{i-1}, L_1 = \min(f_p)$, $R_M = \max(f_p) + \varepsilon$, i.e., $S_i = \{(f_p, f_d) \mid L_i \leq f_p < R_i\}$.

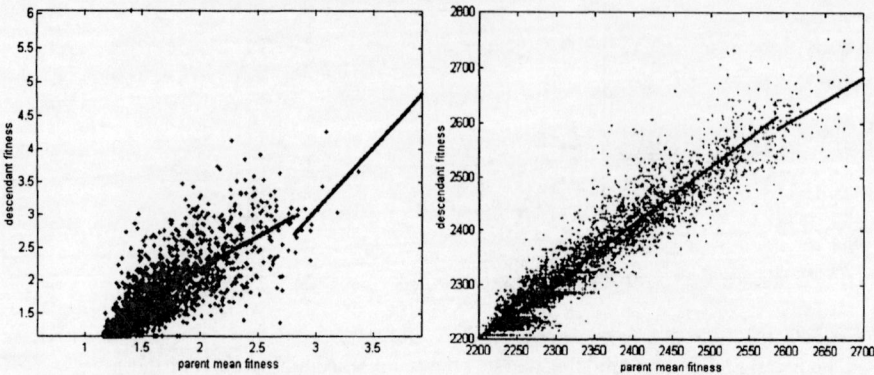

Fig. 4. Linear models for UX crossover in the brachystochrone design problem with three intervals (left), and for UCX crossover in a flowshop scheduling problem with three intervals (right).

2. For each subset S_i do $M^i = \text{COMPUTE-MODEL}(S_i)$
3. Output the complete model $M(\omega) = \{M^1, \cdots, M^M\}$

where COMPUTE-MODEL is the procedure described in the previous section.

Subsequently, the VGA determines in runtime which submodel must be used in each operator, according to the corresponding parents' fitness. In this work, the application interval for each submodel is determined by an equidistant segmentation of the X-axis. Extending the notation introduced in Section 2, this VGA model will be denoted as L-VGA$_M$, where M is the number of submodels as mentioned above.

4 Experimental Results

This section provides some examples of the application of the L-VGA model for several test problems. Except where otherwise noted, experiments have been done with an elitist generational GA/VGA ($popsize = 100, p_X = .9, p_m = 0.01, maxevals = 10000$) using ranking selection. All results correspond to series of 50 runs.

First of all, we focus on the quantitative goodness of the predictions. Fig. 5 (left) shows the results for a 15-machine 30-task flowshop scheduling problem (taken from the OR-Library [3]) using the UCX recombination operator [5]. Notice that while L-VGA$_1$ deviates from the real GA at the end of the run, L-VGA$_2$ provides a good resemblance of the GA behavior. A clearer example is shown in Fig. 5 (right), corresponding to the Fletcher-Powell function (cf. [2]) with $n = 8$, and using SPX crossover. As it can be seen, the accuracy of L-VGA$_2$, L-VGA$_4$, and L-VGA$_8$ is increasingly better.

Focusing now in the qualitative goodness of predictions, some results are provided in Fig. 6. The goal here is estimating how a particular GA will perform with respect to another particular GA. For example, let A and B be two configurations for GA/L-VGA; it is pursued to obtain the same relative properties from a head-to-head

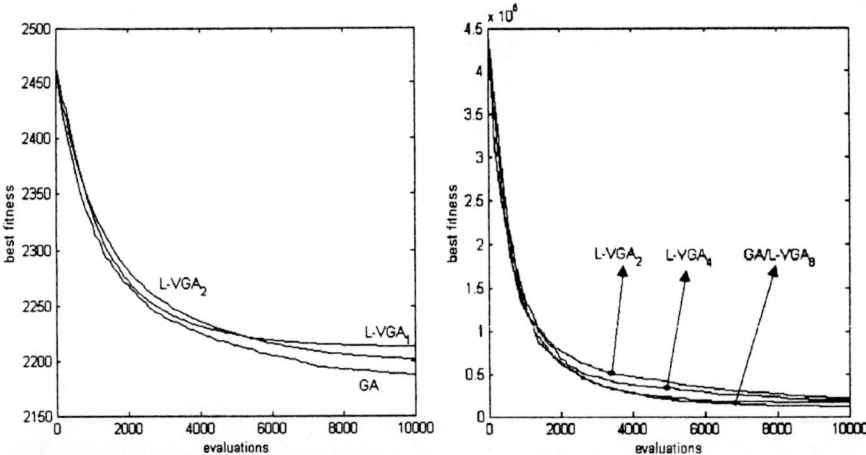

Fig. 5. Comparison of GA vs. L-VGA on a flowshop scheduling problem (left), and on the Fletcher-Powell function (right).

comparison between the VGA^A and the VGA^B than those from a GA^A vs. GA^B comparison. This is shown for GA/L-VGA$_2$ in the Rastrigin function ($n = 8$, $length = 128$ bits) with three configurations: $A = \{popsize = 100, p_X = .9, p_m = 0.01\}$, $B = \{popsize = 50, p_X = .5, p_m = 0.1\}$, and $C = \{popsize = 80, p_X = .75, p_m = 0.001\}$.

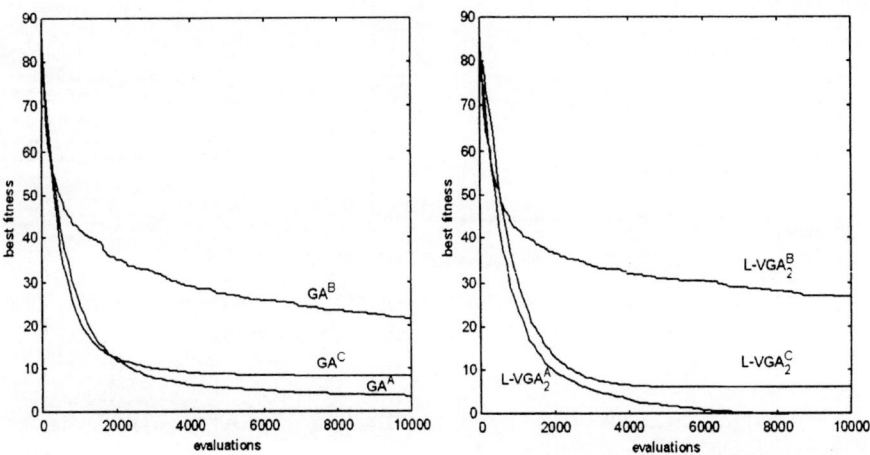

Fig. 6. Comparison of several configurations for the Rastrigin function using SPX crossover. (Left) Results of the GA. (Right) Results of the L-VGA$_2$.

As it can be seen, the relative behavior of the different L-VGAs accurately matches the relative behavior of the real GA. This provides some support to the use of this model to analyze several configurations, using the real GA only upon selection of the best one. It must be noted that the computational cost of the L-VGA is a small fraction of the cost of the real GA. Furthermore, the larger the problem size, the larger the speed-up (this clearly follows from the fact that the L-VGA is not sensitive to the problem size, since it just simulates the effects of genetic operators using the statistical model).

The last set of experiments has been done using island-model parallel genetic algorithms [15,4] as the target algorithm whose performance is to be estimated. To be precise, we have considered a ring-topology, being the number of islands, the frequency of migrations, and the type of migrant selection (random selection vs. best-of-population) additional parameters to be adjusted. Fig. 7 shows the results for a PGA/L-PVGA$_2$ on a flowshop scheduling problem. In this case, each island is a generational GA ($p_X = .9, p_m = 0.01, maxevals = 20000$), the migration frequency is 500 evaluations, and the random migrant selection is used. The free parameter is the number of islands, which in turn determines the population size for each island (due to the fact that a total number of 400 individuals is kept for the PGA). The configurations studied are: $A = \{GAs = 1\}, B = \{GAs = 2\}, C = \{GAs = 4\}$ and $D = \{GAs = 8\}$.

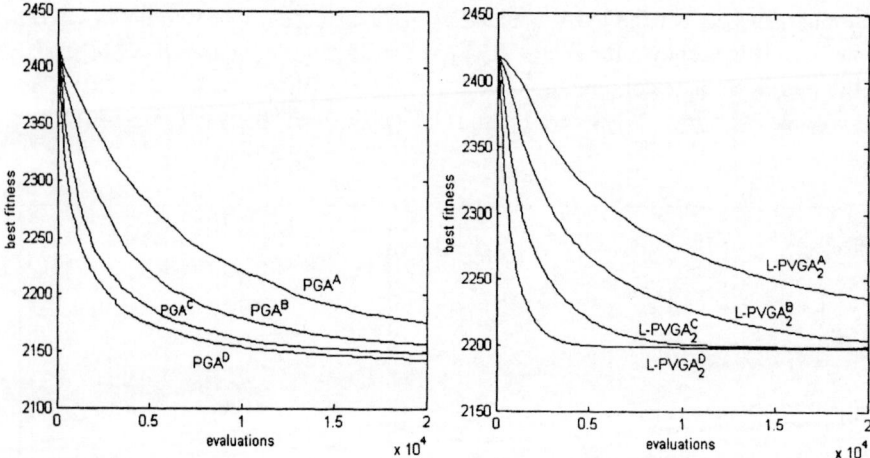

Fig. 7. Comparison of several configurations on a flowshop scheduling problem using UCX crossover. (Left) Results of the PGA. (Right) Results of the L-PVGA$_2$.

Notice how the relative behavior of the different L-PVGAs accurately matches the relationship between real PGAs[2]. Further results are shown in Fig. 8. In this case, each island is a generational GA with $p_X = .9$, $p_m = 0.01$, and $maxevals = 20000$. The free parameters are the number of islands (for a total number of 300 individuals), the migration frequency and the type of migrant selection. To be precise, the configurations considered are: $A = \{GAs = 2, migfreq = 1000, migrate - best\}, B =$

---

[2] The convergence of the L-VGAs in the long term is a side effect of the use of upper/lower limits for simulated values. The utilization of these limits is very useful for avoiding the L-VGA from dropping quickly to low fitness values though.

$\{$GAs $= 2$, $migfreq = 300$, $migrate - random\}$ and $C= \{$GAs $= 4$, $migfreq = 500$, $migrate - random\}$.

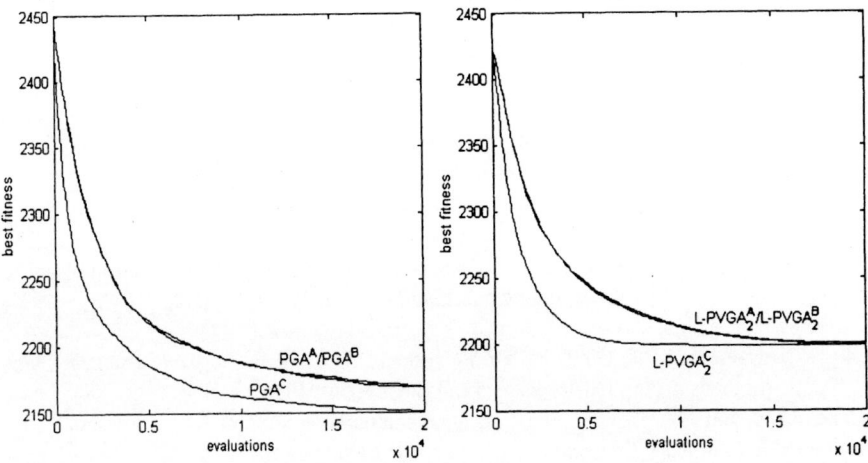

Fig. 8. Comparison of several configurations on a flowshop scheduling problem using UCX crossover. (Left) Results of the PGA. (Right) Results of the L-PVGA$_2$.

Again –and despite the final convergence in the long term–, the L-PVGAs provide a very good indication of the relative performance of the different PGAs: configurations A and B are virtually identical in both cases, and configuration C results in faster advance to low fitness regions.

5 Conclusions

This work has focused on the utilization of statistical techniques based on fitness distributions of genetic operators for predicting GA performance. To this end, some extensions of the VGA model have been presented and evaluated on a benchmark composed of non-trivial test problems. To be precise, these extensions are grounded on the partitioning of the dataset so as to obtain several prediction models of restricted applicability but higher accuracy. The combined used of these models has allowed improving the quality of the predictive capabilities of the VGA approach.

This improvement notwithstanding, it is clear that there is still room for further enhancements of these tools. We are currently trying to apply these techniques to the design of fuzzy-logic controllers [1]. While the qualitative results we are obtaining are indicative of the relative performance of several parameterizations of the GP algorithm, we are not satisfied with the quantitative results. The nature of this particular fitness function is certainly an important factor hindering fitness estimation. Since this is something we have to cope with, we are trying to improve the tools used for the statistical analysis. In this sense, machine learning techniques such as artificial neural networks are appealing candidates.

Acknowledgments

Carlos Cotta is partially supported by the Spanish *Comisión Interministerial de Ciencia y Tecnología* (CICYT) under grant TIC99-0754-C03-03.

References

1. E. Alba, C. Cotta, and J.M. Troya. Evolutionary design of fuzzy logic controllers using strongly-typed GP. *Mathware & Soft Computing*, 6:109–124, 1999.
2. Th. Bäck. *Evolutionary Algorithms in Theory and Practice*. Oxford University Press, New York, 1996.
3. J.E. Beasley. OR-library: Distributing test problems by electronic mail. *Journal of the Operational Research Society*, 41(11):1069–1072, 1990. http://mscmga.ms.ic.ac.uk/info.html.
4. E. Cantú-Paz and D.E. Goldberg. *Parallel genetic algorithms: theory and practice*. Computer Methods in Applied Mechanics and Engineering. Elsevier, New York, 2000.
5. C. Cotta and J.M. Troya. Genetic forma recombination in permutation flowshop problems. *Evolutionary Computation*, 6 (1):25–44, 1998.
6. J.J. Grefenstette. Predictive models using fitness distributions of genetic operators. In L.D. Whitley and M.D. Vose, editors, *Foundations of Genetic Algorithms III*, pages 139–161, San Mateo, CA, 1995. Morgan Kaufmann.
7. J.J. Grefenstette. Virtual genetic algorithms: First results. Technical Report AIC-95-013, Navy Center for Applied Research in Artificial Intelligence, 1995.
8. J.H. Holland. *Adaptation in Natural and Artificial Systems*. University of Michigan Press, Ann Harbor, 1975.
9. T.C. Jones. *Evolutionary Algorithms, Fitness Landscapes and Search*. PhD thesis, University of New Mexico, 1995.
10. P. Merz and B. Freisleben. Fitness landscapes and memetic algorithm design. In D. Corne, M. Dorigo, and F. Glover, editors, *New Ideas in Optimization*, pages 245–260. McGraw-Hill, 1999.
11. R. Nogueras and C. Cotta. Using statistical techniques to predict GA performance. In J. Mira and A. Prieto, editors, *Connectionist Models of Neurons, Learning Processes, and Artificial Intelligence*, volume 2084 of *Lecture Notes in Computer Science*, pages 709–716. Springer-Verlag, 2001.
12. N.J. Radcliffe. Equivalence class analysis of genetic algorithms. *Complex Systems*, 5:183–205, 1991.
13. N.J. Radcliffe. Fitness variance of formae and performance prediction. In L.D. Whitley and M.D. Vose, editors, *Foundations of Genetic Algorithms 3*, pages 51–72, San Mateo CA, 1994. Morgan Kauffman.
14. H.J. Sussmann. From the brachystochrone to the maximum principle. In *Proceedings of the 35th IEEE Conference on Decision and Control*, pages 1588–1594, New York NY, 1996. IEEE Publications.
15. R. Tanese. Distributed genetic algorithms. In J.D. Schaffer, editor, *Proceedings of the Third International Conference on Genetic Algorithms*, pages 434–439, San Mateo, CA, 1989. Morgan Kaufmann.
16. D.H. Wolpert and W.G. Macready. No free lunch theorems for optimization. *IEEE Transactions on Evolutionary Computation*, 1(1):67–82, 1997.

Part VII

Prediction, Design and Diagnosis

A Comparative Assessment on the Application of Knowledge-based Networks and Genetic Algorithms to the Design of Fuzzy Diagnosis Systems for Rotating Machinery

Jesús Manuel Fernández Salido[1] and Shuta Murakami[1],

[1]Kyushu Institute of Technology, Department of Computer Science, 1-1 Sensui-Cho, Tobata-Ku, Kitakyushu-Shi, 804-8550 Japan

Abstract. In this paper, we approach the problem of automatically designing fuzzy diagnosis rules for rotating machinery, which can give an appropriate evaluation of the vibration data measured in the target machines. In particular, we explain the implementation to this aim and analyse the advantages and drawbacks of two Soft Computing techniques: Knowledge-Based Networks (KBN) and Genetic Algorithms (GA). An application of both techniques is evaluated on the same case study, giving special emphasis to their performance in terms of classification success and computation time.

1 Introduction

Automatic fault diagnosis of industrial machinery is an area of research that has experimented a considerable growth in the last years. The main reason for this is certainly the great economic importance that adequate maintenance policies have in industry. This can be illustrated by the case example that bearing company SKF gives in a recent report [9]. Here, it is explained how the successful implantation of a monitoring program for early fault detection in a Swedish Paper mill has resulted in an estimated savings in potential lost production of USD 4 million in two years.

In the case of large, industrial rotating machinery, Vibration Analysis-based diagnosis is one of the most widely applied maintenance strategies. Many years of experience in this area have proved how vibration signals measured at critical points of a rotating machine can provide abundant information that can be used for the detection and diagnosis of its most common faults [10]. New advances in Signal Processing algorithms (including Envelope Analysis, Wavelet transforms, Wigner-Ville distributions, Short Fourier Transforms) [7] and novel schemes for Feature Extraction have dramatically increased the diagnosis accuracy that can be obtained from the study of machinery vibration, but, at the same time, a higher level of expertise is also required for the interpretation of this data. As a consequence, this field has recently become an attractive ground for the application of several Soft Computing techniques. The use of Neural Networks [7] or Fuzzy Rules-based systems[5][11], with their ability to process noisy and imprecise data, can be specially adequate for the resolution of this problem.

If a rotating machine is suffering some kind of fault, several of the vibration parameters measured in it will experiment significant changes from their reference level. These changes will conform a pattern that can be called the *Fault Signature* of the machine. Usually, the Fault Signature will be different for every machine, even in the case of machines of the same type and working under similar working conditions, and are not always easy to detect. Neural Networks, with their innate ability to classify non-linearly separable objects, have been used early on for the detection and classification of Fault Signatures. The main inconvenience for their application is the need for a fairly large and representative amount of Vibration Data of the different fault states, in order to train the network. In the case of large, critical rotating machinery, this data will usually be scarce or not available at all, as the machine will not be allowed to operate in an unstable mode. Another well-known drawback of this paradigm is the "black box" nature of the most common type of networks, which cannot offer any explanation as to the diagnosis conclusions that have been reached.

On the other end, Rules-based Systems like Fuzzy Logic Systems can be programmed using only the Vibration Analysis Expert's knowledge. Although these diagnosis systems can be easily implemented in linguistic terms, they have less diagnosis accuracy than Neural Networks, and in some cases can only offer a general diagnosis, as they cannot always detect the specific Fault Signature of a machine. However, this type of Knowledge-Based systems may be the only available alternative when Fault Data cannot be collected.

Among the different Fuzzy Diagnosis Systems proposals [5][11], in [4], we suggested the implementation of a Fuzzy Logic based diagnosis system for rotating machinery using Fuzzy Pattern Matching (FPM) techniques, based on the papers of Dubois, Prade and Testemale [1]. The system could cope with several sources of uncertainty generated during the measurement process due to variations of the operating conditions of the machinery, and could diagnose the most general faults. In its early implementation, the application of Genetic Algorithms for tuning some of the system's parameters was also studied [4]. Although only vibration data measured in the machine in good state of operation was initially necessary for the implementation of the system, we also realized that it would be highly desirable if new diagnosis rules could be generated from vibration data once some kind of fault had occurred. In this way, the diagnosis system's knowledge-base would gradually adapt itself to the real fault characteristics of the target machinery, as fault data was being accumulated.

It was in more recent papers [2][3], that we explored the abilities of Knowledge-Based Networks [6] for the generation of new diagnosis rules out of the measured vibration data. These networks have the advantage, compared to other fuzzy-rules learning systems like neurofuzzy techniques or other types of fuzzy neural networks, that the learning is concentrated on the structure of the rules and the nature of the connectives, and not so much on the membership functions. We felt that, as in Industrial Diagnosis Systems there is usually plenty of data available for the machinery in its normal state of operation, these membership functions could be constructed using this data with statistical techniques, and that the learning process should be concentrated on the structure of the rules and the nature of the connectives.

Although the diagnosis results obtained using Knowledge-Based Networks were satisfactory, we realized that an adequate implementation of Genetic Algorithms could also have been used for the same purpose. This is the contribution of this paper, in which we discuss and compare both strategies on this framework, with which we hope to provide a better understanding as to the utility of applying connectionist or evolutionary approaches in technical diagnosis.

To do this, after reviewing the basics of the Fuzzy Diagnosis System that was proposed in previous papers, we shall study the characteristics and the implementation difficulties of both techniques, stressing their different advantages and drawbacks towards the extraction of fuzzy diagnosis rules. As an application example, we shall also discuss the specific implementation of both strategies in the same case study and examine their performance in terms of diagnosis accuracy, computational cost and complexity of the extracted knowledge. The paper will close with some concluding remarks, in which a comparative summary of the obtained results is compiled.

2 Structure of the Fuzzy Diagnosis System

In this section, we shall briefly discuss the basics of the Fuzzy Diagnosis System that we proposed in [4]. The structure of the Fuzzy Diagnosis rules to be generated are also overviewed.

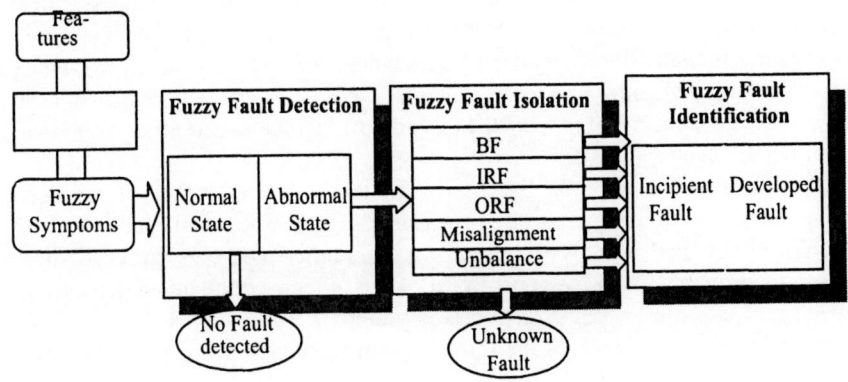

Fig.1. Structure of the Fuzzy Diagnosis System

The basic diagnosis process has been expressed graphically in Fig. 1. Here, the set of features measured in the machine's vibration signal, F', is fuzzified and transformed into a set of *measured* fuzzy symptoms, S' (with triangular shape membership functions). Diagnosis will consist in determining whether set S', matches the knowledge previously implemented in the diagnosis system (described by the fuzzy variables S) about the evolution of the different faults under study. This is to be done using the three classic steps of diagnosis:

1. **Fault Detection:** Determining if something is wrong in the system.
2. **Fault Isolation:** Establishing which fault has been detected.
3. **Fault Identification:** Measuring the size of the fault.

All three processes are implemented through a Fuzzy Pattern Matching (FPM) technique, that is, the set of measured fuzzy symptoms is matched against linguistic patterns that describe the possible state of the measured symptoms. There will be patterns describing the machine in normal or abnormal condition (for fault detection), the different faults that may be diagnosed (for fault isolation), and different degree of development of these faults (for fault identification).

In general, every pattern will be a set of requirement (rules), expressed in one of the following ways:

1. Not Quantified Requirements:
Req. j: $(S_k$ is $A_a)$ [OR $(S_l$ is $A_b)$ [OR $(S_m$ is $A_c)...$]]   (1)
Example:
(Vibration Level (5-40 KHz) is increased) OR *(Vibration Level (0-1 KHz) is increased)*

2. Quantified Requirements:
Req. j: $(Q_k\,S_k$ are $A_a)$ [OR $(Q_l S_l$ are $A_b)...$]]   (2)
Example:
*(**At least Some** Rotating Frequency Harmonics (Spectrum) are Markedly Increased)*

Here, the $S_k, S_l, S_m ...$ are symptom variables, while $A_a, A_b, A_c ...$ express different fuzzy attributes for these symptoms (*normal, increased, slightly increased...*), (which are modeled by triangular or trapezoidal membership functions), and Q_k, $Q_l...$ are Fuzzy Quantifiers (*some, at least some...*). Every requirement will also be assigned a weight that reflects the importance that the expert gives to it towards the final diagnosis.

The set of measured Fuzzy Symptoms, S', is matched with each of the requirements of a pattern using possibility measures. A global matching index for every pattern is obtained by aggregating all the possibility measures corresponding to every requirement using the generalized means, which can be tuned depending on whether a sensible or robust diagnosis is preferred. The patterns with the highest matching indexes for each of the three steps of the global process will be the output of diagnosis.

In this way, the mathematical structure of FPM can be used to model the expert's knowledge as a set of linguistic patterns and the knowledge base can be implemented in a simple fashion.

3   KBN for the construction of new Diagnosis Rules

Among the different hybrid paradigms for the construction of fuzzy rules using neural learning, Knowledge-Based Networks can be of great utility when the learning of the rules structure is preeminent over other factors.

In this paper, we use the term *Knowledge-Based Networks* to describe those Neural Networks whose neurons are not standard lineal or sigmoid functions, but weighted Fuzzy Connectives. Basically, these can be OR neurons and AND neurons, although when compensated connectives like the generalized mean are used, they will have an in-between meaning. This term was originally used in this sense in the fuzzy literature by Pedrycz and Hirota [6], who suggested their utility for modeling real life Decision Making processes. Other authors also refer to these networks under the denomination of *Hybrid Networks*.

The selection of the fuzzy aggregators to be used in these networks can be very flexible. The only condition to be imposed in this respect is that they should be derivable, if the network is to be trained using a backpropagation-like algorithm. The delta rule can then be modified to include the nature of the connectives.

An example of a Knowledge-Based Network, which we shall use to learn the requirements of the linguistic patterns described by eq.1 and eq. 2 is shown in Fig. 2. The numeric values that shall be the inputs to this kind of network will be the indexes obtained after matching, using possibility measures, each measured fuzzy Symptom S_j' with the corresponding expressions "S_j is A_k" that are initially considered. Thus, if at first n S_j' have been computed, whose measured state may be expressed in a partition of m attributes, then the network will have an input of n x m elements. The membership functions that define these m attributes are initially designed according to the statistical distributions of the corresponding symptoms measured under nominal operating conditions.

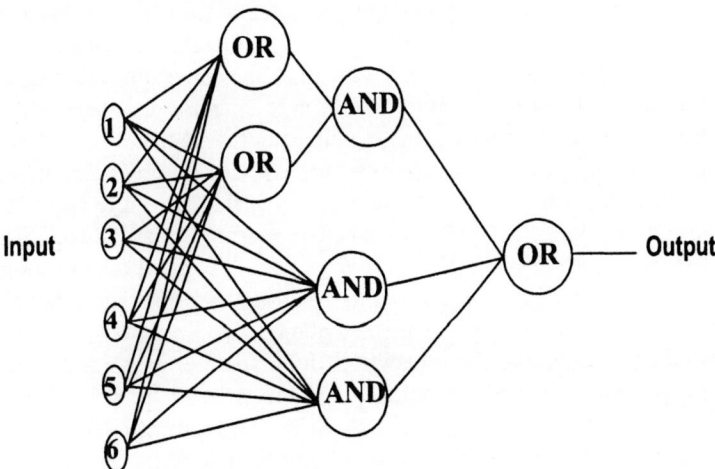

Fig.2. Example of a Knowledge-Based Network

As to the choice of operators for the implementation of the fuzzy OR and AND connectives, we have opted for Hurwicz's compensatory operator, whose output is a simple linear combination of the weighted minimum and maximum aggregators:

$$G(\bar{x},\bar{w}) = a \cdot Max(\bar{x},\bar{w}) + (1-a) \cdot Min(\bar{x},\bar{w}) \qquad (3)$$

Here, a will be a parameter used to tune the nature of the connective. Its value is set in the interval [0.5, 1] for OR-like neurons or in [0, 0.5] for AND-like neurons. $\bar{x}$ is a vector of input values, while $\bar{w}$ represents their associated weights.

The weighted versions of the minimum and maximum operators can be implemented through the weighted generalized means and its De Morgan's dual:

$$Max(\overline{x}, \overline{w}) = \left(\frac{\sum_i w_i^p x_i^p}{\sum_i w_i^p} \right)^{1/p} \tag{4}$$

$$Min(\overline{x}, \overline{w}) = 1 - \left(\frac{\sum_i w_i^p (1 - x_i)^p}{\sum_i w_i^p} \right)^{1/p} \tag{5}$$

Parameter p is constant and given a high enough value, for example $p=1000$, in order to assure the convergence of these operators to their non weighted counterparts in case that $w_1 = w_2 = \ldots = w_n = 1/n$.

It can be easily seen that neuron connectives defined in this way have continuous derivatives in respect to x_i, w_i and a. Therefore, they can be used in backpropagation and other algorithms based on gradient-descent As the value of a is also adapted through learning, the nature of the connectives will change during the training process. If the network is designed using neurons of a predefined type (AND-like or OR-like), as is the case of Fig. 1, the learning algorithm will have to check that a remains in the corresponding value interval for that neuron. It should be noted as well that, in most classification tasks a could take a constant value of 1 for OR neurons and 0 for AND neurons, without being subject to learning.

It must be stressed that the learning rate of every parameter is very dependent on the nature of its corresponding aggregation neuron. As, unlike in networks composed of sigmoid neurons, every neuron possesses different characteristics, the optimal learning rate for every parameter will usually be different. For this reason, the introduction of a system of adaptive learning rates, as suggested by Pedrycz, Lam an Rocha [8] is highly recommended. All of these factors contribute for a significant increase in complexity of the learning algorithm and bring about a higher computational cost for their implementation.

The advantage of applying these networks, however, lies in the fact that, after the training process has been completed and the network is pruned of those branches with low weights, the resulting structure can be interpreted as a collection of fuzzy rules. Depending on the difficulty of the diagnosis case and the type of training data, the complexity of the extracted fuzzy diagnosis rules will vary. However, the application of gradient-descent based learning algorithms for a sufficiently long time will result in many branches of rules with limited diagnosis capacity attaining very low values. Thus, a network trained in such a way will in many cases represent a simple set of fuzzy rules from which knowledge can be easily extracted. The number of rules obtained will depend on the complexity of the training data, and of the quantity of neurons which have initially been set in the network.

In regards to the problems associated with the learning algorithm, these are essentially the same as in other gradient descent–based networks. The optimal structure that gives the smallest classification error will rarely be obtained, as the network's learning will usually be trapped at some point in a local optimum. This effect can be alleviated to a certain extent if techniques like simulated annealing are incorporated to the training algorithm, with an additional augmentation of implementation complexity. Another strategy to cope with this problem would be to increase the number of neurons of the network, which gives the learning algorithm extra space to learn new rules even when some of its neurons have been held at some local minima. Our experience dictates that a good number of neurons for a network can lie between 4 or 5 times the expected number of fuzzy rules. However, the decision for the optimal number of neurons still remains mainly a trial and error process.

From these arguments, it can be inferred that the implementation of Knowledge-Based Network for diagnosis is not simple, and that the tuning of its parameters can be a hard and time consuming work. In the next section, we shall show how the application of evolutionary strategies can greatly simplify the implementation process, at the expense of a higher computational cost.

4   GA for the construction of new Diagnosis Rules

Due to the constant improvement of computational hardware, a growing number of applications of evolutionary computation are claiming their place in a wide range of optimization problems. In particular, Genetic Algorithms with standard selection, crossover and mutation operators can usually be implemented in many situations in a direct manner, without much effort spent into parameter tuning. In this section, we shall attempt to show how this technique can be applied for the construction of fuzzy diagnosis rules, and analyze some of its main characteristics. As the mechanisms of the standard Genetic Algorithms implementations are well known, we shall concentrate on how the structure of our Fuzzy Pattern Matching rules and the connectives' parameters encoded for GA optimization. Some insights will also be given as to the nature of the constructed rules using this technique, and how evaluation functions should be used.

A first, straightforward approach to this would be to apply the learning mechanism of GA to the weight structure of the equivalent Knowledge-Based Network, with each weight being codified as a gene. Every chromosome of the GA population would then be composed of all the genes representing the KBN's weights and those genes codifying the a parameters used in equation (3), that are necessary to model the nature of the neurons. If we consider a basic KBN with a first input layer of l components, a second layer of m OR-like neurons, a third layer of n AND-like neurons and a fourth layer composed of a single OR-like neuron, then every chromosome would be composed of $(l \times m + m \times n + n)$ genes encoding the weights structure and $(m + n + 1)$ genes that represent the a_i parameters for the connectives, as is expressed in Fig. 3(a).

The obvious problem of this strategy is that in Vibration Analysis the number of features that can be obtained from the measured signals is usually very high, as many harmonics and sidebands can be extracted from the related signals' spectra at different frequencies of interest. Even with the inclusion of an·input feature reduction algorithm (As we did in [3], with the application of Fuzzy Rough Sets theory), after the matching process of every feature with its corresponding fuzzy attributes, 100-200 input values to the network can be had. The chromosome's length would then be extremely long and the required computation times for the algorithm to attain convergence would make this approach unfeasible. In order to avoid this, the number of weights that attain a non null value in the input layer can be limited to a fixed quantity, N, for every neuron of the second layer. The N weights of the input layer to be selected can be identified by an index, so that, for this layer, only the values of N indexes and their respective weight values have to implemented as genes in every chromosome. This shortens substantially the chromosome's length to $(N \times m + m \times n + n) + (N \times m) + (m + n + 1)$ genes. Of course, the bit length will also depend on the discretization rate adopted, which, for the sake of computational efficiency, cannot be set to high. This reduced approach for gene codification is represented in Fig.3 (b) for a case of $N=4$.

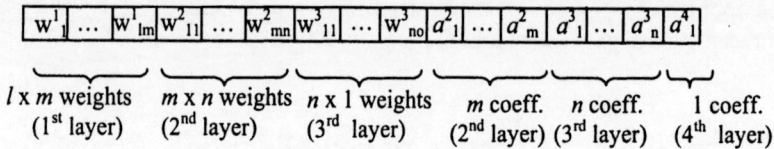

$l \times m$ weights $m \times n$ weights $n \times 1$ weights m coeff. n coeff.   1 coeff.
(1st layer)   (2nd layer)   (3rd layer)   (2nd layer) (3rd layer) (4th layer)

(a) Direct Approach

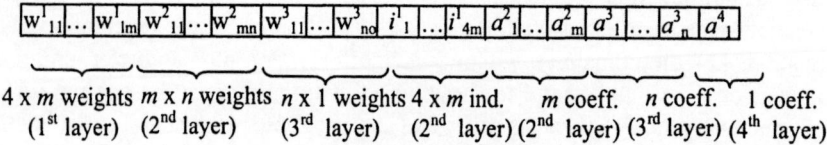

$4 \times m$ weights $m \times n$ weights $n \times 1$ weights $4 \times m$ ind. m coeff. n coeff.   1 coeff.
(1st layer)   (2nd layer)   (3rd layer)   (2nd layer) (2nd layer) (3rd layer) (4th layer)

(b) Reduced Approach

Fig.3. Two approaches for Gene Codification

As GA performs a global search, it is much less vulnerable than KBN to stagnate its learning process at a local optimum, and will give better classification results at the cost of a much higher computation time. In the experiments that we have conducted, this is usually 3 to 6 times higher (depending on the number of samples) than the computation time that Knowledge-Based Network requires. Furthermore, this will escalate with the number of samples available for training, as all of them must be evaluated for every chromosome in the population. Therefore, if the number of training samples is very high, these should have to be reduced to a representative group, running the risk of discarding some important information. In this aspect, the performance of GA is weaker than that of KBN.

Although an evolutionary search usually results in a better classification rate, there are also other disadvantages associated with the global character of it. One of these is that, unlike gradient-descent based techniques, the weights related with features with limited diagnostic capability do not have to attain a low value after the learning process has been completed, as the search is being done in the whole parameter space. A consequence of this is the extraction of fuzzy rules that can be much more complex than in the KBN case, with the appearance of many symptoms that do not make a significant contribution towards the diagnosis decision. Knowledge extraction from such rules will be a more difficult process and the diagnosis knowledge base will become less intuitive. As the application example will make clear, this can be alleviated with a modification of the evaluation function (which, originally, is usually the classification error). In it, a term that measures the simplicity of the extracted rules can be incorporated.

5 Application Example

In order to measure the characteristics and performance of both techniques in the same benchmark, we have chosen to apply them in the vibration data measured at the transmission system of a pump in a pumping station at The Netherlands. The main advantage of using these signals is that they have been made public by Delft University of Technology, and, therefore, are available for comparison with other techniques. This dataset was acquired in the Delft "Machine diagnostics by neural networks"-project with help from TechnoFysica B.V, The Netherlands, and can be downloaded freely at the following web-address:

http://www.ph.tn.tudelft.nl/~ypma/mechanical.html

At this point, we want to express our gratitude to the owners of this dataset.

The machine measured in the data files is a pump driven by an electromotor. The incoming shaft is reduced in speed by two delaying gear-combinations, in which a progressed pitting in both gears has been produced. Channel 3 from the available signals, which correspond to the machine in good state of operation and the machine with the gear fault, was used. Furthermore, the signals have been obtained in for both states in two modes: minimum and maximum load mode.

From this dataset, we have constructed 304 vibration samples for the gear mechanism in good state of operation, and other 304 samples for the mechanism with the gear flaw. Out of these, half are meant to be used for training, and the remaining half is for evaluation. All samples are 512 points long, and both load states have been included, in order to increase the complexity of the problem and to obtain a flexible diagnosis towards changes in the measurement conditions.

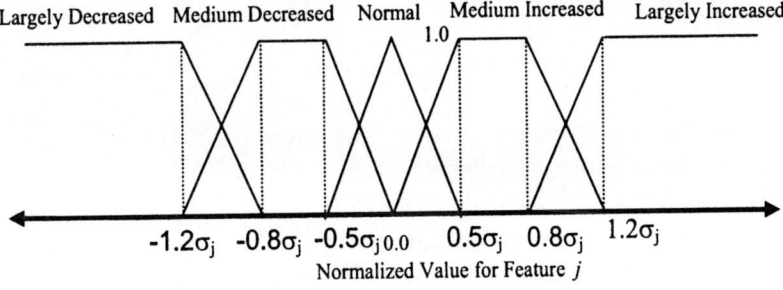

Fig.4. Fuzzy Partition for each feature

In each sample, 28 vibration features were estimated in the time and frequency domain. These were normalized relative to their mean in low load, no fault condition. They were then fuzzified and represented by degrees of membership to a partition of 5 linguistic labels designed using each feature's standard deviation in low load, no fault condition (Fig. 4). Thus, each sample has $28 \times 5 = 140$ values.

In what follows, we present the application results in this dataset using the techniques that are object of this paper, and point out their implementation details.

5.1 Knowledge-Based Networks

In order to train a Knowledge-Based Network with this dataset, the input features were divided into three groups: a group of 13 features measured in the time domain (Kurtosis, Crest Factor, Peaks Mean, etc), another group of 5 features measured in the frequency spectrum (frequency mean, frequency standard deviation, vibration levels...) and a third one of 10 features, also extracted from the frequency spectrum, but composed exclusively of peaks measured at certain frequencies of interest. The separation of features in groups was done in order to prevent the formation of rules containing symptoms of unrelated domains.

These three feature groups become three input layers (with 13x5, 5x5 and 10x5 inputs) to a KBN structure like the one represented in Fig. 2: its input layers are connected to an initial OR layer (12 neurons, 4 for each input group) and AND layer (also 12 neurons). The OR neurons of this first layer are further connected to a second layer of 4 AND neurons. The output of the network is given by a single OR neuron, which collects the results of the first and second AND layers. The nature of these neurons is fixed ($a=1$ for OR neurons and $a=0$ for AND neurons).

The network was trained with a backpropagation-like algorithm (which included the derivatives of our OR and AND neurons, expressed by Eqs. 3, 4, and 5), with momentum for weights training (momentum constant = 0.2). The learning rates were initially set to 0.005 for weights training. A mechanism for learning rates adaptation, as suggested by Pedrycz et al [8] was also implemented.

Learning for the KBN converged after approximately 20 epochs, each one requiring a calculation time of 120 seconds in an Athlon 700 MHz PC. After training, the network was normalized and pruned of those branches with weights lower than 0.4. It could be reduced to the following diagnosis rules:

[[Peaks Mean is *largely decreased* (weight: 1) OR
        Peaks Max is *largely increased* (weight: 0.533)]   (weight: 1)
AND
        [Vib. Level (10 Hz-5Khz) is *markedly decreased*]   (weight: 0.971)
AND
        [Freq. Mean is *markedly increased*   (weight: 0.853) OR
        Vib. Level (10 Hz-1Khz) is *mark. decreased* (weight: 1)] (weight: 0.969)]
OR   [Peak at 218 KHz is *markedly decreased*]   (weight: 0.501)]
THEN   A Gear Fault has been detected.

Due to the position of the selected sensor, the vibration levels were actually lower in the faulty samples than in the samples of the pump in good condition. The constructed rules are useful to point this fact.

These diagnosis rules were used in the 304 samples reserved for evaluation, with 286 samples correctly classified, giving a success ratio of 94.07%.

5.2 Genetic Algorithms

The weights of a KBN with the same structure as above were codified using the reduced coding approach developed in Section 4 (four weights for first layer neurons). With this, 160 three-bits genes were used for the weights codification, and 96 six-bit genes were needed for the index codifications. As to the evaluation function, the classification error was adopted. A population of 101 chromosomes was let to evolve in a basic GA scheme, with a mutation rate of 0.0077, and crossover rate of 0.77. After 70 generations (each one 160 seconds), the algorithm converged. The optimal gene codified a complex diagnosis system, which is not reproduced. It classified correctly 296 out of the 304 evaluation samples (97.36%).

Obviously, this improvement also implied a considerable increase in the complexity of the knowledge base. In order to alleviate this, the following evaluation function, which penalizes the formation of long rules, was introduced:

$$f(chromosome) = -\frac{N_S^{CC}(chrom.)}{N_S} + 0.15\frac{N_F^{LL}(chrom.)}{N_F^{LL}} \qquad (4)$$

Here, N_S stands for the number of evaluation samples and N_S^{CC} is the number of correctly classified samples. N_F^{LL} is the maximum number of features in the last layer, and $N_F^{LL}(chrom)$ is the number of features in the last layer of the chromosome's corresponding KBN. With this new learning approach, a much more simple diagnosis scheme of 1 rule including 3 symptoms was obtained:

IF   [St. Deviation is *markedly decreased* (weight: 0.571)]   AND
     [Vib. Level (10 Hz-5Khz) is *markedly decreased*] (weight: 1)   AND
     [Crest Factor is *markedly decreased*] (weight: 0.428)
THEN   A Gear Fault has been detected.

The classification success rate was 96.38% (293 correctly classified samples).

Concluding Remarks

In this paper, we have dealt with the issue of extracting fuzzy diagnosis knowledge from vibration data measured in rotating machinery. For this, two different Soft Computing strategies, one based on gradient-descent learning and another one that uses the concept of evolutionary adaptation, have been applied. From our theoretical study and the results of the application example, this comparative analysis can be summarized in the following points:

- **Implementation**: Undoubtedly, the implementation of KBN requires more programming effort than GA (for which many software packages are available) and the tuning of its parameters can be a difficult task.
- **Computing Time**: The computing time required by GA will be 5-7 times higher than when KBN are applied. This factor augments sensibly with the number of training samples

- **Complexity of the Extracted Knowledge:** The rules extracted by a KBN paradigm tend to be much more simple than the ones learned by GA. In order to simplify the structure of the extracted rules, the evaluation function (originally based on the classification error) must be modified with terms that penalize the formation of complex rules

- **Diagnosis Performance:** The better search capabilities of GA will result in a more accurate diagnosis compared to KBN. However, if limits on rule complexity are imposed, this will be handicapped.

These results lead us to believe that, in most cases, KBN will give an acceptable performance for the extraction of fuzzy knowledge. The use of GA is more appropriate when diagnosis accuracy is paramount and when the related computational cost can be undertaken.

References

1. Dubois D., Prade H. and Testemale C. (1988) Weighted Fuzzy Pattern Matching. Fuzzy Sets and Systems **28**, 313-331
2. Fernández J.M. and Murakami S. (2000) Construction of Fuzzy Diagnosis Rules for Rotating Machinery using Vibration Analysis and Knowledge-Based Networks. 6th International Conference on Soft Computing IIZUKA 2000, Iizuka, Japan,
3. Fernández J.M. and Murakami S. (2000) A Proposal for the Automatic Design of Fuzzy Diagnosis Systems for Rotating Machinery. Proceedings of COMADEM'2000, 13th International Congress on Condition Monitoring and Diagnostic Engineering Management, Houston, Texas, USA, 755-764
4. Fernández J.M. and Murakami S. (2001) Application of Fuzzy Pattern Matching and Genetic Algorithms to Rotating Machinery Diagnosis. Practical Applications of Soft Computing in Industry, edited by Sung Bae-Cho, World Scientific, Singapore, 255-285.
5. Frank, P.M. and Marcu, T. (1999) Fuzzy Techniques in Fault Detection, Isolation and Diagnosis. Fuzzy Logic Control-Advances in Applications, edited by H.B. Verbruggen and R.Babuska, World Scientific, Singapore 239-258.
6. Hirota, K. and Pedrycz, W. (1993) Knowledge-based networks in classification problems. Fuzzy Sets and Systems **59**, 271-279.
7. Paya, B.A., Esat, I.I. and Badi, M.N.M. (1997) Artificial Neural Network Based Fault Diagnosis of Rotating Machinery using Wavelet Transforms as a Preprocessor. Mechanical Systems and Signal Processing **11**(5), 751-765.
8. Pedrycz, W., Lam, P.C.F and Rocha, A. (1995) , Distributed Fuzzy System Modeling. ,IEEE Transactions on Systems, Man and Cybernetics **25**, 769-780.
9. SKF Group (2000) Paper mills gains from condition monitoring. Evolution: Business and Technology Magazine from SKF. March Issue.
10. Renwick J.T. and Babson P.E. (1985) Vibration Analysis –A proven Technique as a Predictive Maintenance Tool. IEEE Transactions on Industry Applications **21**(2), 324-330
11. Ulieru, U. and Iserman, I. (1993) Design of a Fuzzy-Logic based Diagnostic Model for Technical Processess. Fuzzy Sets and Systems **58**, 249-259.

Evolution of a Tactile Wall-Following Behavior in Real Time

Frank Hoffmann[1] and Juan C. S. Zagal Montealegre[1]

Royal Institute of Technology, NADA/CVAP
10044 Stockholm, Sweden
E-mail: {hoffmann,juan}@nada.kth.se,
WWW home page: http://www.nada.kth.se/{~hoffmann,~juan}

Abstract. This paper describes the evolution of a simple, reactive wall-following behavior. The robot perceives its environment by means of tactile sensors and is supposed to travel along a looping maze as smoothly as possible. A genetic algorithm learns the mapping from sensory information to motor actions. Candidate behaviors are executed on a LEGO Mindstorm robot and the distance covered during the trial serves as a fitness measure.

1 Introduction

Evolutionary robotics is a design paradigm to adapt robotic behaviors by means of simulated evolution[12]. The basic idea it to use evolutionary computation in order to automatically synthesizes robot controllers that exhibit a useful behavior in a complex environments. The evolutionary algorithm searches through space of possible mappings from sensory perceptions to control actions. A population of candidate controllers is processed from one generation to the next by means of selection, recombination and mutation. A scalar fitness function describes the performance of the controller with respect to the task assigned to the robot.

In practice, the evaluation of the controller often takes place in a simulator of the robot and the environment [9]. The underlying assumption of this approach is that the simulation is authentic enough so that controllers once transfered to the real robot exhibit a similar performance. It has been demonstrated in previous work that controllers are successful on the real robot, if a carefully constructed and validated simulation contains the right amount of noise [13]. JAKOBI proposed a a framework called "minimal simulation" for the construction of fast-running robotic simulators [10]. The basic idea is to model only those aspects that are highly relevant to the behavior in mind and thereby to assure that the evolved controllers are more robust against uncertainty and incomplete information inherent to real world situations.

The alternative approach is to evolve behaviors directly on the robot itself as the world is its best own model [3]. Numerous authors report experiments in which the evolutionary learning takes place on the real robot [4,2]. In [5], the authors describe the evolution of a wall-following behavior in which a miniature robot perceives its environment by means of infra-red proximity sensors. In [2], the authors design an obstacle avoidance controller for the same miniature robot by means of genetic programming. Their real-time learning approach overcomes the problem of limited evaluation time, by allowing the system to learn from past experiences stored in memory. Our approach is different in that the robot possesses no remote sensing capabilities, but only relies on tactile sensors.

2 Genetic Algorithm

Evolutionary algorithms provide a universal optimization technique that mimics the type of genetic adaptation that occurs in natural evolution [7]. Unlike specialized methods designed for particular types of optimization tasks, they require no particular knowledge about the problem structure other than the objective function itself.

A population of competing candidate solutions evolves over time by means of genetic operators such as mutation, recombination and selection. Genetic algorithms operate on binary strings $s = \{s_1, \ldots, s_N\}$ that are mapped into potential solutions to the optimization problem. A scalar objective function $f(\{s_1, \ldots, s_N\})$ evaluates the quality of a candidate string $\{s_1, \ldots, s_N\}$. The role of selection is to exploit good candidate solutions by allowing those strings that achieve a high fitness to reproduce offspring. Crossover cuts and splices two parent strings in order to generate new variants. Mutation randomly alters individual and thereby maintains the genetic diversity in the population. The interplay of exploiting good solutions via selection and exploring the search space in form of crossover and mutation constitutes the fundamental theme in evolutionary optimization. The quality of solutions improves gradually as a result of this basic cycle of selection, reproduction, recombination and mutation.

3 Robotic Architecture

The experiments described in this paper use the LEGO Mindstorm robotic platform. The tactile wall-following behavior was down-loaded to the Robotic Controller X (RCX), a a programmable micro-controller packaged in a palm-size LEGO brick. The RCX can receive input from up to three touch, light or rotation sensors and is able to control up to three outputs connected to electric motors. A resistor network proposed in [6] mimics a simple digital to analog converter which allows it to simultaneously connect three touch sensors to a single RCX input. This three touch sensor expander allows the RCX to monitor six touch sensors and a rotation sensor.

The RCX communicates with a remote PC via an infrared serial link. This feature is used in our experiments to down-load the behavior parameters from the PC to the RCX and to up-load the scalar fitness value from the RCX to the PC upon execution of a behavior.

Figure 1 shows a top-view of the robot used in the experiments. The robot has two flexible antennas, two touch sensors per antenna detect whether the antenna bends to the left or right. In addition a left, central and right front bumper are connected to two touch sensors that detect collisions with object. The left and right motor of the robot drive the front and rear wheel at both sides. The motor direction

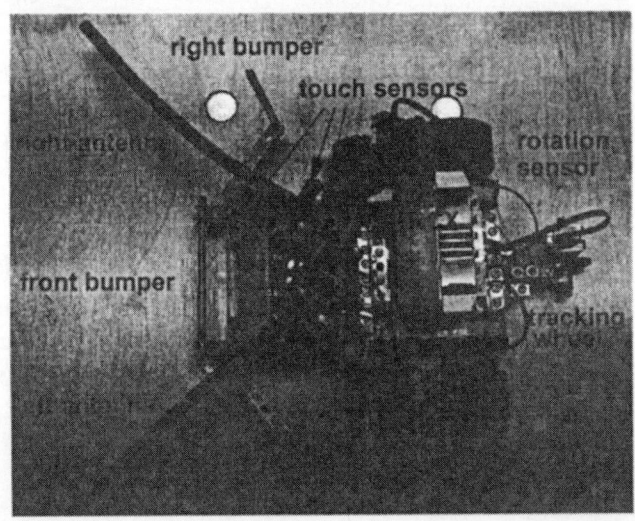

Fig. 1. Robot with antennas, bumpers, touch sensors and rotation sensor.

of the left and right wheel pair can be set to forward, reverse or floating mode during which the motor operates with zero torque.

By counting the rotations of a tracking wheel in the robot's tail distance traveled by the robot can be computed.

The robot behavior is almost purely reactive, the RCX reads the sensor states and based on these the controller sends a motor command. This motor command is executed for a specified time period until the controller captures a new sensor state. In other words, changes in the sensor states are ignored during the execution of the current motor command. The duration of the motor command can be different for left and right motor. This feature enables the robot to rotate by an amount that is proportional to the time delay between left and right motor action.

4 Learning a Tactile Wall-Following Behavior

4.1 Behavior Representation

The robot controller obtains input from six binary touch sensors, which means that in principle the perception space contains $2^6 = 64$ different sensory states. However, as the two antennas can only actuate either its left or right touch sensor, the actual number of possible sensory states is 32. Due to the small number of perceptual states the robot is often unable to distinguish between situations that require different control actions. It would be possible to augment the state space, by considering not only the current but also previous sensor readings. This would at least partially solve the problem of perceptual aliasing as the robot can better distinguish between different world states. The RCX imposes a severe technical restriction, in that it only provides 32 storage locations for variables.

The actual controller only discriminates among nine different sensor perceptions $S^1,\ldots,S^9$. Despite the small number of states the robot demonstrates a meaningful, albeit clumsy and sub-optimal wall-following behavior. Video captures of the evolved behavior are available online at [8].

In state S_1, no touch sensor is active, this corresponds to a situation in which the robot has no contact with walls or obstacles. A default action of both motors running in forward direction is assumed for this state. Imposing this default action greatly reduces the risk of evaluating a large number of inferior behaviors in the early generations. The states S_2, S_3, S_4 capture the bumper sensors and correspond to frontal bumper pressed, left bumper pressed and right bumper pressed. The role of the antennas is to provide some remote sensing capability, they are therefore only considered in case no bumper sensor is active [1] The states S_5, S_6, S_7, S_8 reflect situations in which the left respectively right antenna is bend inwards or outward. Finally, the state S_9 captures the situation in which both antennas are bend outwards.

As left and right motor can operate in forward, reverse or floating mode there are in principle nine possible motor actions available. However, setting left and right motor to floating mode causes the robot to stop. As the sensors would remain in their current state, the robot would never start moving again for the rest of trial. Therefore, only eight motor actions are considered which are encoded by three bits $s_1,\ldots,s_3$ per state S_i.

Five additional bits $s_4,\ldots,s_8$ per state specify the durations $\Delta t_l, \Delta t_r \in [0 - 70ms]$ of the left and right motor action. The longer a motor command is executed, the larger the distance traveled by the robot. In case both motors operates in reverse for example, the parameters $\Delta t_l, \Delta t_r$ determine how far the robot backs up. By choosing different values for $\Delta t_l, \Delta t_r$ the robot can execute more complex motion patterns within a single control action, such as backing up straight and then turning left.

The six bits encode the duration of motor actions in milliseconds in the following way.

$$\Delta t_l = s_4 \times 40 + s_5 \times 20 + s_6 \times 10$$
$$\Delta t_r = s_4 \times 40 + s_7 \times 20 + s_8 \times 10 \qquad (1)$$

The action associated to a particular state S_i is encoded by single byte. The behavior distinguishes between nine perceptual states, but since state S_1 has a default forward motor command, only the chromosome only encodes actions of states $S_2,\ldots,S_9$. Therefore, the entire chromosome has a length of $8x8 = 64$ bits.

The genetic algorithm employed fitness-proportionate selection with linear scaling, no elitism scheme, two-point crossover with a crossover probability of $p_c = 0.7$ and mutation with a mutation rate of $p_m = 0.01$ per bit. The population size is 10 individuals evolved over a course of 20 generations.

---

[1] In recent experiments not reported in this paper the robot is able to perceive two additional states corresponding to simultaneous contact with the antennas and the bumper. This allows the robot to distinguish between direct contacts with the bumpers and those activations caused by the antennas touching the vertical bars mounted at the tip of the side bumpers.

4.2 Fitness Function

The objective is to evolve a controller that maximizes the forward motion of the robot measured by the freely spinning wheel in the tail. Each candidate behavior down-loaded onto the robot and evaluated over a period of 50 seconds. Including the time needed for communication between the robot and the host computer that runs the genetic algorithm, a single fitness evaluation takes about one minute. A longer evaluation period would provide a less noisy and more accurate estimate of the behavior performance. The overall experimentation time is limited by the capacity of the batteries which last for about four hours.

After 50 seconds of operation the overall number of rotations of the tracking wheel is returned as a fitness value. Notice, that moving backward reduces fitness as the rotation counter decrements. When the robot turns on the spot the spinning wheel slides rather than rotates and therefore the rotation counter usually neither increments nor decrements during such a motion. The theoretical maximum fitness of about 1000 rotations is obtained if the robot moves straight forward in an obstacle-free environment over the entire trial. However, as the robot possesses no remote sensing capability the theoretical maximal fitness is never achieved as collisions with walls and obstacles in the maze force the robot to turn and back-up. The robot is able to minimize evasive manoeuvres by assigning appropriate motor actions to the sensor states. In addition the robot is partially able to determine which future sensory information it receives as the selected action effects its position in the environment and thereby indirectly its perception of the environment as well.

As mentioned previously, the limited number of perceptual states results in a substantial amount of ambiguity over the actual situation. The robot is unable to distinguish between world states that although generating the same perception require different control actions. This problem is well known in the robotic literature as *perceptual aliasing*. The problem of perceptual aliasing can be partially resolved by selecting control actions that minimize the likelihood of entering ambiguous world states. Another solution is to choose control actions that result in disambiguous future perceptions, a strategy known as active perception [1]. The evolved behavior seems to incorporate active perception to some extent, as the robot wiggles back and forth in an apparently erratic fashion when it gets stuck at an obstacle.

5 Experiments

A genetic algorithm evolved the 64-bit strings that encoded the mapping from eight possible sensor states to motor actions as described in section 4.1 with a population size of ten individuals evolved over twenty generations.

Each behavior is down-loaded on the LEGO Mindstorm robot and its performance is evaluated on the robot traveling through the race track of size $160x120cm$ shown in figure 2. In an earlier experiment the track contained only side walls but no other obstacles. Due to the homogeneous geometry of the original track the robot was able to successfully traverse the track using only the bumper sensors while ignoring the antennas.

Fig. 2. Track used in the experiments.

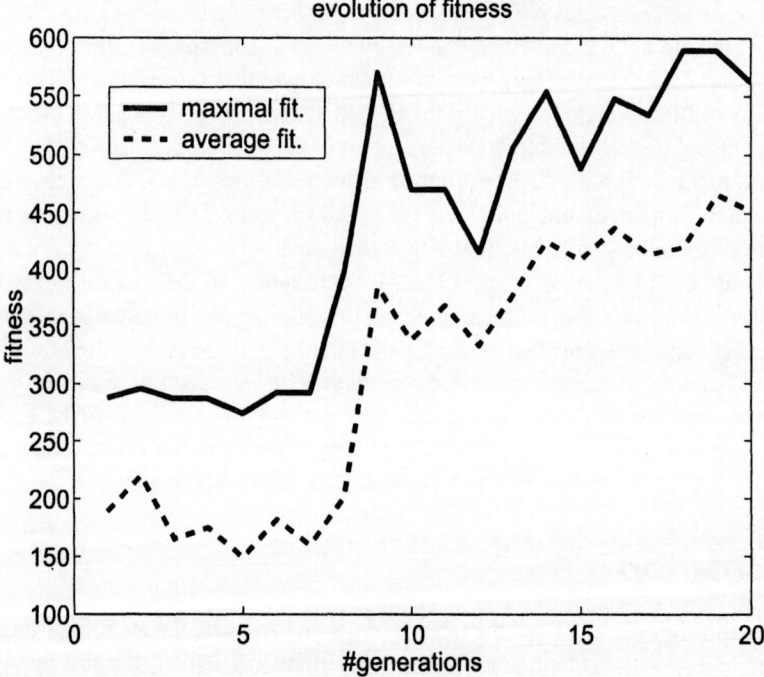

Fig. 3. Evolution of maximal and average fitness over time.

In order to provide the learning algorithm with a more challenging task, we increased the complexity of the environment by adding additional obstacles. Cans mounted at the sides narrowed the track making it more difficult for the robot to pass these segments. With the additional obstacles the robot gets stuck more easily and a successful wall-following behavior requires more complex pattern of motions.

Each behavior is executed on the robot over a period of 50 seconds after which the RCX transmits the number of rotations accumulated during the trial as the fitness to the genetic algorithm running on the host computer. Figure 3 shows the evolution of the average and maximal fitness with the number of generations. The best individual in the initial population achieved a a fitness of 288 rotations, the best overall behavior emerged in 18th generation and achieved 589 rotations, which corresponds to a traveled distance of about $3.5m$.

The mapping from sensor states $S_1, \ldots, S_9$ to motor actions of the best individual is shown in table 1. Notice, that in case of a left bumper collision the robot turns right (left forward, right motor in reverse) and vice versa turns right in case the right bumper is pressed. The robot reacts similar when it perceives obstacles with its tactile antennas.

We evaluated the manual and the evolutionary designed behavior in four trials of five minutes each in order to obtain a less noisy assessment of their actual quality. The manually designed wall-following behavior achieved an average performance of about 2202 rotations per five minute trial, the evolved behavior was slightly better achieving on average 2287 rotations per trial.

Movies that show the evolved behavior on the real robot are available at
http:\\www.nada.kth.se\~hoffmann\lego.html

| sensor state | left motor | | right motor | |
	direction	delay Δt_l	direction	delay Δt_r
S_1 :no contact	fwd	0 ms	fwd	0 ms
S_2 :front bumper	rev	50 ms	rev	50 ms
S_3 :left bumper	fwd	40 ms	rev	70 ms
S_4 :right bumper	rev	30 ms	fwd	30 ms
S_5 :left antenna outward	fwd	60 ms	fwd	60 ms
S_6 :left antenna inward	float	20 ms	rev	30 ms
S_7 :right antenna inward	rev	60 ms	fwd	70 ms
S_8 :right antenna outward	rev	70 ms	fwd	40 ms
S_9 :left & right antenna outward	fwd	20 ms	rev	10 ms

Table 1. Mapping from sensor states $S_1, \ldots, S_9$ to motor actions for the best behavior.

6 Conclusion and Future Work

This paper presented an evolutionary learning approach for the design of a tactile wall-following behavior. The robotic behavior evolved on the real robot demonstrated the same performance as the manually designed control scheme.

In evolutionary robotics evaluating the fitness on the real robot is extremely time-consuming, compared to which the computational effort associated with selection, crossover and mutation can be neglected. In the future we are going to investigate how a fitness model build from observations on the real robot can be used speed-up the learning process [2,11]. Such a model would allow the genetic algorithm to estimate the fitness of an individual prior to evaluation and to allocate new trials only to promising candidate behaviors.

Acknowledgments

This research presented in this paper has been sponsored by the Swedish Foundation for Strategic Research. We thank the reviewers for valuable comments on how to improve the paper.

References

1. R. Bajcsy. Active perception. *Proceedings of the IEEE*, 76(8):996–1005, 1988.
2. W. Banzhaf and P. Nordin. An on-line method to evolve behavior and to control a miniature robot in real time with genetic programming. *Adaptive Behaviour*, 5(2):107–40, 1997.
3. R.A. Brooks. New approaches to robotics. *Science*, 254:1227–1232, 1991.
4. D. Floreano and F. Mondada. Evolution of homing navigation in a real mobile robot. *IEEE Transactions on Systems, Man, and Cybernetics-Part B: Cybernetics*, 26(3):396–407, 1996.
5. D. Floreano and F. Mondada. Evolutionary neurocontrollers for autonomous mobile robots. *Neural Networks*, 11:1461–1478, 1998.
6. M. Gasperi. Mindstorms rcx sensor input page. http://www.plazaearth.com/usr/gasperi/lego.htm, 1999.
7. D.E. Goldberg. *Genetic Algorithms in Search, Optimization, and Machine Learning*. Addison-Wesley, 1989.
8. F. Hoffmann and J. C. S. Zagal Montealegre. Evolution of a tactile wall-following behavior in real time. http://www.nada.kth.se/ hoffmann/lego.html, 2001.
9. F. Hoffmann and G. Pfister. Evolutionary design of a fuzzy knowledge base for a mobile robot. *International Journal of Approximate Reasoning*, 17(4):447–469, 1997.
10. N. Jabobi. Half-baked, ad-hoc and noisy: Minimal simulations for evolutionary robotics. In *Proceedings of the Fourth European Conference on Artificial Life*, pages 348–357. MIT Press, 1997.
11. P.B. Nair M.A. El-Beltagy and A.J. Keane. Metamodeling techniques for evolutionary optimization of computationally expensive problems: Promises and limitations. In *GECCO-99 : Proceedings of the Genetic and Evolutionary Computation Conference*. Morgan Kaufmann Publishers, 1999.

12. S. Nolfi and D. Floreano. *Evolutionary Robotics – The Biology, Intelligence, and Technology of Self-Organizing Machines*. Intelligent Robotics and Autonomous Agents. MIT Press, 2000.

13. O. Miglino S. Nolfi, D. Floreano and F. Mondada. How to evolve autonomous robots: Different approaches in evolutionary robotics. In R. Brooks and P. Maes, editors, *Artificial Life IV*, pages 190–197. MIT Press/Bradford Books, 1994.

Fuzzy-Memetic Approach for Prediction of Chaotic Time Series and Nonlinear Identification

Leandro dos Santos Coelho, Marcelo Rudek, and Osiris Canciglieri Junior

Pontifícia Universidade Católica do Paraná, LAS/CCET/PUCPR
Rua Imaculada Conceição, 1155, Prado Velho
80210.390 Curitiba, PR, Brazil, e-mail: {lscoelho,rudek,osiris}@rla01.pucpr.br

Abstract. This paper presents the configuration of a fuzzy system of Takagi-Sugeno-Kang type based on optimization through a memetic algorithm. The memetic algorithm is composed by a fast evolutionary programming combined with simulated annealing algorithm. The fuzzy-memetic system is evaluated for two case studies: (i) prediction of chaotic system with maps involving non-differentiable functions called Lozi map, and (ii) identification of a system composed of a nonlinear continuous stirred tank reactor. Simulations deal the estimation and validation procedures of dynamic model based on fuzzy-memetic system. The performance of the fuzzy-memetic system for case studies in time series prediction and nonlinear identification are presented and discussed.

1 Introduction

Different approaches to time series prediction and nonlinear identification problems have been investigated over the years, such as evolutionary computation [11], neural networks [4], fuzzy systems [10], [14], and hybrid intelligent systems [12]. Recently, a considerable amount of research activity has been directed toward developing fuzzy models for identification and control of complex dynamical systems. Generally, a fuzzy model should have two major attributes: high nonlinearity ad simple structure. The former is required for representing the highly nonlinear behavior of the complex system white the latter is required for easy analysis and design of the system.

The development and use of dynamic models based on fuzzy inference systems is an adequate methodology for the treatment of problems that present complex dynamical, such as noise, nonlinearities, and chaotic behavior. Fuzzy systems viewed as nonlinear systems are potential candidates for modeling and control of general nonlinear systems. Many applications can be found in the literature. They include those in water purification process, liquid level rigs, chemical process control, power systems, robotics, manufacturing, control systems, and others [5],[6],[7],[8].

This paper presents the configuration of a fuzzy inference system of Takagi-Sugeno-Kang type based on optimization through a memetic algorithm for time series prediction and nonlinear identification. The proposed memetic algorithm is constituted by a fast evolutionary programming for global search combined with simulated annealing algorithm for local search.

The Takagi-Sugeno-Kang model exhibits both high nonlinearity and simple structure. As reported in the literature, it is capable of approximating a complex system using fewer fuzzy rules compared to conventional fuzzy models of Mamdani's type. The identification problem in Takagi-Sugeno-Kang (*TSK*) modeling consists of two major parts, the structure identification and the parameter identification. Furthermore, the *TSK* system comprises the premise part identification adn the consequent part identification. Identification of the premise parte cosnsite of determining the premise space partitionk and extracting the number of rules. The consequent part identification consists of determing the structure of the rules' output parts. Finaly, the parameter learnig task consists of determining the system parameters so that a performance measure based on the output errors is minimised. In this paper the structure identification of the premise and the consequent part are separately performed. The structure identification is realized based memetic algorithm for premise part optimization and the consequent part optimization is realized by least mean square method.

The fuzzy-memetic system is evaluated for two case studies: (i) prediction of chaotic system wtih maps involving non-differentiable functions called Lozi map, and (ii) identification of a system composed of a nonlinear continuous stirred tank reactor. Simulations deal the estimation and validation procedures of dynamic model based on fuzzy-memetic system. The performance of the fuzzy system for case studies in time series prediction are presented and discussed. The results indicate that the proposed fuzzy-memetic approach is attractive to applications of production systems prediction, financial market applications ?nd nonlinear identification of processes industrial.

The remainder of this paper is organized as follows. The fuzzy-memetic approach is presented in section 2. Section 3 presents the description of two case studies and the analysis of obtained simulation results. The conclusion and future works are provided in section 4.

2 Fuzzy-Memetic approach

The fuzzy systems were originally introduced as a way of formally describing and manipulating linguistic information. As fuzzy theory is developed, the research on fuzzy modeling, which describes a real system very successfully with its nonlinear property, is conducted actively. The fuzzy models have advantages of excellent capability to describe a given system and intuitive persuasion toward human operators over linear models [10].

The Takagi-Sugeno-Kang fuzzy model is based on rules in which the consequent is not a linguistic variable, as in the Mamdani-type fuzzy model, but a function of the input variables. Recently, the Takagi-Sugeno-Kang fuzzy reasoning has become a most significant topic in several applications of fuzzy modeling and control. The identification of a Takagi-Sugeno-Kang fuzzy model involves two primary tasks: parameter tuning and structure optimization. The parameter tuning procedure deals with the estimation of a feasible set of parameters for a given structure. The structure optimization procedure aims to find the optimal structure of the local models, the relevant premise variables and a suitable partition of the premise space. In the section follow the fundamentals of Takagi-Sugeno-Kang system based on memetic optimization are presented.

2.1 Fuzzy system of Takagi-Sugeno-Kang (*TSK*) type

The *TSK* models consist of linguistic IF-THEN rules that can be represented by the following general form:

$$R^{(j)}: \quad IF \ z_1 \ IS \ A_1^j \ AND \dots AND \ z_m \ THEN \ g_j = w_0^j + w_1^j \ u_1^j + \dots + w_{qj}^j u_{qj}^j \quad (1)$$

The IF preconditioned statements define the premise part while the THEN rule functions constitute the consequent part of the fuzzy system; $\underline{z} = [z_1, \dots, z_m]^T$ is the input vector of the premise p, A_i^j and   are labels of fuzzy sets.

The parameters $\underline{u} = [u_1^j, \dots, u_{qj}^j]^T$ represents the input vector to the consequent part of $R^{(j)}$ that comprising q_j terms; $g_j = g_j(\underline{u}^j)$ denotes the *j*-th rule output which is a linear polynomial of the consequent input terms u_i^j, and $\underline{w}^j = [w_0^j, w_1^j \dots, w_{qj}^j]^T$ are the polynomial coefficients that form the consequent parameter set. Each linguistic label A_i^j is associated with a membership function, $\mu_{A_i^j}(z_i)$, which described by

$$\mu_{A_i^j}(z_i) = exp\left[-\frac{1}{2} \frac{(z_i - m_{ij})^2}{\sigma_{ij}^2} \right] \quad (2)$$

where m_{ij} and σ_{ij} are the mean value and the standard deviations of the Gaussian type membership function, respectively. The union of all these parameters formulates the premise parameter set. The firing strength of rule $R^{(j)}$ represents its excitation level and it is given by:

$$\mu_j(\underline{z}) = \mu_{A_1^j}(z_1) \cdot \mu_{A_2^j}(z_2) \cdots \mu_{A_m^j}(z_m) \quad (3)$$

The fuzzy sets pertaining to a rule form a fuzzy region (cluster) within the premise space, $A_1^j \times A_2^j \times \cdots \times A_m^j$, with a membership distribution described by equation (3). Given the input vectors $\underline{z}$ and $\underline{u}^j$, $j = 1, \dots, M$, the final output of the fuzzy system is inferred by taking the weighted average of the local outputs $g_j(\underline{u}^j)$ that is given by

$$y = \sum_{j=1}^{M} v_j(\underline{z}) \cdot g_j(\underline{u}^j) \tag{4}$$

where M denotes the number of rules and $v_j(\underline{z})$ is the normalized firing strength of $R^{(j)}$, which is defined as

$$v_j(\underline{z}) = \frac{\mu_j(\underline{z})}{\sum\limits_{j=1}^{M} \mu_j(\underline{z})} \tag{5}$$

The structure identification of *TSK* system is realized based memetic algorithm for premise part optimization and the consequent part optimization is realized by least mean squares method [13]. The procedure of fuzzy-memetic approach are presented in figure 1.

2.2 Memetic optimization of *TSK* fuzzy system

The memetic algorithm proposed in this paper consists of a fast evolutionary programming combined with simulated annealing algorithm for improved convergence rates and better performance of fuzzy design optimization. This approach combines local and global searches that characterize a form of Lamarckian evolution. Hybrid algorithms can combine global search using evolutionary algorithms and local search using individual learning algorithms using individual learning algorithms. Hybrid evolutionary algorithms can exploit learning either actively via Lamarckian inheritance or passively via the Baldwin effect [17]. In the 19th century, Jean Baptiste Lamarck, who proposed that environmental changes throughout an organism's life cause structural changes that are transmitted to offspring, challenged Darwin's theory. This theory lets organisms pass along the knowledge and experience that they acquire in their lifetime [9],[17].

This is analogous to Lamarckian inheritance in evolutionary theory, whereby characters acquired during a parent's lifetime are passed on their offspring. Evolutionary programming produces the offspring of the next generation through the evaluation function to determine its fitness. One of the disadvantages of evolutionary programming in solving some high dimensional optimization problem is its slow convergence to a good near optimum. This procedure can benefit from the advantages of Lamarckian theory. By letting some of the organism's "experiences" be passed along to future organisms. Following a Lamarckian approach, first would try inject some "smarts" into the offspring organism before returning it be evaluated. A traditional hill-climbing routine could use the offspring organism as a starting point and perform quick, localized optimization. The hill-climbing procedure is realized by simulated annealing approach. This optimization procedure is presented in figure 2.

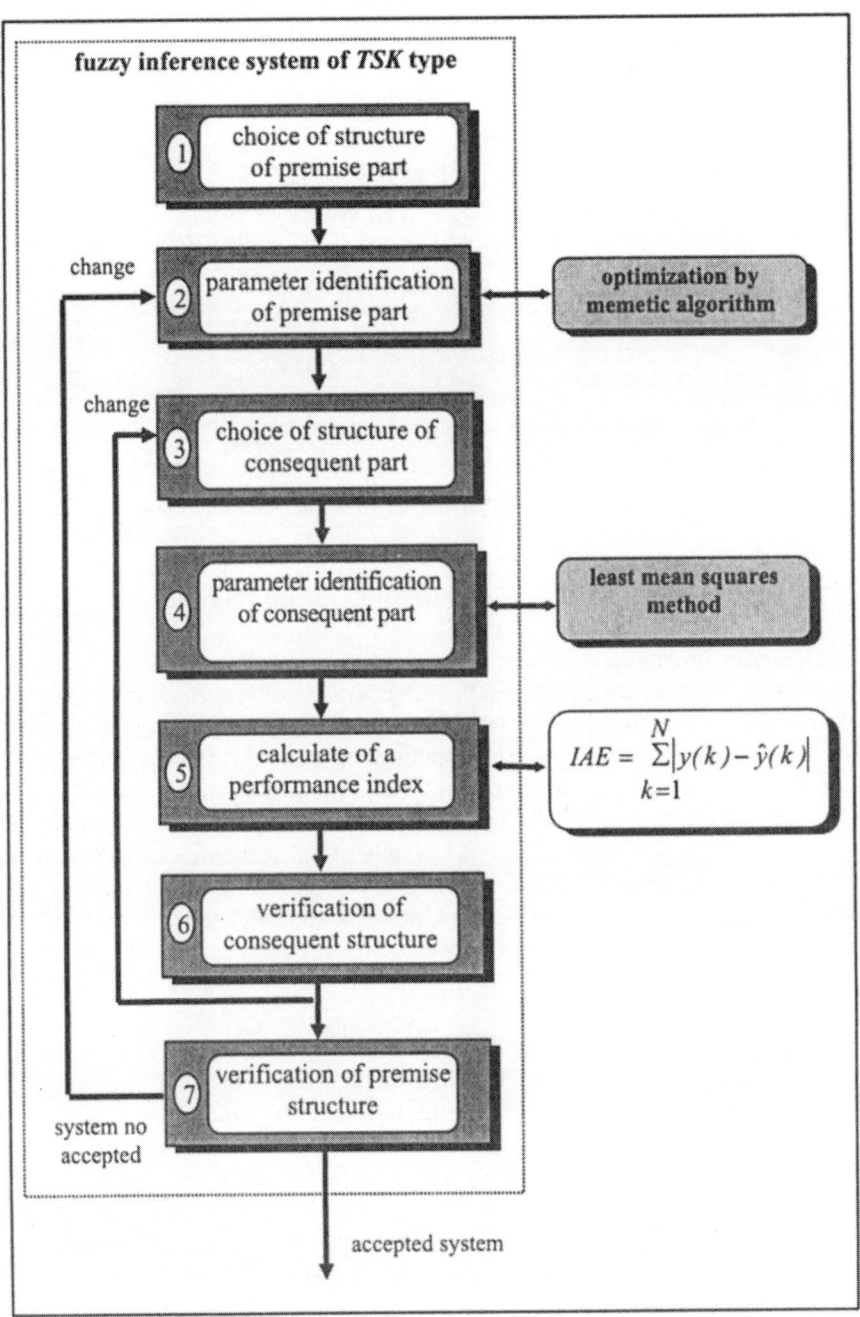

Fig. 1. *TSK* fuzzy system with optimization procedure based on memetic algorithm and least mean squares method.

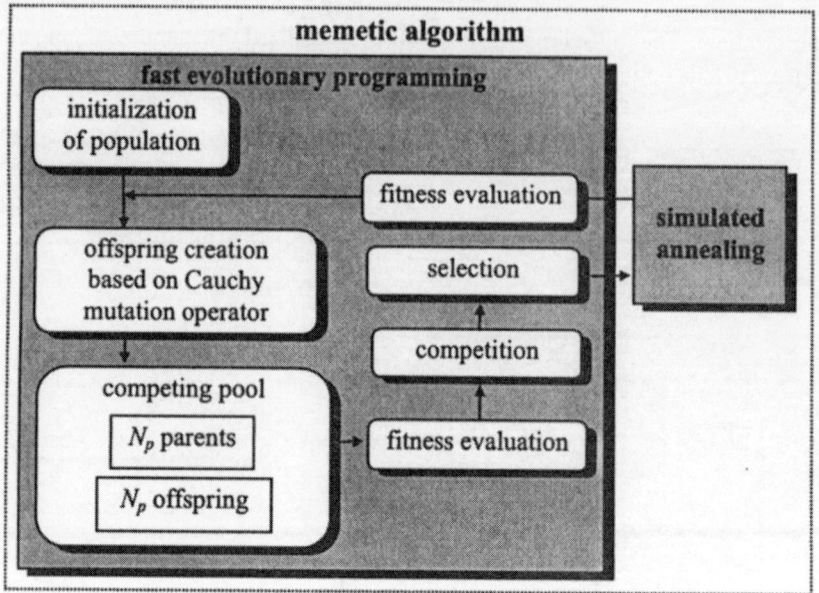

Fig. 2. Optimization procedure based on memetic algorithm.

2.2.1 Basic concepts of fast evolutionary programming

Evolutionary programming is a stochastic optimization strategy similar to genetic algorithms, which placed emphasis the behavioral link between an individual and its offspring, rather than seeking to emulate specific genetic operators as observed in nature. The evolutionary programming operates as follows: The initial population is selected at random and is scored with respect to a given cost function. Offspring are created from these parents through random mutations, i.e., each component is usually perturbed by a Gaussian random variable with mean zero and an adaptable standard deviation term. It uses probability transition rules to select generations. Selection is based on a probabilistic tournament where each individual competes with other individuals in a combined population of the old generation and the mutated old generation. The competition results are valued using a probabilistic rule. The winners individuals in the old generation constitute the next generation. A relevant detail of evolutionary programming is that the individuals present floating point representation.

The fast evolutionary programming uses Cauchy mutations [18] and is employed in this work. The relationship between the classical evolutionary programming using Gaussian mutation and the fast evolutionary programming using Cauchy mutation is analogous to that between classical simulated annealing and fast simulated annealing. Evolutionary programming using Gaussian mutation is conducted by Box-Muller algorithm [15] for generation of random values with normal distribution, where

$$p(y)dy = \frac{1}{\sqrt{2\pi}}e^{-y^2/2}dy \tag{6}$$

The mutation operator of evolutionary programming with Cauchy distribution presents a one-dimensional Cauchy density function centered at the origin and defined by

$$f_t = \frac{1}{\pi} \frac{t}{t^2 + x^2}, \qquad -\infty < x < \infty \tag{7}$$

where $t > 0$, is the scale parameter. The corresponding distribution function is

$$F_t(x) = \frac{1}{2} + \frac{1}{\pi} arctan\left(\frac{x}{t}\right) \tag{8}$$

The variance of the Cauchy distribution is infinite. Many studies have indicated the benefits of variance are increased in Monte-Carlo algorithms [18].

2.2.2 Basic concepts of simulated annealing

In the proposed approach of memetic algorithm, a local hill climbing methodology — simulated annealing — is used to improve the obtained solution by fast evolutionary programming. Simulated annealing is an optimization technique inspired from Monte Carlo methods in statistical mechanics. It attempts to avoid local optimal by probabilistically taking non-locally optimal steps in the search space. The probability of taking such steps decreases with the "temperature" of the system, which in turn decreases with time. Simulated annealing algorithm has the advantage of asymptotically producing the global optimal solution. The proposed memetic algorithm is based on hybrid genetic algorithm with local search based on simulated annealing. The annealing procedure is applied to the evolutionary programming at each generation for fine tune of the individual values with low temperature factor and fast factor of annealing [16].

3 Case studies

3.1 Case study 1: Lozi map

The Lozi map [2] consists of a time series that present maps involving non-diffentiable functions and it is described by equation:

$$y(k+1) = -P \cdot |y(k)| + Q \cdot y(k-1) + 1 \tag{9}$$

where k is the number of sample and y is the system output. The Lozi presents a chaotic attractor when $P = 1.8$ and $Q = 0.4$. The evaluated number of samples in this case is 200 samples, where 100 samples are utilized for the estimation stage (learning procedure) of fuzzy-memetic system, and others 100 samples are utilized for the validation stage of obtained fuzzy model. The adopted initial conditions are $y(0) = 1$ e $y(1) = 0$. The obtained results with fuzzy-memetic approach are

764

presented in figure 3. The adopted performance index, *IAE*, is the sum of absolute error among the estimated output, $\hat{y}(k+1)$, and the real output, $y(k)$, of Lozi map. The input signals of *TSK* fuzzy system were $\{y(k-1), y(k)\}$ and the output of fuzzy system is given by $\hat{y}(k+1)$. The adopted parameters for the memetic algorithm of optimization of premise part (parameters of membership functions) were: population = 30 individuals, generations (stopping criterion) = 300, percentage of population that is applied of simulated annealing procedure in each generation = 30%.

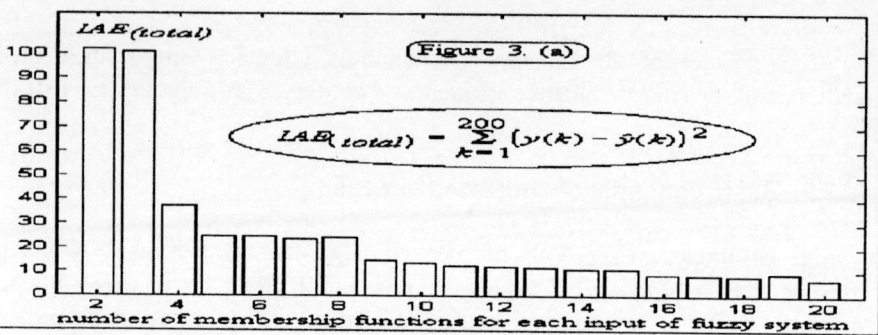

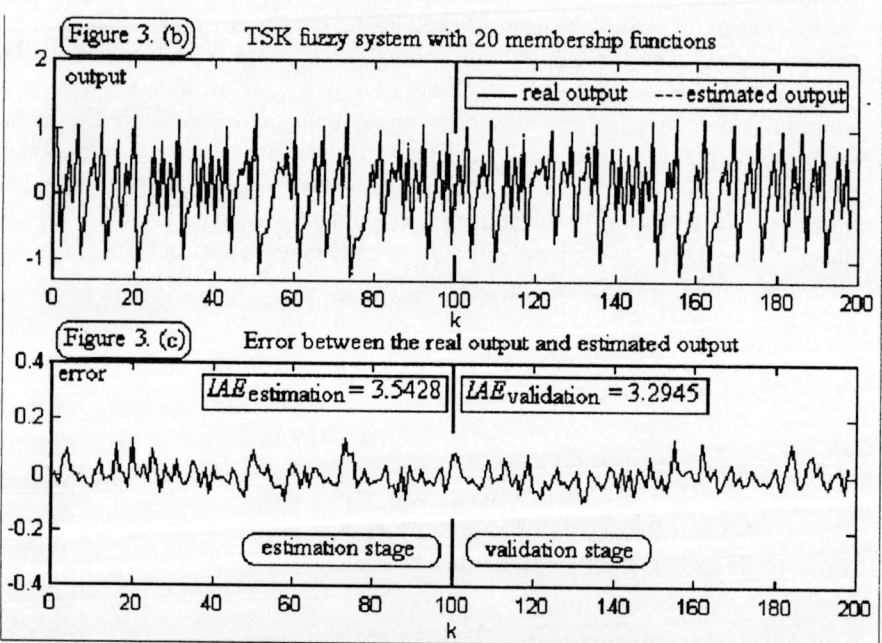

Fig. 3. (a) Histogram: *IAE* (total, 200 samples) = *IAE*(estimation, 100 samples) + *IAE*(validation, 100 samples) versus number of membership functions for each input of *TSK* fuzzy system; (b) real and estimated of fuzzy-memetic system; (c) error among the real output and estimated of Lozi map obtained by fuzzy-memetic system.

3.2 Case study 2: Continuous Stirred Tank Reactor (*CSTR*)

In this preliminary study the case study of *CSTR* is presented. The dynamic equations of *CSTR* represent the nonlinear process are given by:

$$\frac{dx_1}{dt} = -x_1 + D_a(1-x_1)e^{\frac{x_2}{1+x_2/\varphi}} \tag{10}$$

$$\frac{dx_2}{dt} = -(1-\beta)x_2 + BD_a(1-x_1)e^{\frac{x_2}{1+x_2/\varphi}} + \beta u \tag{11}$$

$$y = x_2 \tag{12}$$

where x_1 and x_2 represent the reagents concentration (dimensionless) e reactor temperature, respectively. The control input, u, is the temperature of the cooling jacket surrounding the reactor. The physical constants are D_a, φ, B and β which represent the Damköhler number, the activation energy, reaction heat and the heat transfer coefficient, respectively. The nominal parameters of the system are $D_a = 0.072$, $\varphi = 20$, $B = 8$ and $\beta = 0.3$. In this case the process exhibit the open-loop unstable behavior [3]. The simulation of dynamic behavior of reactor is realized by conversion of the equation (10), (11) and (12) for the discrete equations system, by the utilization of Euler method [1]. Consequently, the resultants equations are:

$$x_1(k+1) = x_1(k) + T_s\left[-x_1(k) + D_a(1-x_1(k))e^{\frac{x_2(k)}{1+x_2(k)/\varphi}}\right] \tag{13}$$

$$x_2(k+1) = x_2(k) + T_s\left[-(1-\beta)x_2(k) + BD_a(1-x_1(k))e^{\frac{x_2(k)}{1+x_2(k)/\varphi}} + \beta u(k)\right] \tag{14}$$

$$y(k) = x_2(k) \tag{15}$$

where T_s denotes the sampling time (adopted $T_s = 200\ ms$) and k is the k-th control step. The relation between the system output, y, and control input, u, can be obtained by substitution of the equation (6) in the equation (7) with $k = k+1$:

$$y(k+1) = x_2(k) + T_s\left[-(1-\beta)x_2(k) + BD_a(1-x_1(k))e^{\frac{x_2(k)}{1+x_2(k)/\varphi}} + \beta u(k)\right] \tag{16}$$

The input signal, $u(k)$, is generated random with uniform distribution with amplitude, $u(k) \in [-1,4;\ 1,4]$. In response of input signal, u, the output signals

766

vector of process, y, is obtained. In figure 1, the preliminary results obtained with *TSK* for the estimated output, , and real output of *CSTR*, $\hat{y}(k+1)$, are presented. The two inputs utilized in the TSK model are $\{u(k), y(k)\}$, and the output of *TSK* fuzzy model is given by $\hat{y}(k+1)$.

The simulation results with fuzzy-memetic approach are presented in figure 4. The adopted performance index, *IAE*, is the sum of absolute error among the estimated output, $\hat{y}(k+1)$, and the real output, $y(k)$, of *CSTR* process. The input signals of *TSK* fuzzy system were $\{y(k-1), y(k)\}$ and the output of fuzzy system is given by $\hat{y}(k+1)$. The adopted parameters for the memetic algorithm of optimization of premise part (parameters of membership functions) were: population of 30 individuals, stopping criterion of 100 generations, percentage of population that is applied of simulated annealing in each generation is 30%.

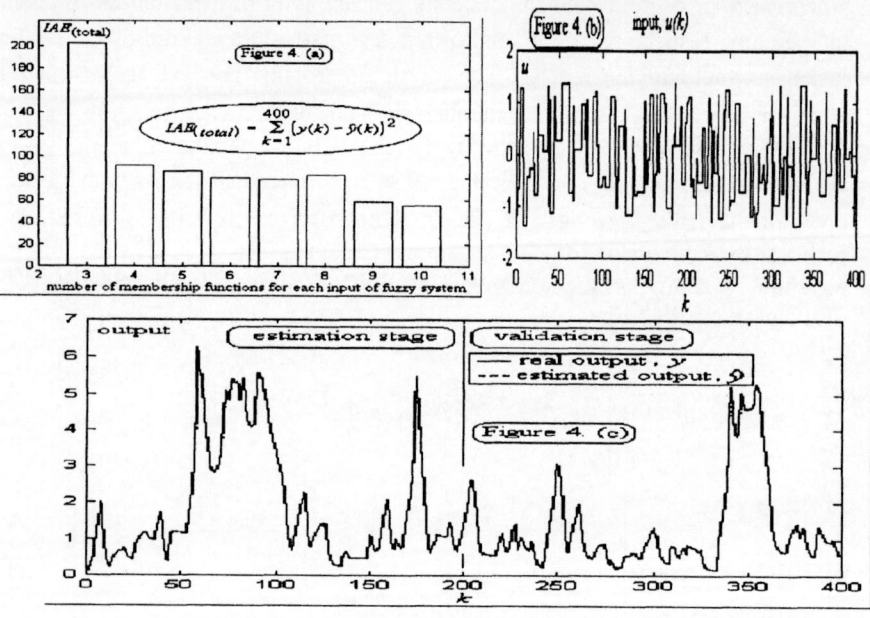

Fig. 4. (a) Histogram: *IAE* (total, 400 samples) = *IAE*(estimation, 200 samples) + *IAE*(validation, 200 samples) versus number of membership functions for each input of fuzzy system; (b) input of *CSTR*; (c) real and estimated of *CSTR*.

The obtained results show the robustness and efficiency of the *TSK* fuzzy system based on memetic optimization for prediction of Lozi map and identification of *CSTR* nonlinear process. However, the problem of structural selection occurs with the *TSK* fuzzy system that depends of centers and spread choices for a adequate design. These choices present a compromise among number of selected centers and the approximation quality of *TSK* fuzzy system. For problems with low complexity are needed a small number of centers that for a complex problem.

767

4   Conclusion and future developments

This paper presents a memetic approach for premise part optimization of a *TSK* fuzzy system. The fuzzy-memetic system is evaluated for two case studies: (i) prediction of chaotic system with maps involving non-differentiable functions called Lozi map, and (ii) identification of a system composed of a nonlinear continuous stirred tank reactor.

Simulations deal the estimation and validation procedures of dynamic model based on fuzzy-memetic system. The simulations indicate the good performance for fuzzy-memetic approach for prediction and identification tasks. This characteristic confirms the usefulness and robustness of the proposed method. The future developments of this work treat several approaches of fuzzy design for predictive and adaptive control applications of nonlinear multivariable processes.

References

1.   Åström K. J., Wittenmark B. (1984). Computer controlled system: theory and design, Prentice-Hall, NJ
2.   Chen G., Chen Y., Ogmen H. (1997). Identifying chaotic systems via a Wiener-type cascade model, IEEE Control Systems, **17**(5), 29-36
3.   Chen C.-T., Peng, S.-T. (1997). A nonlinear control strategy based on using a shape tunable neural controller. Journal of Chemical Engineering of Japan, **30**(4), 637-646
4.   Farmer J.D., Sidorowich J.J. (1987). Predicting chaotic time series, Physical Review Letters, **59**(8), 845-848
5.   Fukuda T., Shimojima K. (1995). Fusion of fuzzy, nn, ga to the intelligent robotics. *Proceedings of IEEE International Conference on Systems, Man and Cybernetics*, Vancouver, British Columbia, Canada, 2892-2897
6.   Ghezelayagh H., Lee K. Y. (2000). Application of neuro-fuzzy identifier for a fossil fuel boiler system. *IEEE Power Engineering Society Winter Meeting*, Syngapore, **2**, 1135-1139
7.   Hirota K., Sugeno M. (1995). Industrial applications of fuzzy technology in the world, advances in fuzzy systems — applications and theory, **2**, World Scientific: Syngapore
8.   Huaguang Z., Bien Z. (1998). Fuzzy system identification and predictive control of load system in power plant. IEEE World Congress on Computational Intelligence, Conference on Fuzzy Systems, **1**, 342-347
9.   Kennedy S. A. (1993). Five ways to a smarter genetic algorithm. *AI Expert*, December, 35-38
10.   Kim E., Park M., Ji S., Park M. (1997). A new approach to fuzzy modeling, IEEE Transactions on Fuzzy Systems, **5**(3), 328-337
11.   Kim I., Lee S.-R. (1999). A fuzzy time series prediction method based on consecutive values. Proceedings of IEEE International Fuzzy Systems Conference, Seoul, Korea, **2**, 703-707
12.   Lee D.-W., Sim K.-B. (1999). Evolving chaotic neural systems for time series prediction. Congress on Evolutionary Computation, Washington, DC, **1**, 310-316
13.   Ljung L. (1987). System identification: theory for the user, Prentice-Hall, NY.
14.   Osawa K., Watanabe T., Kanke M. (1997). Fuzzy auto-regressive model and its applications, First International Conference on Knowledge-Based Intelligent Electronic Systems, Adelaide, Australia, **I**, 112-116

768

15. Press W.H., Teukolsky S.A., Vetterling W.T., Flannery B.P. (1994). Numerical recipes in c: the art of scientific computing, Cambridge Press
16. Tan K.C., Li Y., Murray-Smith D.J., Sharman K.C. (1995). System identification and linearisation using genetic algorithms with simulated annealing, Proceedings of IEE/IEEE GALESIA, Sheffield, UK, 164-169
17. Whitley D., Gordon S., Mathias K. (1994). Lamarckian evolution, the Baldwin effect and function optimization. Parallel Problem Solving for Nature, Springer-Verlag, Berlin, 6-15
18. Yao X., Liu Y. (1996). Fast evolutionary programming. In: Fogel L.J., Angeline P.J., Bäck T. (eds) Proceedings of the 5th Annual Conference on Evolutionary Programming, San Diego, CA, MIT Press, 451-460

Finite-Element Mesh Adaptation via Time Series Prediction Using Neural Networks

Larry Manevitz[1], Akram Bitar[1], and Dan Givoli[2]

[1] Department of Computer Science, University of Haifa, Haifa, Israel
manevitz@cs.haifa.ac.il   akram@il.ibm.com
[2] Faculty of Aerospace Engineering, Technion-Israel Institute of Technology, Haifa, Israel
givolid@aerodyne.technion.ac.il

Abstract. In this paper, basic learning algorithms and the neural network model are applied to the problem of mesh adaptation for the finite-element method for solving time-dependent partial differential equations. Time series prediction via the neural network methodology is used to *predict* the areas of "interest" in order to obtain an effective mesh refinement at the appropriate times. This allows for increased numerical accuracy with the same computational resources as compared with more "traditional" methods.

Keywords: finite-element method, neural networks, time-dependent PDEs, time series prediction, mesh adaptation.

1 Introduction and Background

The finite-element method (FEM) [7,1] is a computationally intensive method for the numerical solution of partial differential equations (PDEs). It is a widely used tool and in many cases is the method of choice. Basically, the FEM works by deciding, *a priori* on a certain kind of simple approximation to the solution, by dividing up the domain of solution into a finite number of nonoverlapping elements (i.e. subdomains), and by allowing the parameters of the simple approximation to vary from element to element. This collection of elements and the connections between them constitutes the finite-element mesh. The requirement that the individual local solutions remain consistent with each other and with the boundary conditions results in linear constraints on the parameters. These are then solved by standard linear algebra techniques.

The quality of the numerical results thus depends on the geometry; i.e. the domain of solution and its division to small elements as well as the kind of approximation taken (see [8]).

In addition, time is typically treated differently than spatial dimensions in the solution phase of time-dependent partial differential equations. That is typically time is not treated as simply another dimension, but instead time is *simulated* i.e. the equation is repeatedly solved for different constant times; using the previous solution as the starting condition for the next one.

However, in time-dependent PDEs [7,5], this implies that one should not use the same mesh at different times since the areas of "interest" (i.e. areas where the simple approximations are inherently less accurate) are of course changing with time.

For example, when solution of hyperbolic problems involves a shock wave, which propagates through the mesh the location of the shock vicinity keeps changing in time. Thus one wants to have the mesh more refined around the area of the shock vicinity and less refined elsewhere. Another example is the problem of fluid flow in a cavity, where flow cells are generated and undergo continuous changes in their shapes and size as time proceeds.

This means that the mesh adaptation is a crucial part for the efficient computation of the numerical method. Thus, in order to achieve an optimal mesh (one which the solution error is low relative to the number of nodes in the mesh), the mesh choice should be dynamic and varying with time.

In current usage, the method is to use indicators (e.g. gradients) from the solution at current time to identify where the mesh should be modified (i.e. where it should be refined and where it can be made coarser) at the next time stage. However, this suffers from the obvious defect that one is always operating one step behind. In other words, if the areas of interest are propagated, then one may be always refining behind the most interesting phenomena.

In this paper, we present a new approach for solving the mesh adaptation problem . Our approach looks at this as special instance of a control problem and uses the neural network to solve it in a similar way that such networks have been used to predict time series (see [9]).

The neural network is a universal approximator that learns from the past to predict the future values. It receives, in some form, as input the "areas of interest" at recent times and predicts the "areas of interest" at the next time stage. Using this predictor we can forecast the position of the "action" and refine the mesh accordingly.

1.1 Background of Time Series Neural Networks

Neural networks (NNs) [6] is a biologically inspired model, which tries to simulate the network of neurons, or the nervous systems, in the human brain. The artificial neural networks consist of simple calculation elements, called neurons, and weighted connections between them. A neural network can be trained to perform complex functions by adjusting the values of the connections (weights) between the elements (neurons) according to one of several algorithms. In general, neural network tasks may be divided into four main types of distinct applications: classifications, associations, codification and simulations.

For our purposes, the neural network may be considered simply as a data processing technique that maps, or relates, some type of stream information to an output stream of data. For example, to classify input vectors by dividing the input space into two regions: one for input vectors which are above and to the left of a decision boundary line, and the other for vectors which are below and to the right of the decision boundary line. However, and importantly, NNs can also determine non-linear decision boundaries.

The most common NN model is the supervised-learning, feed-forward network. Typical the feed-forward network contains three types of processing units, input units, output units and hidden units, organized in a hierarchy of layers, as demonstrated in Fig. 1.

The training algorithms such as back-propagation or Newton-Gauss consist of two phases: feed-forward propagation and backward propagation. In forward propagation, the input units send the input signals forward through the network to produce an output. Then, the difference between the actual and desired outputs produces error signals which are sent backwards through the network to modify the weights between neurons. The forward and backward propagation are executed iteratively over the training set until convergence occurs (when the average squared error between the network outputs and the desired outputs reaches an acceptable value).

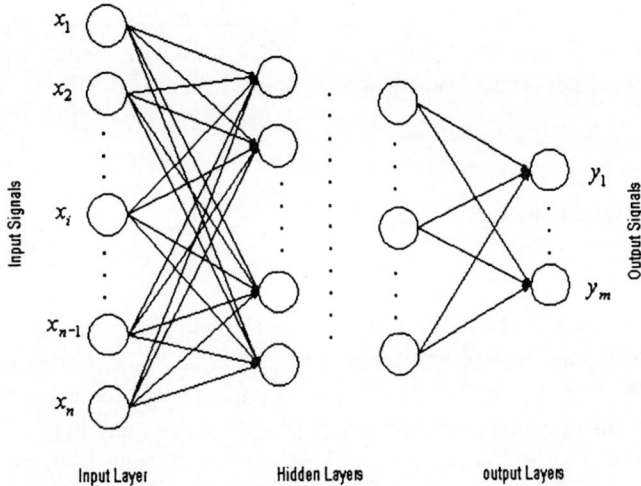

Fig. 1. Feed-forward network architecture

Time series are well suited for data where past values in the series may influence future values. In this case, a future value is a nonlinear function of its past m values:

$$x(n) = f(x(n-1), x(n-2), ..., x(n-m)) \qquad (1)$$

This means, that it is necessary to fit a function x through its past values in order to extrapolate this function to the near future.

Since a feed-forward network can approximate any function after a suitable amount of training [3] it can be applied to this problem, by submitting discrete values of this function to the network. The net is then expected to learn the function rule by the training algorithm. The behavior of the network is changed by modifying the values of the weights.

Therefore, we can use the back-propagation network as nonlinear model that can be trained to map past and future values of a time series. This method is called time series prediction with neural networks and is used in the forecasting of financial markets [2] (e.g. to predict whether stock market rates will rise or fill). The following section describes how we use this method to predict the mesh refinement placement for time-dependent PDEs.

1.2 Time-dependent PDEs

PDEs arise in modelling numerous phenomena in science and engineering. The time-dependent PDEs tend to be divided into two categories: hyperbolic and parabolic. The hyperbolic PDE is used for transient and harmonic wave propagation in acoustics and electromagnetic, and for transverse motions of membranes; the basic prototype of the hyperbolic PDE is the wave equations . The parabolic PDE is used for unsteady heat transfer in solids, flow in porous media and diffusion problems; the basic prototype parabolic PDEs is the heat equations .

2 Applying NNs to Time-dependent PDEs

For many PDEs critical regions should be subject to local mesh refinement. The critical regions are the regions for which the local gradient shows bigger changes. In order to meet this problem, the FEM adaptation process makes a local refinement in those areas, thus the ensuing mesh may be more gross in the other areas. In time-dependent problems, the mesh refinements should be dynamic and depends on the error estimation in each time stage.

In current usage, most of the error estimate methods take into account the solution gradient; in this work we developed a new approach based on predicting the future gradient value of the solution and we used this as the refinement criteria.

In dynamic systems, e.g. hyperbolic equations , the areas of interest, i.e. the areas with high gradient are propagated through the domain. Therefore, for each mesh element the future gradient value is influenced by the past gradient values of the element and of its direct neighbors. In other words, the future gradient value can be considered as a nonlinear function of its past values. This is a proper time series problem and the time series neural networks can be used to predict the future gradient values. Fig. 2 illustrates this concept.

Thus there are two steps to our methodology: (a) training the neural network to predict the indicators (at least) one step in advance (b) applying the indicators to the refinement in the FEM solution.

In our experiments we examine the numerical results of applying this procedure using the FEM on different time-dependent PDE problems using different parameters for the NN algorithm and comparing this with (i) FEM with no adaptation and (ii) FEM using the "standard" adaptation via the current gradient indicator.

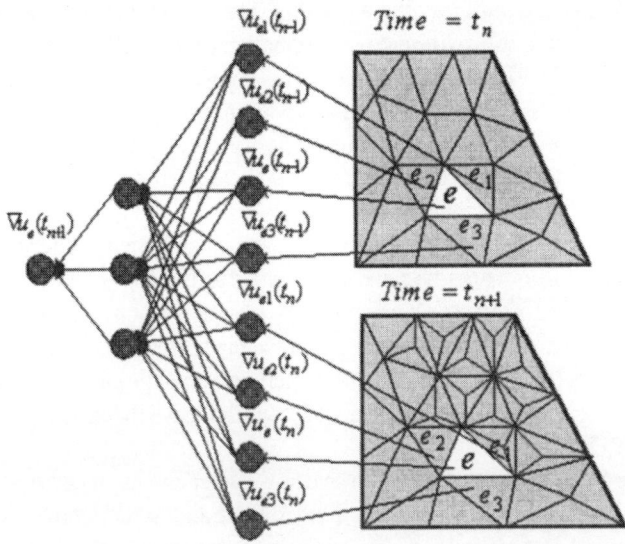

Fig. 2. Using Neural Networks to forecast future FEM gradient values. Gradient values for two previous times and all neighboring elements are used as input to the network

2.1 Measures of Solution Quality

To measure the quality of FEM solution, we calculate both of the L^2 and L^∞ error norm per value in each time stage. (Here u is the analytic solution and u_h is the numerically computed solution).

$$L^2 error/value = \frac{\sum_{nodes} |u(node) - u_h(node)|_2}{\sum_{nodes} |u(node)|_2}, \tag{2}$$

$$L^\infty error/value = \frac{\sum_{nodes} |u(node) - u_h(node)|_\infty}{\sum_{nodes} |u(node)|_\infty}, \tag{3}$$

The analytic solution is very important in order to measure the precise error in the solution. When the analytic solution of a PDE is not available, we do one calculation with a very small time step and a very fine constant mesh, and then we use this solution as a reference to the analytic one. In all cases, we report at the end of each experiment the average of L^2 and L^∞ error per value over all the time space.

2.2 Neural Network Architecture and Training

In this study, the MATLAB's Neural Network Toolbox [4] was used for designing and training the neural network; and the MATLAB's Partial Differential Equation Toolbox was used for defining, and solving the two dimension PDEs problems. For one dimension problems, we used a FEM solver which we developed especially for our research needs. In the examples tested so far the results are fairly dramatic.

First, using the Levenberg-Marquardt training algorithm [10] (a variant of Gauss-Newton) the training was both quite swift and exceptionally accurate. (See Fig. 3.) Second, the improvement in the FEM numerical results (as compared with the "standard" gradient adaptive method) reached as high as 25% on some examples; and never fell significantly below the standard method. (The variance in the improvement depends on the shape of the wave; and is to be expected. That is, for some waves it is more important to predict the gradient than others.)

We used two different networks, one for boundary elements and one for interior elements. The architecture of networks was six input units for boundary elements network, and eight input units for interior elements networks (corresponding to the value of the gradient of the element and its two neighbors in the current and previous times); and for both networks six hidden units (with hyperbolic tan-sigmoid transfer function), and one output unit (with linear transfer function) that gave the prediction of the output value. See Fig. 2.

In order to make the training more efficient: (a) we normalized the input and output data between the values 0 and 1; and (b) we divided the training data into two disjoint subsets: training set and testing set. The training set is used for computing the gradient and updating the network weights and biases. The testing on the validation set is monitored during the training process; as long as the error decreases, training continues. When the error begins to increase, the net begins to overfit the data and loses it's ability to generalize; at this point the training is stopped.

To generate training data: (a) we calculate the solution on the initial non-dynamic mesh over all the given time space; (b) we choose random time stages, and build training examples for all the elements in these time stages. The training data, consists of more than 600 examples (about 400 for training set and 200 for testing set).

2.3   One Dimension Wave Equations

We have run the NN modifier over a variety of initial conditions for the wave equation. In all cases, the NN predictor was extremely accurate. Fig. 3 shows the results of a typical prediction test for interior and boundary elements. Training took about 117 epochs to converge to extremely small error (about 0.00024) in the interior elements prediction. Results for the boundary elements were similar. When applying this modifier to the FEM mesh, the numerical improvement over the "standard" gradient modifier varied from no significant improvement to an improvement of more than 25% (both in the L^2 norm and in the L^∞ norm).

In this paper we present one sample example, where the initial condition of the wave is a Gaussian. See Example 2 in Table 1.

The analytical solution is well known for these types of problems and it depends on the initial and boundary conditions. The wave splits into two waves (with the same width but half the height) that travels to the left and to the right with speed $c = 1$. When such a travelling triangle reaches the edge it turns over and returns upside down (see Fig. 4). The NN modified solution and the "standard" gradient modifier are displayed in Fig. 5. Compare the graphs in Fig. 5. Observing the areas indicated in the figure, one can see that the NN has chosen to place its resources

Prediction for Boundary Elements

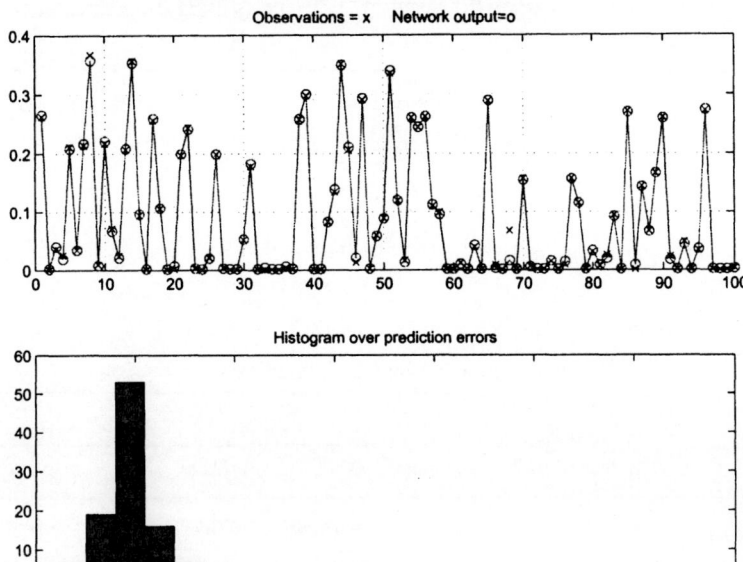

Prediction for Interior Elements

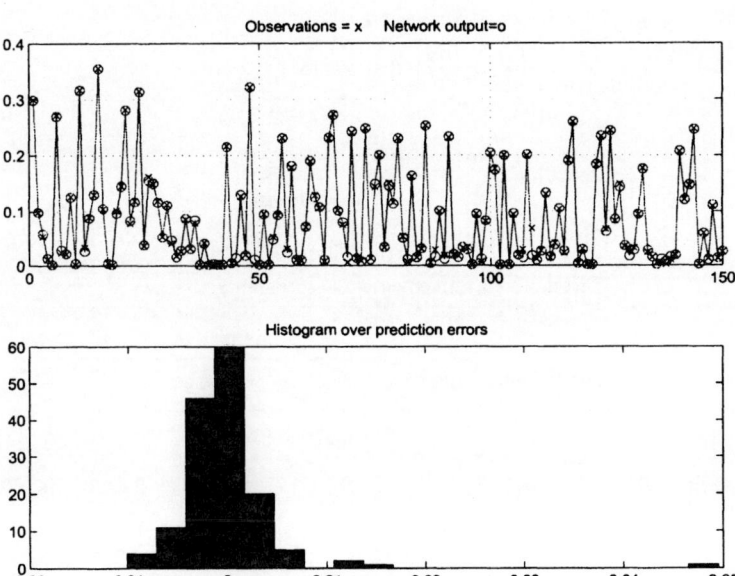

Fig. 3. Time Series Prediction Test for the 1D Wave Equation. (See Text.) Blue (x) indicates test values; red (o) the network response

Table 1. One dimension examples. Comparison between FEMs run with (i) The neural network predictor of the gradient measure. (ii)"Standard" refinements using the gradient measure. (iii) No adaptation.

Example 1

$$\frac{\partial^2 u}{\partial t^2} = \frac{\partial^2 u}{\partial x^2}, \ 0 \leq x \leq 10$$

$$u(0,t) = 0 \quad and \quad u(10,t) = 0$$

$$u(x,0) = \begin{cases} 1 - |1-x|, & 1 \leq x \leq 2 \\ 0, & otherwise \end{cases}$$

Number of Initial Elements:10, Time:12, Time Step:0.08

Threshold for refinement = 0.08 (gradient)

Method	Number of Refined Elements	L^2 Norm Max	Average	L^∞ Norm Max	Average
NN Modifier	70	0.15756	0.0869	0.1653	0.1056
Standard Modifier	70	0.1826	0.1022	0.1914	0.1222
No Adaptation	0	2.5332	1.0053	3.4708	1.0643

Improvement: L^2 error norm = 15%, L^∞ error norm = 13.6%

Example 2

$$\frac{\partial^2 u}{\partial t^2} = \frac{\partial^2 u}{\partial x^2}, \ 0 \leq x \leq 25$$

$$u(0,t) = 0 \quad and \quad u(25,t) = 0$$

$$u(x,0) = \begin{cases} \exp(\frac{-(x-5)^2}{2}), & 0 \leq x \leq 10 \\ 0, & otherwise \end{cases}$$

Number of Initial Elements:15, Time:25, Time Step:0.12

Threshold for refinement = 0.2 (gradient)

Method	Number of Refined Elements	L^2 Norm Max	Average	L^∞ Norm Max	Average
NN Modifier	98	0.4423	0.1928	0.5190	0.2342
Standard Modifier	91	0.6671	0.2686	0.8230	0.3142
No Adaptation	0	1.4622	0.6985	1.6288	0.6456

Improvement: L^2 error norm = 28%, L^∞ error norm = 25%

in the correct places. Looking at the refinement markings (in red or dots on the x-axis); one can see that, as suggested by our theory, the NN is keeping pace with the development of the solution, whereas the "standard" method is always one-step behind, which at critical locations causes increased numerical error.

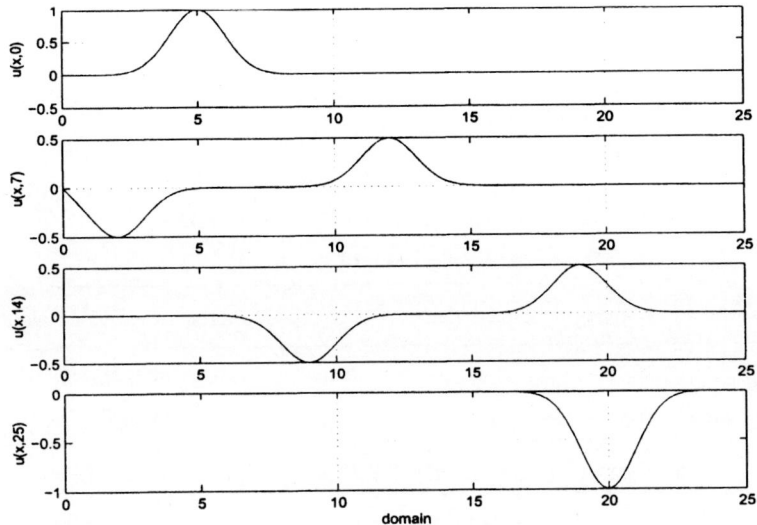

Fig. 4. Analytic Solution

Since these examples have analytic solutions, we can keep track of the actual numerical errors of each of the methods. In Fig. 6 we track the errors (both in L^2 norm and in the L^∞ norm).

2.4 Two Dimensional Waves

So far we have done a few of initial experiments. In the examples done, see Table 2,(i) the prediction of the gradient was very accurate (see Fig. 7 and Fig. 8); and (ii) the improvement in the FEM numerical results were around 10% over the standard gradient methods.

3 Summary

We have implemented a version of a NN modifier for the FEM mesh; designed to adaptively change the mesh based on a *prediction* of the gradient. In experimental work, we have shown that the NN can accurately predict the gradient and applying this mesh results in a substantial numerical improvement.

Acknowledgment

Supported in part by the *HIACS* Research Center, the University of Haifa.

Table 2. Two dimension examples. Comparison between FEMs run with (i) The neural network predictor of the gradient measure. (ii)"Standard" refinements using the gradient measure.

Example 1

$$\frac{\partial^2 u}{\partial t^2} = \frac{\partial^2 u}{\partial x^2} + \frac{\partial^2 u}{\partial y^2}, \quad -1 \le x \le 1, -1 \le y \le 1$$

$$u(-1,y,t) = 0 \quad and \quad u(1,y,t) = 0 \quad for \quad -1 \le y \le 1$$

$$u(x,-1,t) = 0 \quad and \quad u(x,-1,t) = 0 \quad for \quad -1 \le x \le 1$$

$$u(x,y,0) = \begin{cases} 15x(x+1)y(y+1), & -1 \le x \le 0, \ -1 \le y \le 0 \\ 0, & otherwise \end{cases}$$

Number of Initial Elements:28, Time:3, Time Step:0.05

Threshold for refinement = 1 (gradient)

Method	Number of Refined Elements	Average L^2 Norm	Average L^∞ Norm
NN Modifier	803	0.4057	0.4846
Standard Modifier	803	0.4314	0.5029

Improvement: L^2 error norm = 6%, L^∞ error norm = 3.6%

Example 2

$$\frac{\partial^2 u}{\partial t^2} = \frac{\partial^2 u}{\partial x^2} + \frac{\partial^2 u}{\partial y^2}, \quad -1 \le x \le 1, -1 \le y \le 1$$

$$u(-1,y,t) = 0 \quad and \quad u(1,y,t) = 0 \quad for \quad -1 \le y \le 1$$

$$u(x,-1,t) = 0 \quad and \quad u(x,-1,t) = 0 \quad for \quad -1 \le x \le 1$$

$$u(x,y,0) = \arctan(\cos(\tfrac{\pi}{2x}))$$

$$\frac{\partial^2 u}{\partial t^2}(x,y,0) = 3\sin(\pi x)\exp(\sin(\tfrac{\pi}{2y}))$$

Number of Initial Elements:28, Time:3, Time Step:0.08

Threshold for refinement = 2.2 (gradient)

Method	Number of Refined Elements	Average L^2 Norm	Average L^∞ Norm
NN Modifier	246	0.2962	0.3359
Standard Modifier	232	0.3256	0.3807

Improvement: L^2 error norm = 9%, L^∞ error norm = 11%

Neural Network Predictor   "Standard" Gradient Indicator

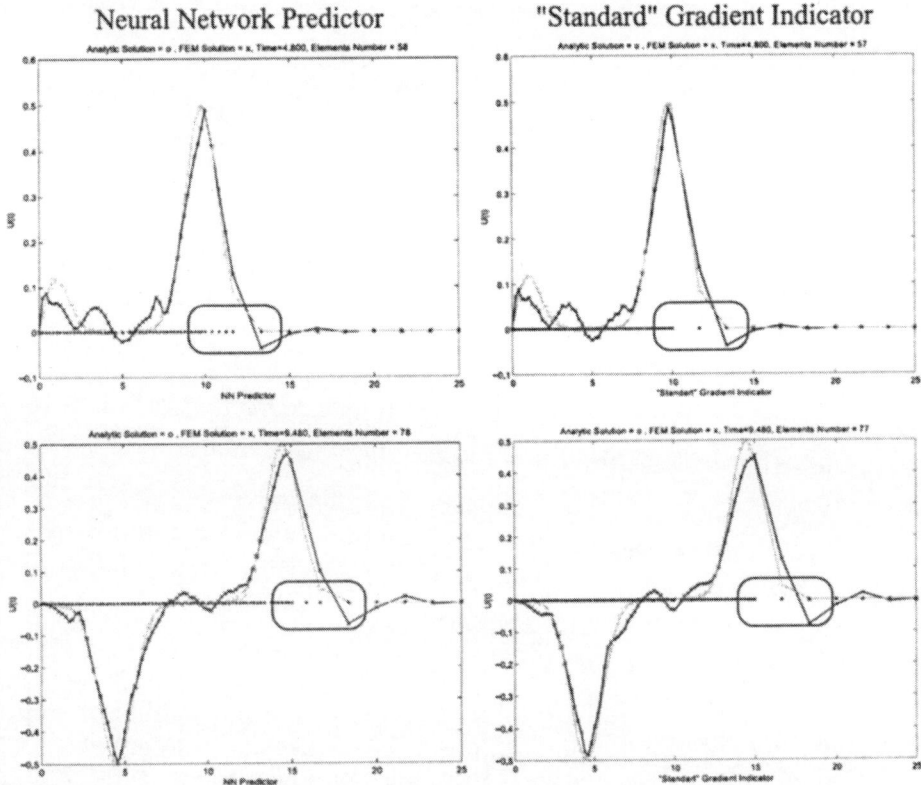

Fig. 5. Results from FEM on 1D Wave Equation. The left figures are refined with the NN predictor. Also indicated is the analytic solution. The right figures are refined with the "standard" gradient indicator. Compare the segments of the curves on the left (enclosed rectangles) with the corresponding ones on the right to see how the NN predictor focuses the resources in the correct places

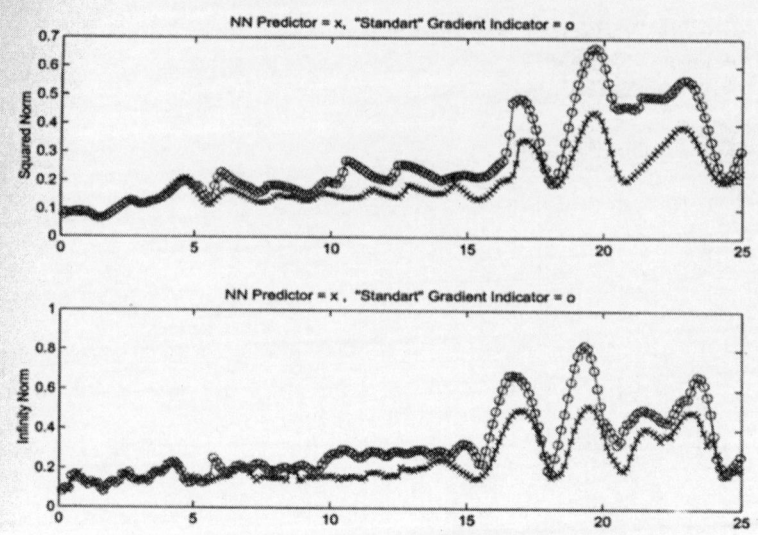

Fig. 6. L^2 error norm and L^∞ error norm over time

Prediction for Interior Elements

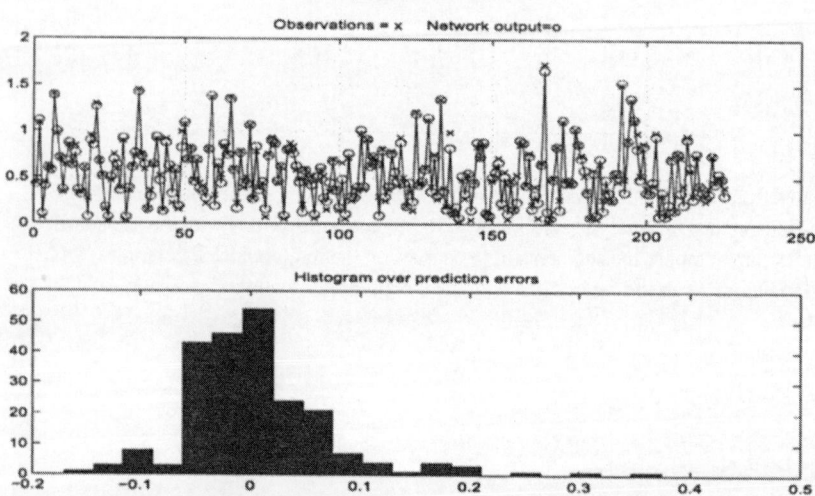

Fig. 7. Time Series Prediction Test for the 2D Wave Equation. Blue (x) indicates test values; red (o) the network response

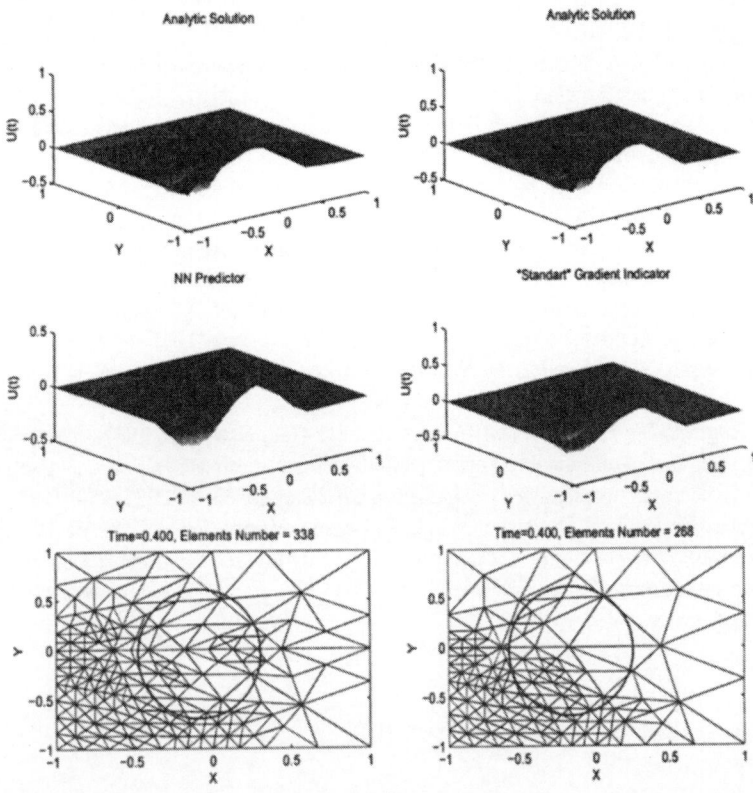

Fig. 8. FEM result; 2D Wave Equation Example 1 (see Table 2). The left figures are refined with the NN predictor. The right figures are refined with the "standard" gradient indicator

References

1. O. Axelsson and V. A. Barker. *Finite Element Solution Of Boundary Value Problems.* Academic Press, Inc., London, 1984.
2. E. Michael Azoff. *Neural Network Time Series Forecasting Of Financial Markets.* John Wiley & Sons Ltd, England, 1994.
3. G. Cybenko. Approximation by superpositions of a sigmoidal function. *Math. Control Signal System,* 2:303–314, 1989.
4. Howard Demuth and Mark Beale. *Neural Network Toolbox User's Guide.* The Math Works Inc., Natick, 2000.
5. K. Eriksson, D. Estep, P. Hansbo, and C. Johnson. *Computational Differential Equations.* Springer-Verlag,, London, 1996.
6. Laurene Fausett. *Fundamentals Of Neural Networks.* Prentice Hall, Inc., New Jersey, 1994.
7. T. J. R. Hughes. *The Finite Element Method.* Prentice-Hall, New York, 1987.
8. L. Manevitz, D. Givoli, and M. Yousef. Finite-element mesh generation using self-organizing neural networks. *Microcomputers in Civil Engineering,* 12:233–250, 1997.
9. M. Norgaard, O. Ravn, N. K. Poulsen, and L. K. Hansen. *Neural Networks for Modelling and Control of Dynamic Systems.* Cambridge University Press, Cambridge, 2000.
10. B. D. Ripley. *Pattern Recognition and Neural Networks.* Cambridge University Press, Cambridge, 1996.

Multi-Objective Genetic Algorithms on Financial Ratio Selection for Earning Forecast

Ping-Chen Lin[1,2] and Jiah-Shing Chen[2]

[1] Department of Information Management, Van Nung Institute of Technology, Jungli, Tao-Yuan, Taiwan 320, R.O.C
[2] Department of Information Management, National Central University, Jungli, Tao-Yuan, Taiwan 320, R.O.C

Abstract. Conventionally, the linear and univariate time-series models are broadly used in the earning forecast. However, the forecasting accuracy will be limited seriously without considering sufficient important factors. On the other hand, it's not guaranteed to obtain better forecasting accuracy by using more variables. In addition, considering more variables will cause inefficiency. The Multi-Objective Genetic Algorithms (MOGA) is guaranteed to select the variable set with population diversity and performs efficient searches in large space. Simultaneously, the multiple regression model can efficiently evaluate the predicting accuracy by using the least SSE. Therefore, in this paper, combining both the advantages of multiple regression and MOGA, we propose a new efficient forecasting mechanism where the forecasting accuracy is maximized with minimal number of financial ratios. Furthermore, this mechanism also includes the SWMR (Sliding Window Multiple Regression) mechanism, which can predict more accurate EPS by using sliding windows concept.

1 Introduction

Most investors have great difficulty in the stock selection decisions of their investment behavior due to their cognitive, informational or psychological limitations. The company's Earning Per Share (EPS) is broadly used by investors to help with their stock selection decisions. It means that the correct or near correct EPS plays an important role in the stock selection policy. Currently, the univariate time series models [1,2] and multiple regression models [3] are mostly used to predict future EPS.

In the univariate time series models, such as the simple linear trend model, the simple exponential model, the simple auto regressive model and etc., they mainly predict future EPS from previous EPS. It is not reasonable in the real business because the real EPS not only depends on the previous EPS but also on the various financial ratios. That is, the forecasting accuracy will be seriously limited without considering sufficient necessary conditions.

In the other way, although the multiple regression model uses multiple predictor variables to predict EPS, it is difficult to select a suitable variable subset. The commonly used selection methods are the stepwise method, the forward selection

784

method, the backward elimination method, and etc. [3]. Among them, the linear selection characteristic is the kernel methodology, because they select variables according to the specific sequence. That is, the final selected variable subset and the predicted EPS depend on the variables sequence. Each method may only produce suboptimal solution. Therefore, an efficient forecasting near-optimal EPS accuracy with a smallest variables set mechanism will be proposed in this paper.

Genetic algorithm (GA) which was introduced by Holland, is a well-known efficient nonlinear search method in large space [4]. Many relative variable selection problems can be efficiently solved by using GA [5–13]. These GA-based variable selection algorithms combine many different optimization targets into one single cost function. Among them, it is very difficult to find various possible variable sets when considering the trade-off between forecasting accuracy and variable selection.

On the other hand, using the Multi-Objective Genetic Algorithms (MOGA) to select the variable set maintains the population diversity while performing efficient searches in large space. The previous empirical results [14–18] show that the predicting accuracy of multi-objective GA-based selection is better. In this paper, combining both the advantages in multiple regression and MOGA, we will propose a new efficient forecasting mechanism where the performance forecasting accuracy is maximized under selecting the minimum subset of financial ratios. Except that, this mechanism also includes the SWMR (Sliding Window Multiple Regression) mechanism, which can predict more accurate EPS by using the sliding window concept. Following the sliding window moving trajectory, the trend of each financial ratio is predicted more accurately.

The rest of this paper is organized as follows. Section 2 reviews the Genetic Algorithm. Section 3 explains the proposed system architecture in details and designs a proper fitness function. In section 4, some performance evaluation results are shown. Finally, Section 5 gives our conclusions.

2 Genetic Algorithms

Genetic algorithm is a search method based on evolution and genetics. It combines survival of the fittest with a structured and randomized information exchange mechanism. It is a simple but powerful computation tool and makes no restrictive assumptions about the searching space [4,19–21].

To solve a problem with genetic algorithms, an encoding mechanism must first be designed to represent each possible solution of the problem by a fixed length binary string or individual. Each individual will be evaluated by a fitness function for its goodness. Genetic algorithms use a population, which is simply a set of binary strings, to search the solution space. During each generation, three genetic operators are applied to the population: selection, crossover, and mutation.

Selection operator picks individuals in the population based on their fitness. Each pair of individuals or parents undergo crossover at random by exchanging their information with each other to generate new individuals or offspring. Each bit is randomly mutated (flipped) with a small mutation rate. The process continues until the termination criterion or the predetermined generation number is reached.

3   System Architecture

The genetic algorithm is a population based search method. It uses a fixed length binary string to represent a possible solution or individual for a problem domain. In this study, we propose an efficient SWMR MOGA-based (Sliding Window and Multiple Regression model based on Multi-Objective Genetic Algorithm) architecture shown in Figure 1 to predict the most accurate EPS. In this architecture , 65 financial ratios, such as R308 (Book Value Per Share), R408 (Total Growth Rate), R432 (Operating Income Growth Rate), R612 (Fixed Asset Turnover), R835 (Operation Income Per Employee) and so on, are selected to compose the predictor variable set. Then, the GA mechanism generates the next-generation population from the parent population under considering its fitness. The SSE (Error Sums of Square) of each individual in the newer population will be evaluated by the SWMR mechanism. Finally, the fitness function is evaluated from the SSE and the number of selected predictor variables. The following subsections will detail explain these components in Figure 1.

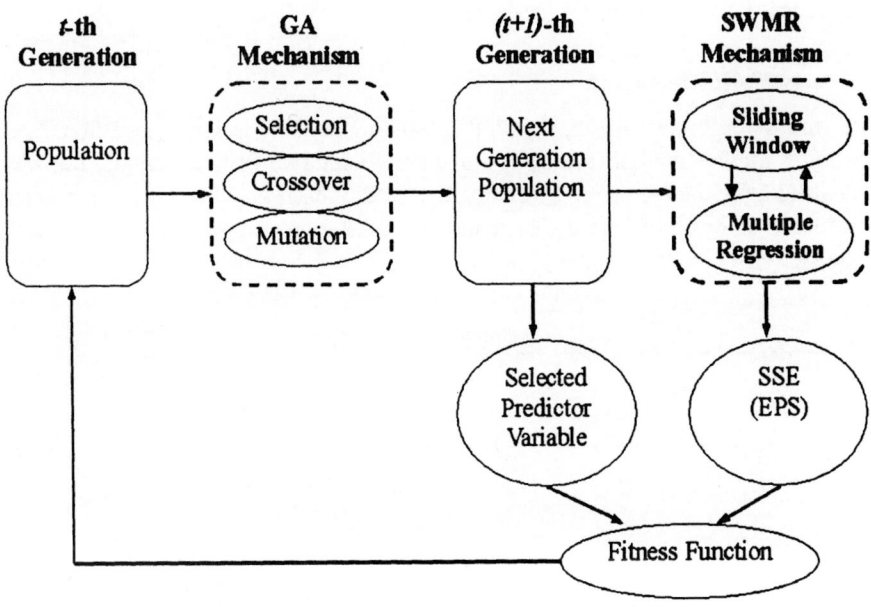

* *SWMR: Sliding Window Multiple Regression*

Fig. 1. SWMR MOGA-based Architecture

3.1 Encoding and Fitness Function

To encode a financial predictor variable in GA, it is only necessary to represent the ratio serial numbers for each string. In this paper, the total number of variables T is 65 because 65 financial ratios are considered. Thus, the search space of the predictor variables for genetic algorithms is 2^{65}. Each individual is represented by a T-bit string. In each bit, '1' or '0' means present or absent of the corresponding financial ratio predictor variable, respectively. The fitness of each individual will be evaluated by a fitness function.

In this paper, the most accurate predicting EPS under considering less variables are preferred. Thus, both the predicting accuracy criterion and the number of predictor variables criterion are considered simultaneously. Let F(SSE(EPS), E(rp)) denote the fitness function which is defined as follows.

$$F(SSE(EPS), E(rp)) = \lambda * SSE(EPS) + \gamma * [T - E(rp)/T]$$

Where SSE(EPS) denotes the error sums of square of forecasting EPS by SWMR mechanism, T is totally number of feature in entire set, E(rp) is the function depending on the number of selected variables, and λ, γ is some particular real number meaning the weight of them. In the SSE(EPS) function, the higher accuracy obtains higher score, and vice versa. Similarly, the less number of selected variables produces higher score in T-E(rp)/T function.

3.2 SWMR Mechanism

In the SWMR mechanism, the sliding window concept is introduced first. Each sliding window consists of one Training Period and one Testing Period. Following the sliding window moving trajectory, the trend of each financial ratio is predicted, which is shown in Figure 2. The detail explain sliding window simulation process in Section 4.

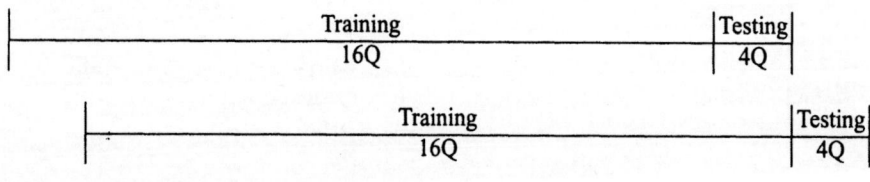

Fig. 2. Sliding windows simulation process

In each sliding window, the selected predictor variables in the individual of the newer generation population are $X_1, X_2, ..., X_m$ in the estimated multiple regression function, which is shown as below.

$\hat{Y}=b_0 + b_1X_1 + b_2X_2+ \ldots\ldots+b_mX_m$

The $\hat{Y}$ denotes the estimated EPS value in the corresponding sliding window when considering the minimum SSE in this paper.

4 Performances Evaluation

In this section, we specify the simulation method and illustrate the simulation results by using the proposed mechanism in Figure 3 - 4 and Table 1. The simulation program related our proposed mechanism was written in Borland C++ Builder 5.0 and library of Sugal 2.1. The stepwise method was simulated in SPSS 8.0 package. The used data sources are based on the Microsoft SQL-2000 server database management system. All of them were run in the Microsoft Windows 2000 server environment.

The parameters used in our GA runs are set as follows. The population size is 50 and chromosome length is 65. The number of generations is set to 2000. The selection method is roulette wheel and the crossover method is one-point crossover. The crossover rate is set to 0.6 and mutation rate is set to 0.001 because both the searching efficiency and results are best [21].

For comparing our proposed mechanism with stepwise method, five electronic businesses are randomly selected to be our simulation target which are shown as follow:

- 2303, United Micro Electronics Co., Ltd. (UMC)
- 2308, Delta Electronics Inc. (Delta)
- 2313, Compeq Manufacturing Co., Ltd. (Compeq)
- 2315, MITAC International Corp. (MITAC)
- 2316, WUS Printed Circuit Co., Ltd. (WUS)

These companies enter the stock market within 1985-1991 and has larger market value among all the companies listed and traded in the Taiwan Stock Exchange (TSE). The related sets of data are from Taiwan Economic Journal Data Bank (TEJ). The data encompass the entire period from the first quarter of 1991 to the last quarter of 1999. Each cycle in Figure 2 consists of a training period with 16 quarters and a testing period with 4 quarters.

From the selected financial ratios and generated coefficients, the forecasting EPS are obtained from the stepwise method and SWMR MOGA mechanism and shown in Table 1. In this table, we denote the $F_{EPS}G$ by forecasts earning in SWMR MOGA mechanism and the $F_{EPS}S$ denote by forecasts earning in Stepwise method. In Table 1, it is obvious that the forecasting accuracies of our proposed mechanism are better in each business. The overall mean differences between them are 0.7231 and 2.2137. It also demonstrates that our proposed mechanism provides more accurate EPS.

To reinforce the robustness and stability of our proposed mechanism, Figure 3 and Figure 4 shows that the superior results ($F_{EPS}G$) are obtained from randomly selected five companies.

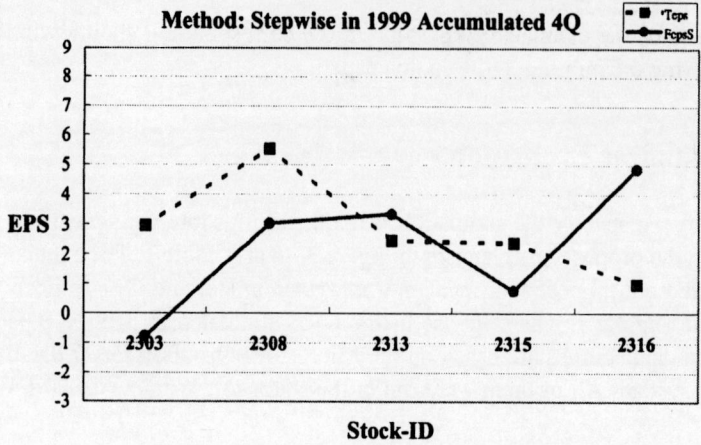

Fig. 3. T_{EPS} and F_{EPS}S of Stepwise in 1999 Accumulated 4Q

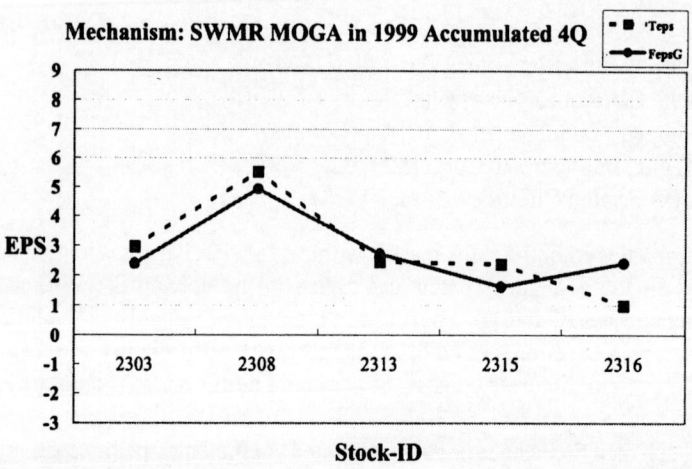

Fig. 4. T_{EPS} and F_{EPS}G of SWMR MOGA in 1999 Accumulated 4Q

Table 1. The different from T_{EPS}, $F_{EPS}G$ and $F_{EPS}S$

T_{EPS}		$F_{EPS}G$		$F_{EPS}S$	
1999		SWMR MOGA		Stepwise	
Stock-ID	value	Value	Diff.	Value	Diff.
2303	2.9700	2.4002	0.5698	-0.7696	2.2004
2308	5.5400	4.9315	0.6085	3.0329	2.5071
2313	2.4700	2.7262	0.2562	3.3497	0.8797
2315	2.3900	1.6257	0.7643	0.7724	1.6176
2316	1.0000	2.4169	1.4169	4.8637	3.8637
Mean Diff.		0.7231		2.2137	

5 Conclusions

A company's EPS is very helpful in the stock selection decision. Conventionally, the forecasting accuracy will be seriously limited whether using linear univariate time series model or multiple regression model, because insufficient necessary conditions are considered or the final forecasting results are influenced by the searching sequence. On the other hand, it's not guaranteed to obtain better forecasting accuracy if considering more variables. Therefore, an efficient forecasting near-optimal EPS accuracy with a smallest variables subset mechanism (SWMR MOGA-based Mechanism) is proposed in this paper.

The proposed mechanism combines both the advantages of Multiple Regression model and Multi-Objective Genetic Algorithm to forecast the more accurate EPS. Besides, this proposed mechanism also efficiently combines the characteristics of sliding window which can further improves the predicting accuracy and stability. From the results on 5 randomly selected electronic companies listed and traded in the TSE, The feasibility of our proposed mechanism is clearly demonstrated as shown in Figure 3 and 4.

References

1. Frank, J. F. (1999) Investment Management, Prentice Hall.
2. Brown, L. D., Hagerman, R. L., Griffin, P. A., Zmijewski, M. E. (1987) Security Analyst Superiority Relative to Univariate Time-Series Models in Forecasting Quarterly Earnings, 61–87
3. Johnson, D. E. (1998) Applied Multivariate Methods for Data Analysts, Duxbury Press.
4. Holland, J. (1975) Adaptation in Natural and Artificial Systems: An Introductory Analysis with Applications to Biology, Control and Artificial Intelligence, MIT Press.
5. Aluallim, H., Dietterich, T. G. (1992) Efficient Algorithms for Identifying Relevant Features, Proceeding of 9th Canadian Conference on Artificial Intelligence, 38–45.
6. John, G. H., Kohavi, R., Pfleger, K. (1994) Irrelevant Features and the Subset Selection Problem, Proceeding of 9th International Conference on Machine Learning, 121–129.

7. Jain, A., Zongker, D. (1997) Feature Selection: Evaluation, Application, and Small Sample Performance, IEEE Transaction on Pattern Analysis and Machine Intelligence, 153–158.

8. Kira, K., Rendell, L. A. (1992) The Feature Selection Problem: Traditional Methods and a New Algorithm, Proceeding of 10th Nat'l Conference on Artificial Intelligence, 129–134

9. Liu, H., Setiono, R. (1996) A Probabilistic Approach to Feature Selection: A Filter Solution, Proceeding of 13th International Conference on Machine Learning, 319–327.

10. Pei, M., Goodman, E. D., Punch, W. F., Ying, D. (1993) Genetic Algorithm for Classification and Feature Extraction Proceeding of 5th International Conference on Genetic Algorithms, 557–564.

11. Siedleck, W., Sklansky, J. (1988) On Automatic Feature Selection, International Journal Pattern Recognition and Artificial Intelligence, 197–220.

12. Vafaie, H., De Jong, K. A. (1993) Robust Feature Selection Algorithm, Proceeding of IEEE International Conference Tools with Artificial Intelligence, 356–363.

13. Punch, W. F., Goodman, E. D., Pei, M., Chia-Shun, L., Hovland, P., Enbody, R. (1993) Further Research on Feature Selection and Classification Using Genetic Algorithm, Proceeding of 5th International Conference on Genetic Algorithms, 557–564.

14. Emmanouilidis, C., Hunter, A., MacIntyre, J., Cox, C. (1999) Multiple-Criteria Genetic Algorithms for Feature Selection in Neurofuzzy Modeling, International Joint Conference on Neural Networks, 4387–4392.

15. Emmanouilidis, C., Hunter, A., MacIntyre, J. (2000) A Multiobjective Evolutionary Setting for Feature Selection and a Commonality-Based Crossover Operator, 2000 Congress on Evolutionary Computation, 309–316.

16. Gao, Y., Shi, L., Yao, P. (2000) Study on Multi-Objective Genetic Algorithm, Proceedings of the 3rd World Congress on Intelligent Control and Automation, 646–650.

17. Tamaki, H., Kita, H., Kobayashi, S. (1996) Multi-Objective Optimization by Genetic Algorithms: A Review, Proceedings of IEEE International Conference on Evolutionary Computation, 517–522.

18. Li, M., Kou, J., Dai, L. (2000) GA-Based Multi-Objective Optimization, Proceedings of the 3rd World Congress on Intelligent Control and Automation, 637–640.

19. Goldberg, D. E. (1989) Genetic Algorithm in Search, Optimization and Machine Learning, Addison-Wesley.

20. Mitchell, M. (1996) An Introduction to Genetic Algorithms, MIT Press.

21. Srinivas M., Lalit M. P. (1994) Genetic Algorithms A Survey, IEEE Computer, 18–20.

Hybrid Genetic Algorithm for VLSI Macro Cell Layout

Sathiamoorthy S. and Andaljayalakshmi G.

Department of Computer Science and Engineering,
Thiagarajar College of Engineering, Madurai -625015. INDIA

Abstract. Genetic algorithms have proven to be a well-suited technique for solving selected combinatorial optimization problems. The blindness of the algorithm during the search in the space of encoding must be abandoned, because this space is discrete and the search has to reach feasible points after the application of the genetic operators. This can be achieved by the use of a problem specific genotype encoding, and hybrid, knowledge based techniques, which support the algorithm during the creation of the initial individuals and the following optimization process. In this paper a novel hybrid genetic algorithm, which is used to solve macro-cell placement problem is presented. Two new heuristics are introduced. Due to a tree-structured genotype representation and hybrid, problem- specific operators, the proposed approach is able to show satisfactory performance.

1   Introduction

The design of VLSI *(very large scale integrated)* microchips is a process of many consecutive steps including specification, functional design, circuit design, physical design, and fabrication. Macro-cell layout [2] generation is a task in the *physical design cycle*. The circuit is partitioned and the components are grouped in functional units, the *macro-cells*. These cells can be described as rectangular blocks with *terminals* (pins) along their borders. These terminals have to be connected by *signal nets*, along which power or signals (e. g., clock ticks) are transmitted between the various units of the chip. A net can connect two or more terminals, and some nets must be routed to *pads* at the outer border of the layout, since they are involved in the I/O of the chip. The layout defines the positions of the cells (Figure 1).

The major objectives are chip area minimization and interconnection wire length minimization. Since the number of possible placements increases explosively with the number of blocks, even subsets of the problem have been shown to be NP-complete or NP-hard [7]. In this article, a hybrid genetic algorithm with two new heuristics is introduced for this problem. A genotype representation based on binary trees is used, and the genetic operators work directly on this tree structure.

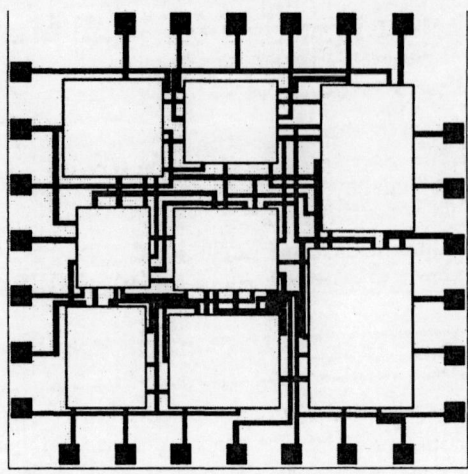

Fig. 1. The schematic representation of a VLSI macro-cell layout, which shows the position of eight cells, the routes for the signal'nets, and the I/O pads.

2 Problem Description

Inputs of the placement problem are
• a set of blocks with fixed geometries and   fixed pin positions
• a set of nets specifying the interconnections between pins of blocks
• a set of pads (external pins) with fixed positions
• a set of user constraints, e.g., block positions/orientations, critical nets, if any

Given the inputs, the objective of the problem is to find the positions and orientations of each block, so that the chip area and interconnection wire length between blocks are minimized while satisfying all the given constraints. We take wire length into account simultaneously in the optimization process. Since it is impossible to calculate the exact wire length at this stage where detailed routing has not yet been carried out, we estimate the length of each net as one-half of the perimeter of the bounding box of the net.

The objective function, which measures the quality of the resulting placement, can be expressed as follows,

$E = 1/(C_1 \text{ ChipArea} + C_2 \text{ WireLength})$

where C_1, C_2 are the corresponding weights.

3 Genetic Layout Optimization

3.1 The Hybrid Genetic Algorithm

A Hybrid Genetic Algorithm is designed to use heuristics for improvement of offspring produced by crossover. Initial population is randomly generated. The offspring is obtained by crossover between two parents selected randomly. The layout improvement heuristics RemoveSharp and LocalOpt are used to bring the offspring to a local maximum. If fitness of the layout of the offspring thus obtained is greater than the fitness of the layout of any one of the parents then the parent with lower fitness is removed from the population and the offspring is added to the population. If the fitness of the layout of the offspring is lesser than that of both of its parent then it is discarded. For mutation a random number is generated within one and if it is less than the specified probability of the mutation operator a layout is randomly selected and removed from the population. Its layout is randomized and then added to the population. The algorithm works as below:

Step 1 :
  Initialize population randomly
Step 2 :
  Apply **RemoveSharp** algorithm to all layouts in the initial population
  Apply **LocalOpt** algorithm to all layouts in the initial population
Step 3 :
  Select two parents randomly
    Apply **Crossover** between parents and generate an offspring
    Apply **RemoveSharp** algorithm to offspring
    Apply **LocalOpt** algorithm to offspring
If Fitness(offspring) > Fitness (any one of the parents) then replace the weaker parent by the offspring
Step 4 :
  Mutate any one randomly selected layout from population
Step 5 :
  Repeat steps 3 and 4 until end of specified number of iterations.

3.2 Genotype Representation

The phenotypic representation for the placement problems is basically the pattern that describes the position of the blocks. Binary slicing trees are well suited to represent placement patterns and have already been used in genetic algorithms [6]. During recombination, partial arrangements of blocks are transmitted from parents to offspring. The corresponding operation is the inheritance of subtrees from the parents. Encoding the tree in a string complicates this operation, since the string needs to be decoded into the slicing tree to execute the recombination, then recoded into an offspring chromosome afterwards. There is no reason for using a string encoding except for the analogy to the natural evolution process, where the genetic information is encoded in a DNA string.

794

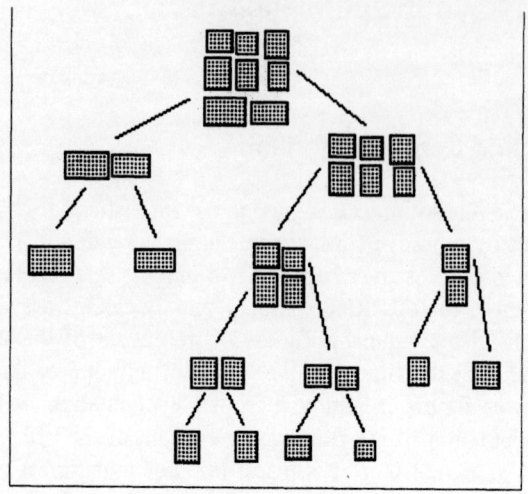

Fig. 2. The genotype

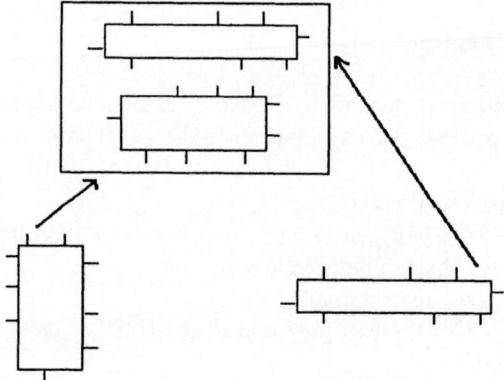

Fig. 3. The composition of a meta-block

When directly using the slicing tree as the genotype representation, further de-coding or encoding the tree when applying genetic operators is avoided. The geno-type is encoded as a binary slicing tree, which defines the relative placement of the cells (fig.2). It is composed in a bottom-up fashion. In each inner node two *blocks* (in the lowest level these are single cells) are joined to a *meta-block* (partial placement). In each meta-block the orientations of the combined blocks are fixed (fig.3). Therefore every tree describes several possible shapes for the correspond-ing layout, which enormously improves the performance of the GA. Blocks or sub-patterns in a tree defining a layout, is always stacked vertically upon each other. The pattern characterized by the right successor of an inner tree node is al-ways positioned on top of the pattern characterized by its left successor when combining both parts into a pattern or meta-block.

3.3 Genetic Operators

During the optimization process the placement of the blocks has to be changed. The genetic operators directly work on the tree-structure by combining subtrees of parents (crossover) and modifying the tree of an individual (mutation). The crossover operator takes two individuals (parent) out of which one offspring is composed by combining two subtrees, one from each parents. Unfortunately, these parts usually do not add up to a complete layout. After the combination of the two subtrees the redundant blocks are deleted and the missing blocks have to be added at random positions to the tree to ensure that the offspring finally represents a correct layout. Mutation operator modifies either by exchanging simple blocks or a block (leaf) with a meta-block (subtree) or by exchanging two meta-blocks. These cases represent the exchange of two cells, a cell with a partial layout, and the exchange of two partial layouts on the layout surface.

4 The RemoveSharp Algorithm

The RemoveSharp algorithm removes sharp increase in the wirelength due to a badly positioned macro-cell, which eventually decreases the fitness. The algorithm works as below:

Step 1 : A list (CONNECTIVITY-LIST) containing the m macro-cells highly connected to a selected macro-cell by signal nets is created.

Step 2 : RemoveSharp removes the selected macro-cell from the binary tree.

Step 3 : Now the selected macro-cell is reinserted in the binarytree either as left or right subtree to any one of the macro-cells in CONNECTIVITY-LIST and the fitness of the new layout is calculated for each case.

Step 4 : The genotype, which produces the highest fitness, is selected.

Step 5 : The above steps are repeated for each macro-cell in the binary tree.

4.1 Time Complexity of RemoveSharp

RemoveSharp removes a macro cell, inserts it at 2m (m is CONNECTIVITY-LIST size) different positions, and selects the best one. This is done for n macro cells. Using pointers the time taken for the functions, remove() for removing a block and insert() for inserting next to a specified macro cell is linear. The time complexity of the RemoveSharp algorithm is approximately linear as the basic operations are linear.

5 The LocalOpt Algorithm

The LocalOpt algorithm works as follows:

Step 1: Remove a subtree having q macro-cells from the binary tree.

Step 2: Compute the fitness of all possible arrangements of the n macro-cells (leaves) in the subtree.

Step 3: Find the one with maximum fitness value.

Step 4: Insert the optimized subtree at the position in the binary tree from where it is removed.

5.1 Time Complexity of LocalOpt

When LocalOpt is applied to a subtree only the links between the macro cells (at- ·tached to the leaf nodes) in the subtree and the leaf nodes in the subtree are changed. The binary tree structure of the subtree is not changed. The number of possible ways by which each one of the macro cells in the subtree can be assigned to a leaf node in the subtree is a constant (q!) for a given subtree size. For example it is 24 for subtree having 4 macro cells and 720 for subtree having 6 macro cells. Thus the time taken for applying LocalOpt to all the subtrees of same size is a constant. The Time Complexity of LocalOpt is hence k*O(N), where k is a constant. The constant k varies a little more than linearly with N due to the increase in time taken for finding the subtrees of specified size, when the number of macro cells in the problem (N) increases. Therefore the time taken for LocalOpt increases a little more than linearly for constant subtree size (q).

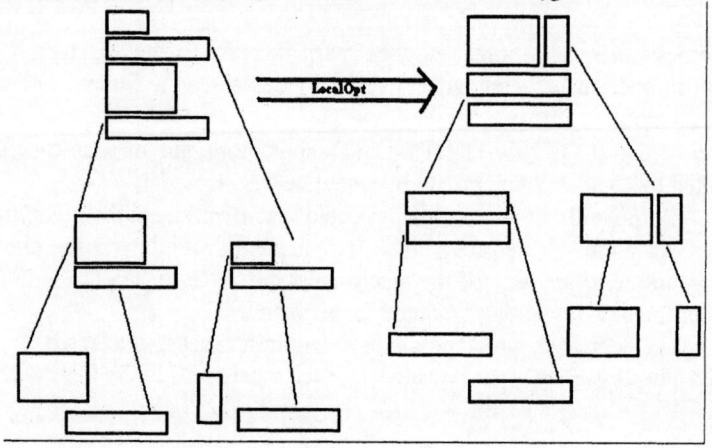

Fig. 4. LocalOpt applied to a subtree.

In fig.4 LocalOpt is applied to a subtree. It is obvious that the one obtained after applying LocalOpt (fig.4 right) results in a placement with higher fitness value.

6 Results

The hybrid genetic algorithm for the layout generation problem was tested on real-life circuits chosen from a benchmark suite that was released for design workshops in the early 90s and is often referenced in the literature as the *MCNC benchmarks*. They were originally maintained by *North Carolina's Microelectronics, Computing and Networking Center*, but are now located at the *CAD Benchmarking Laboratory (CBL)* at North Carolina State University. These benchmarks are standard problems in macro-cell layout, and the characteristics of the circuits are shown in Table 1. All the benchmark problems used in this paper can be downloaded from the URL: www.cse.ucsc.edu/research/surf/GSRC/MCNC/.

Table 1. The benchmark circuits for the macro-cell layout generation problem

Benchmark	Cells
Apte	9
Xerox	10
Hp	11

The algorithm is implemented in C++ and executed on a Pentium 300MHz workstation. The results are obtained within thirty seconds. The effects of various parameters and heuristics are analyzed; the best performance of HGA is found when the values of the parameters are set as below:

RemoveSharp (m)	:	5	LocalOpt(q)	:	6
Probability of			Population size	:	50
Mutation operator	:	0.02	Number of Iterations :		10000

The classical layout problem is introduced in section 2; in which wirelength is also a factor in fitness value. But due to technological progress, technologies like Over-The-Cell routing are now used so there is no need to determine wirelength and to add routing space (through which wires are routed) to the layout [7]. Therefore in the fitness function introduced in section 2, C_1 is assigned one, and C_2 is assigned zero. It is difficult to fairly compare our algorithm with the other approaches reported, because they include routing space in their placements, which is not necessary any more. So the results of our algorithm are compared only with the results in [7].

In Table 2 our algorithm has been compared with Cluster Refinement for Block Placement. In case of Apte and Hp our algorithm has outperformed cluster refinement while in other case same result is obtained. Parallel Genetic Algorithm (PGA) can be designed using the new heuristics and better results can be obtained. It had been already shown that PGA generates better results than the sequential one [1,5].

Table 2. Comparison of HGA with Cluster Refinement Algorithm

Benchmark	HGA	Cluster Refinement
Apte	47.30	48.42
Xerox	20.30	20.30
Hp	9.39	9.575

798

7 Future Research

A parallelization of the genetic algorithm is planned and along with the strategy adaptation given in [5]. It might be ingenious to exchange or move large parts of the layout during the early stage of optimization, and doing only minor changes when the population converges to an optimum. This can become possible by adapting the frequencies of the mutation operator during the optimization. Further a gene-pool recombination operator [6,4] will be implemented which might replace the current crossover operator. For comparison a combinatorial optimization problem like the layout generation, the biologically motivated crossing of two parent chromosomes is likely to be less efficient. The construction of an offspring out of a pool of good building-blocks seems to be more suitable.

8 Conclusion

In this paper an approach has been presented to incorporate domain knowledge into a genetic algorithm, which is supposed to compute near-optimal solutions to VLSI Placement problem. The feasibility of the approach has been demonstrated by presenting performance results for benchmark instances. We find that the implementation of the two newly introduced heuristics result in near optimal solutions in all cases. These heuristics are simple, straightforward and easy to implement when compared to other algorithms. We believe this approach promises to be a useful tool in VLSI Design Automation.

References

1. Lienig J. (1997) A Parallel Genetic Algorithm for Performance Driven VLSI Routing. IEEE Transactions on Evolutionary Computation Vol. I. No.1 :29-39
2. Mazumder P., Rudnick E. (1999) Genetic Algorithm for VLSI Design, Layout and Automation. Addison-Wesley Longman Singapore Pte. Ltd., Singapore.
3. MCNC Benchmarks: www.cse.ucsc.edu/research/surf/GSRC/MCNC/
4. Schnecke V., Vornberger O (1996) A Genetic Algorithm for VLSI Physical Design Automation :In Proceedings of Second Int. Conf. on Adaptive Computing in Engineering Design and Control, ACEDC '96 26-28 Mar 1996, University of Plymouth, U.K., pp 53-58
5. Schnecke V., Vornberger O (1996) An Adaptive Parallel Genetic Algorithm for VLSI-Layout Optimization :In Proceedings of 4th Int. Conf. on Parallel Problem Solving from Nature (PPSN IV) 22-27 Sep 1996, Springer LNCS 1141, pp 859-868
6. Schnecke V., Vornberger O (1997) Hybrid Genetic Algorithms for Constrained Placement Problems. IEEE Transactions on Evolutionary Computation. Vol. I. No.4. :266-277
7. Xu, J., Guo P., et al. (1997) Cluster Refinement for Block Placement: In Proceedings of ACM-DAC California ,paper 47.4

Artificial Neural Networks Modeling as an Industrial Plant Fault Diagnostic Tool

Zvi Boger[1, 2]

[1]Licensing and Safety Division
Israeli Atomic Energy Commission
P.O. Box 7061, Tel-Aviv 61070, Israel
and
[2]OPTIMAL – Industrial Neural Systems Ltd.
P.O. Box 9201, Be'er Sheva 84191, Israel

E-mail: zboger@bgumail.bgu.ac.il

Abstract. Fault diagnosis of a complex industrial plant is an important part of the operator task. Artificial neural networks (ANN) model learns from the plant past behavior and can be used for fault diagnosis. Efficient large-scale ANN modeling algorithms are presented, with two industrial diagnostic cases involving quality control problems in inorganic salt and nitro-cellulose production processes, and an example of process trend classification. Analysis of the trained ANN models revealed process knowledge that helped the operators and increased their confidence in the ANN models

Keywords. Artificial Neural Networks, Industrial Diagnostics, Process Trends, Fault Detection

1. Introduction

The suggested use of artificial neural networks (ANN) modeling techniques in industrial plants, in which the model is learned from data of the plant behavior, arouses strong emotions. "No complicated equations! No man-years of development effort!" cheer the proponents. "No detailed equations? No reliability!" counter the opponents.

One use of such models is on-line fault diagnosis. It was suggested for the chemical industry in the early nineties [1]. The nuclear power industry is interested in this capability, as timely fault diagnosis can prevent costly, and public relations damaging, nuclear power plant (NPP) incidents and shutdowns. The International Atomic Energy Agency has reviewed periodically the feasibility of "intelligent" software in NPP safety systems in Expert Group meetings since

1988, and in most meeting papers were presented on possible ANN-based systems. Several reviews on the use of ANN models in NPP diagnostics were published [2], [3].

In a recent paper [4], two examples of NPP diagnostic using ANN models were given: An example classification of NPP simulator-derived response to three fault situations; An example of actual NPP data, where the cause of an unexplained fault can be found using Genetic Algorithm (GA) analysis of a trained ANN model of this fault behavior.

The nuclear power industry has the unique advantage of having access to full-scale plant simulators used for on-going training of NPP operators, so fault scenarios can be simulated easily, which is not the case in most industrial plants. However the use of ANN modeling in industrial plants is spreading, as other modeling methods are costly, both in resources and time, to fully meets the requirements of fault diagnosis.

An often-cited opposition to the use of ANN modeling in industrial diagnostics is the lack of "explanation" facility, the ability of the operator to understand the basis of the ANN recommendations. In this paper diagnostic examples are presented to show that the "black-box" image of ANN model is misleading, and the trained ANN model can be analyzed to correctly explain the causes of the observed fault symptoms

The structure of the paper is as follows: A review of the PCA-CG algorithms for large-scale ANN modeling, the Causal Index (CI) method of analyzing trained ANN, and the use of the hidden neurons' output values as clustering tool. An example shows the power of the ANN modeling to provide timely information on simulated process trends even in the presence of noise. Two examples of industrial fault diagnosis are given, both analyzing the causes of not meeting a product specification. Because of non-disclosure agreements, the exact details of the second plant, product and variable names are withheld.

2. Large-scale ANN Modeling and Analyzing Issues

2.1 Efficient ANN Modeling and Identification of the Relevant Inputs

One of the difficulties of training large-scale ANN models can be overcome by using the PCA-CG algorithm that starts the training from non-random initial weights. Several papers [5], [6] describe the experience in modeling large-scale systems and extracting some knowledge from the trained ANN models using relatively simple techniques. These are:
- Training large ANN, beginning from non-random initial connection weights;
- Identification of the relevant inputs of a system;
- Optimizing the number of the neurons in the hidden layer;

- Calculation of causal indices (CI), for learning some rules of the relations between each input and output.
- Using the outputs of the hidden layer neurons for fault detection and classification

The first three techniques were developed already in 1990 [7], but the details were published later [6], [8]. With these algorithms ANN models of industrial and other large systems with hundreds of inputs and outputs have been developed on PC computers in a fraction of the time and the number of training epochs usually reported.

2.2 The Causal Index Method of Analyzing ANN Models

The causal index method is an easily, somewhat qualitatively, method for rule extraction [9]. The CI is calculated as the sum of the product of all "pathways" between each input to each output,

$$CI = \sum_{j=1}^{h} W_{kj} * W_{ji} \qquad (1)$$

where there are h hidden neurons, W_{kj} are the connection weights from hidden neuron j to output k, W_{ji} are the connection weights between input i to hidden neuron j.

Plotting the CI for each output as a function of the inputs' number reveals the direction (positive or negative) and the relative magnitude of the relationship of the inputs on the particular output. Although somewhat heuristic, it is more reliable than the local sensitivity checks. Their advantage is that they do not depend on a particular input vector, but on the connection weight set that represents all the training input vectors. This is also one of their limitations, as a local situation may be lost in the global representation.

2.3 The Use of the Hidden Neurons Output for Classification

The PCA-CG algorithm recommends ANN architecture with a small number of neurons in the (single) hidden layer. It leads to efficient clustering, based on the hidden layer's neuron outputs.

The neurons in the hidden layer are supposed to learn concepts, which then should be orthogonal to each other, as in the PCA dimensionality reduction. This led to examining the possibility that a unique binary pattern of the hidden neurons could indicate the correctness of the ANN classification. A similar suggestion was given for calculating the maximum information content of the hidden neuron outputs using entropy concepts [10]. Entropy is calculated as the sum over all hidden neurons of p * ln (p), when p are the hidden neurons outputs. Experience with well-trained classifying ANNs showed this orthogonality by plotting the output of a hidden neuron against the output of another hidden neuron. In most cases the plotted points were on the plot corners and axes. Thus p = [0, 1] decreases the entropy. It was shown that the incorrect binary pattern of the hidden neurons could identify false classifications of chemical substances by an "artificial nose" sensor, when the trained ANN was presented with a highly noisy and drifting input spectra [11].

Once the patterns are clustered in groups, it is easy to calculate the mean value of each attribute for each group. When compared to the overall data attribute means, it is evident which attributes are responsible for the clustering.

2.4 The use of Auto-Associative ANN for Clustering

Auto-associative ANN (AA-ANN) is a ANN model in which the target (output) vectors are identical to the input vectors, (after pre-processing). When the AA-ANN trains to a small error, the previous algorithm of the hidden neurons' output pattern can be used as for non-supervised clustering, as no classes are assumed a-priory. It can also serve as a model of the data consistency. By monitoring the differences between the actual input vector and the predicted AA-ANN output vector, an estimate of a possible plant-model mismatch can made. If only one input differs from the predicted value, most chances are that the actual input is in error, be it from data entry error, or a sensor fault [12]. If several such differences are evident, the probable cause is that the Plant State is different from the states from which the training data was taken. It may be an incipient fault, or a new plant state. In both cases the operator should be alerted. A more detailed description of the use of AA-ANN for clustering is presented in a recent paper [13].

2.5 Identification of the more significant process variables

Once the trained ANN model is obtained, there is still the need to reduce the model complexity. One important motivation is to construct a parsimonious model with an optimal trade-off between fitting accuracy and the number of adjustable parameters in the model. By removing unnecessary structures in the network, over-fitting a limited set of process data with a model containing too many degrees of freedom is avoided. The reduced network model also highlights the dominant underlying process relationships, represented by the remaining inputs, nodes and connections in the reduced network.

When the PCA-CG algorithm is used, the number of nodes in the hidden layer is selected in advance to represent the intrinsic dimensionality of the data set. Inputs that are linearly correlated do not contribute independent information and thus the information content of the correlated inputs can be captured in lower dimensions. The amount of information lost in the projection of the input data into a lower-dimensional space is quantified by the fraction of the variance of the original data not represented in the reduced space. Typically, we specify as many hidden nodes as dimensions needed to capture 70-90% of the variance of the input data set. This immediately places limits on the number of hidden nodes in the initial network, and subsequently it remains only to remove nodes in the hidden layer that represent dimensions of the hidden layer that do not specifically contribute to the output of interest. In our experience, the suggested number of hidden neuron is already very close to the optimal size before the network reduction procedures are applied. Thus, most of the effort was directed at reducing the number of network inputs, (the number of outputs is not a factor since they are fixed by the requirements of the model). The statistical network reduction method does not address removal of connections between nodes retained in the reduced model. Nonetheless, many connections are eliminated when the non-relevant inputs and hidden layer nodes are removed.

A procedure that can avoid the time consuming stages of ANN pruning by the other published techniques was developed [6]. This method is based on elementary statistical considerations involving the variation of the inputs to the hidden layer nodes in the trained network. If the variance of the total (summed) input to a hidden node, calculated over the training set as a whole, is small, then the node contributes little more than a constant bias to the subsequent layer. A similar analysis of variance can determine the importance of network inputs. Inputs that make relatively small contributions to the variance at the input of the hidden layer serve only as biasing constants and can be replaced by fixed biases at the hidden layer nodes. When analyzing a given input, its collective effect on all hidden layer nodes is of interest, rather than the effects on individual hidden nodes, since removing the input de-couples it simultaneously from all hidden nodes.

3. Automatic Process Trend Classification

The data used in this demonstration was donated to a data repository [14]. It consists of 6 classes of process trends, 100 hundred trends in each class, 60 synthetic values in each process time-series vector. The six classes are as follows - normal, cyclic, increasing trend, decreasing trend, upward shift, and downward shift. The start of the deviation from "normal" behavior is at a random point in the time-series vectors. The database was used to identify statistical features that will help to find similarity patterns in time series. However the best success rate was below 50% [15].

The process trend charts were randomly divided to form training and verification sets, on a 70:30 proportion. The data was preprocessed column wise, by subtracting the column mean and dividing by the column standard deviation. The transformed was used as input to the ANN. The output vectors for the classification task had six binary class outputs, with the correct trend class having the value of 0.9, the others having the value of 0.1. The training algorithm suggested four hidden neurons. Thus the classification ANN had the structure 600-4-6.

The classification ANN was trained using the conjugate-gradient optimization [16] in several hundred epochs. The ANN classification was determined by the highest-class output value. All training process trends were classified correctly, and only two out of the 167 verification examples were mis-classified.

To check the robustness of the ANN classification, random normal noise, with increasing amplitude, was added to the original data, and the noisy data was presented to the trained ANN. The results are shown in Table 1. As seen from the results, the addition of large amplitude noise degrade the classification accuracy, but the most troubling mis-classification, of abnormal trend as "normal" error rate remains low until the highest noise level.

Noise level (standard deviations)	Number of mis-classified **Verification** examples	% of mis-classified **Verification** examples	% of mis-classified **Verification** examples as **"normal"**
0	2	1.2%	0%
0.5	6	3.6%	0%
1.0	46	27.5%	1.2%
2.0	54	32.3%	6.0%

Table 1: Robustness of the ANN-based trend classification

4. Diagnosis of the Causes of Inorganic Salt Production Problems

In an inorganic salt production, the properties of the product did not always meet the specifications for two properties. The provided data consisted of 46 examples of the quality control laboratory reports, which gave the size classification of the salt into 5 classes, defined by the sieve mesh number. It also gave three chemical impurities concentrations, A B and C, a production-operating variable D, and the process throughput E. The product properties X and Y were given, too.

The pre-processing was done as in the previous section. The inputs were all the data values, and those were also the output values, after re-scaling to the [0.1-0.9] range. The PCA-CG recommended 4 hidden neurons, thus the AA-ANN architecture was 12-4-12. The training stopping criteria were no error decrease by 0.5% from the previous error, after 10 epochs in which the local minima escape algorithms did not succeed. After training to a small error, the hidden neurons' outputs were rounded to [0,1] values, and grouped according to the resulting binary patterns. The means of each attribute of each group was calculated, and divided by the total data attribute mean. The results are given in Table 2.

Table 2. Relative mean attributes of each salt properties clusters

# of examples	A	X	Y	6 mesh	8 mesh	10 mesh	14 mesh	18 mesh	B	C	D	E
3	0.84	1.1	0.03	0.58	1.16	1.07	1.00	1.00	0.99	0.90	0.93	1.00
4	1.14	1.64	0.09	0.70	0.97	0.97	0.99	1.00	1.00	1.20	1.15	1.00
2	1.41	1.48	0.16	0.92	0.81	0.99	0.99	0.99	1.00	1.04	1.04	1.00
8	0.86	1.18	0.50	0.91	1.07	1.04	1.00	1.00	1.00	0.89	1.02	1.00
6	1.12	1.16	0.39	0.69	0.95	0.97	1.00	1.00	1.00	1.22	1.16	1.00
4	0.89	0.97	0.60	0.83	0.64	0.81	1.00	1.00	1.00	0.96	0.86	1.00
3	1.41	0.73	0.99	0.88	1.05	1.04	1.00	1.00	1.01	1.08	0.92	1.00
10	0.95	0.62	2.05	1.54	1.14	1.09	1.01	1.00	1.01	0.91	0.95	1.01
2	0.92	0.57	2.66	0.58	0.86	0.86	1.00	1.00	0.97	1.09	0.94	0.99
2	0.81	0.52	1.66	1.57	1.27	1.11	1.01	1.00	0.99	0.89	1.04	1.01
2	0.91	0.49	1.94	1.43	1.08	0.95	1.00	1.00	0.99	0.95	0.87	1.00

Table 2 shows that the data can be divided into two groups of clusters. The first group includes clusters of data with better than the average X property, smaller then the average Y property, which are the desired specifications. The second group includes those examples that are not meeting the specifications. It can be seen that the first group tend to have A impurity higher then the second group, smaller proportion of the 6 mesh crystals, and higher than the average production rate D. The other attributes seem to be the same.

5. Quality Control Improvement in a Nitro-cellulose Production Process

A nitration process of cotton wool in a batch reactor is a part of nitro-cellulose production. The quality control specifications of the viscosity and the turbidity of the product were frequently not met in a production facility, and many attempts were made to modify the process recipe in order to overcome the production of off-spec material

An ANN model was taught by from a database of 90 production batches, with 37 recipe variables as inputs and the two quality control measurements as outputs.. The number of the relevant inputs was found to be 12, and the resulting prediction of the viscosity by the reduced ANN gave a good agreement between the prediction and the actual measurement.

806

Causal indices of the reduced ANN are plotted in the Figure, showing, for example, that increasing the nitric acid concentration or the cotton-to-nitric acid ratio increases the product viscosity, and using the "Miluot" cotton as raw material decrease the viscosity. The knowledge acquired from it was in agreement with the plant engineers' experience, or if new, not in contradiction with it or with basic process chemistry.

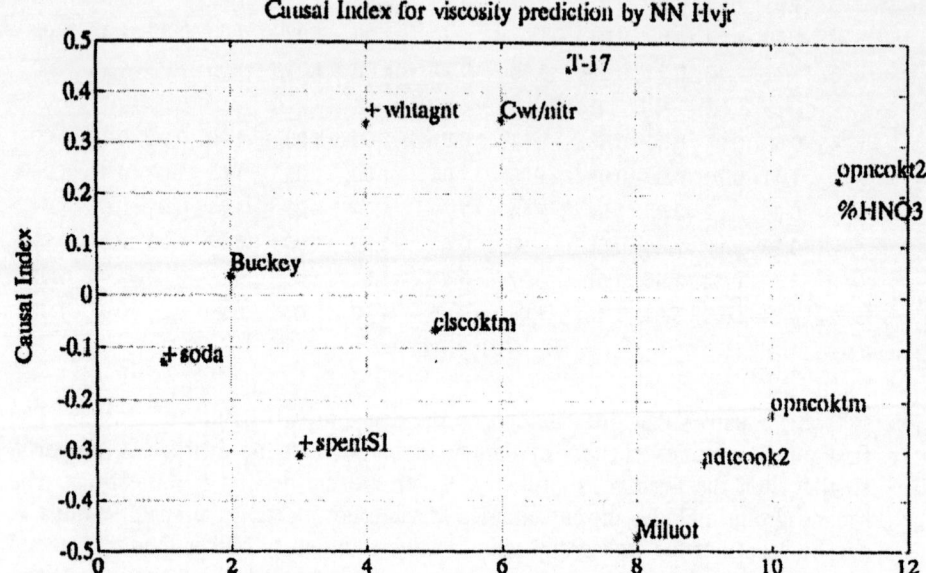

6. Discussion and Conclusions

The examples given in this paper show that the ability to easily train large-scale ANN models can be used for fault classification, provided that enough training data are available. The ANN model may be analyzed by the CI technique, or by the clustering ability of the hidden neurons, to provide a plausible explanation of its recommendation.

The first example given in this paper is a synthetic one, but shows that even when corrupted by noise, the ANN model can classify abnormal trends. The second example shows that the analysis of the clusters identified by the AA-ANN model of the salt production process can point to the causes of the off-spec product. The third example, of the nitro-cellulose production process, demonstrates the usefulness of the causal indices derived from an ANN model as knowledge extraction tool for improving the quality control of the product.

Thus the additional techniques of analyzing ANN models of industrial processes can be useful for increasing the acceptance of these models by the industrial users, and dispel some of their mistaken perception as "black boxes".

References

1. Werbos, P.J., Mc Avoy, T.J. and Su, T. (1992) Neural networks system identification, and control in the chemical process industry. In D.A. White, and D.A. Sofge, (Eds.) Handbook of Intelligent Control, Van Nostrand Reinhold, New York, 283-356
2. Reifman, J. (1997) Survey of artificial intelligence methods for detection and identification of component faults in nuclear power plants. Nuclear Technology 119, 76-97
3. Boger, Z. (1998) Implementing artificial neural networks in nuclear power plant diagnostic systems: Issues and challenges. Proceedings of the IAEA Technical Committee Meeting on Diagnostic Systems in Nuclear Power Plants, Istanbul, Turkey, IAEA-IWG-NPPCI--98/2, 149-160
4. Boger, Z. (2001) Artificial neural networks modeling as a diagnostic and decision making tool. Submitted for publication in D. Ruan and P.F. Fantoni, Power Plant Surveillance and Diagnostics, Springer, and the International Journal of Intelligent Systems, Special issue on Intelligent Systems for Plant Surveillance and Diagnostics
5. Boger, Z. (1997) Experience in industrial plant model development using large-scale artificial neural networks. Information Sciences - Applications 101, 203-212
6. Boger, Z. and Guterman, H. (1997) Knowledge extraction from artificial neural networks models. Proceedings of the IEEE International Conference on Systems Man and Cybernetics, SMC'97, Orlando, Florida, 3030-3035
7. Boger, Z. (1992) Application of neural networks to water and wastewater treatment plants operation. Transactions of the Instrument Society of America 31, 25-33
8. Guterman, H. (1994) Application of principal component analysis to the design of neural networks. Neural, Parallel and Scientific Computing 2, 43-54
9. Baba, K., Enbutu, I., and Yoda, M. (1990) Explicit representation of knowledge acquired from plant historical data using neural network, Proceedings of the International Joint Conference on Neural Networks. San Diego 3, 155-160

10. Kamimura, R. and Nakanishi S. (1995) Hidden information maximization for feature detection and rule discovery, Network Computation in Neural Systems **6**, 577-602

11. Boger, Z., Ratton, L., Kunt, T.A., Mc Avoy, T.J., Cavicchi, R.E. and, Semancik, S. (1997) Robust classification of "artificial Nose" sensor data by artificial neural networks. Proceedings of the IFAC ADCHEM '97 Conference, Banff, Canada

12. Turkcan, E., Ciftcioglu, O. and van der Hagen, T.H.J.J. (1998) Surveillance and fault diagnosis for power plants in the Netherlands: Operational experience. Proceedings of the IAEA Technical Committee Meeting on Diagnostic Systems in Nuclear Power Plants, Istanbul, Turkey, IAEA-IWG-NPPCI--98/2, 53-70

13. Boger, Z. (2001) Hidden neurons as classifiers in artificial neural networks models. Proceedings of the ANNES'01 Conference, Dunedin, New-Zealand

14. Alcock, R.J. (1999) .http://kdd.ics.uci.edu/databases/synthetic_control

15. Alcock, R.J., and Manolopoulos, Y. (1999) Time-series similarity queries employing a feature-based approach. 7th Hellenic Conference on Informatics, Ioannina, Greece

16. Leonard, J. and Kramer, M.A. (1990) Improvement of the backpropagation algorithm for training neural networks. Computers chem. Engng. **14**, 337-341

An Application of Genetic Programming to Electronic Design Automation: from Frequency Specifications to VHDL Code

Roberto Rossi[1], Valentino Liberali[2], and Andrea G. B. Tettamanzi[2]

[1] Università degli Studi di Pavia
Dipartimento di Elettronica
Via Ferrata 1, 27100 Pavia, Italy
Email: roberto.rossi@ele.unipv.it
[2] Università degli Studi di Milano
Dipartimento di Tecnologie dellíInformazione
Via Bramante 65, 26013 Crema, Italy
Email: liberali@dti.unimi.it, tettaman@dsi.unimi.it

Abstract. An evolutionary algorithm is used to design a digital filter with reduced power consumption. The proposed design approach combines genetic optimisation and simulation methodology, to evaluate a multi-objective fitness function which includes both the suitability of the filter transfer function and the transition activity of digital blocks. A client-server computer program allows the user to exploit several CPUís with parallel evolving populations, aiming at the same target specifications but with different fitness functions. Examples on design cases are presented and discussed.

1 Introduction

Electronics design automation must cope with technological trend in silicon integration. Nowadays, the possibility of integrating millions of transistors onto a single silicon chip is demanding for new CAD tools, to bridge the gap between technology capabilities and designer productivity.

Therefore, new branches in computer aided design are expected to emerge in the next future: among them, evolutionary algorithms seem to be very promising, due to their capability to provide solutions to hard design problems. Built on the key concept of Darwinian evolution in biology [1], evolutionary algorithms are a broad class of optimisation methods [2]. In the past years, they have been successfully applied to physical design (partitioning, placement and routing). Now bio-inspired electronic design methods are being considered in a variety of circumstances [3,4]. Evolutionary algorithms for circuit synthesis are a powerful technique, which could provide innovative solutions to several design problems, also when classical decomposition methods may fail [5].

This paper presents an evolutionary approach to the design of digital filters. In a previous work, it was demonstrated that genetic programming can be used to design digital filters starting from mask specifications, and obtaining a synthesizable VHDL code [6]. Now, practical aspects are being considered, aiming at a more efficient design tool, suitable for use in industrial environment.

810

2 Motivation and Problem Statement

Microintegrated technology enables the development of deep submicron CMOS circuits for digital signal processing (DSP) in a broad variety of applications. However, the ever and ever increasing integration scale makes designers to face new problems and requires new design methodologies and tools. Todayís major issues are:

1. the increasing gap between technology capability and designer productivity. Such a scenario is demanding for new design methodologies and innovative CAD tools, as in previous stages of the ìdesign crisisî [7]. Indeed, a remarkable effort is being spent, aiming at the automatic synthesis of electronic design, starting from functional speciÝcations rather than from behavioural description;
2. the increasing power consumption. Circuits with small transistors can be densely packed into a single chip, and size reduction allows frequency to be increased. Consequently, the dynamic power P due to the logic transitions of circuit nodes also increases. It can be expressed as:

$$P = \frac{1}{2}CV^2 f\alpha \tag{1}$$

where C is the total capacitance, V is the power supply voltage, f is the clock frequency, and α is the transition activity, i.e. the average number of logic transitions in a clock period [8,9];
3. product testability. The increasing complexity of integrated systems leads to a huge number of internal nodes which cannot be directly controlled and/or observed. Since manufacturability requires to test every chip to determine whether it is good or bad, it is apparent that more powerful design-for-test techniques are needed.

This paper addresses issues 1 and 2, proposing an approach for the automatic generation of the digital circuits having a reduced transition activity of digital gates. Such a task is not a straightforward one, because digital activity is a non-linear function of input patterns.

3 Genetic Representation of Digital Filters

A description of digital Ýlters can derived from their frequency response in the z-domain [10]. The frequency response of a Ýnite impulse response (FIR) digital Ýlter is:

$$H(z) = \sum_{k=0}^{N-1} h(k)z^{-k} \tag{2}$$

and the canonical direct form of the Ýlter shown in Fig. 1.

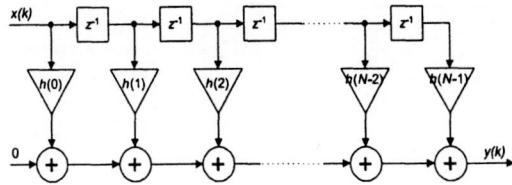

Fig. 1. Canonical direct form of a FIR Ýlter

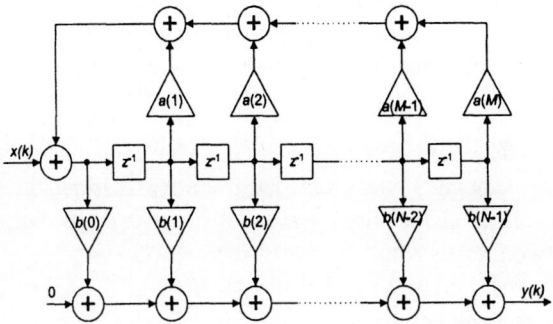

Fig. 2. Canonical direct form of an IIR Ýlter

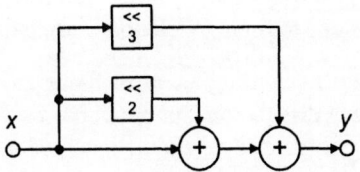

Fig. 3. Multiplication by 13 implemented with shifters and adders

An inÝnite impulse response (IIR) digital Ýlter can be represented with the canonical form shown in Fig. 2, and its frequency response is:

$$H(z) = \frac{B(z)}{A(z)} = \frac{\sum_{k=0}^{N-1} b(k)z^{-k}}{1 - \sum_{j=1}^{M} a(j)z^{-j}} \tag{3}$$

To save area and to reduce power consumption, generic multiplier blocks are often replaced with shifters and adders [11]. As an example, multiplication by 13 can be implemented with two shifts and two additions, as illustrated in Fig. 3, where the block ì $<< n$î means ìleft shift by n positionsî.

The canonical signed digit (CSD) representation [12] assigns a separate sign to each digit: 0, 1 and $\emptyset$(= ñ1). Its goal is to minimise the number of non-zero digits: by encoding the Ýlter coefÝcients with CSD, the Ýlter ouput can be computed using a reduced amount of hardware, since multiplications by zero are simply not implemented. As an example, consider the multiplication by 15: since $15 = 2^3 +$

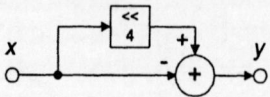

Fig. 4. Multiplication by 15 implemented with one shifter and one subtractor

$2^2 + 2^1 + 2^0 = (001111)_2$, this operation in binary arithmetics would require three shifts and three additions; while using CSD we can write $15 = 2^4 - 2^0 = (01000\bar{1})_2$, and we implement the same operation using only one shifter and one subtractor (Fig. 4).

Starting from these considerations, a digital filter can be described using a very small number of elementary operations. The primitives selected for digital filters are listed in Table 1. Each elementary operation is encoded by its own code (one character) and by two integer numbers, which represent the relative offset (calculated from the current position) of the two operands. When all the offsets are positive (i.e. each block can receive data from previous blocks only), no feedback loop occurs and the resulting structure is a FIR filter. If the relative offsets can be either positive or negative, the resulting filter may have either a FIR or an IIR response, depending on the presence of llops in the signal flow. All primitives include a delay z^{-1}, to avoid possible problems due to timing violations during the synthesis process.

The primitive operators requiring an adder block are more expensive, both in terms of area and of power dissipation. For our purposes, we considered only power consumption: according to gate level simulations, we assigned a relative weight factor to sum, difference and complement.

As an example, the following sequence is made of 6 primitives (6 genes):

(I 0 2) (D 1 3) (L 2 2) (A 2 1) (D 1 0) (S 1 5)

Table 1. Primitives of the genetic algorithm

Name	Code	Op 1	Op 2	Description
Input	I	not used	not used	Copy input: $y_i = x$
Delay	D	n_1	not used	Store value: $y_i = y_{i-n_1} z^{-1}$
Left shift	L	n_1	p	Multiply by 2^p: $y_i = 2^p y_{i-n_1} z^{-1}$
Right shift	R	n_1	p	Divide by 2^p: $y_i = 2^{-p} y_{i-n_1} z^{-1}$
Adder	A	n_1	n_2	Sum: $y_i = (y_{i-n_1} + y_{i-n_2}) z^{-1}$
Subtractor	S	n_1	n_2	Difference: $y_i = (y_{i-n_1} - y_{i-n_2}) z^{-1}$
Complement	C	n_1	not used	Multiply by -1: $y_i = -y_{i-n_1} z^{-1}$

and it is interpreted as follows:

$$y_0 = x$$
$$y_1 = y_0 z^{-1}$$
$$y_2 = 2^2 y_0 z^{-1}$$
$$y_3 = (y_1 + y_2) z^{-1} \tag{4}$$
$$y_4 = y_3 z^{-1}$$
$$y = y_5 = (y_4 - y_0) z^{-1}$$

The last value is the output of the filter. By merging the equations (4), we obtain the transfer function:

$$H(z) = \frac{y}{x} = 5z^{-4} - z^{-1} \tag{5}$$

We note that such representation of a filter has the same essence as a program in a simple imperative programming language, and we can apply genetic programming [13] or, more precisely, Cartesian genetic programming [14] to the task of designing digital filters. Fine granularity of primitives allows a simple genetic encoding, and allows the evolutionary algorithm to perform a better search within the design space.

Efficiency of the designed structure (in term of both area and power consumption) is an important target: it should be optimised under the specification constraints. Evolutionary design have been proven more efficient than structured conventional approaches [15].

It is worth remarking that the proposed genetic representation is different from the evolutionary design of filter coefficients, proposed in previous papers [16,17]. Indeed, the representation of a filter as a sequence of elementary operations is closer to the low-level filter structure, and gives us the opportunity of evaluating directly the power consumption through the switching activity of digital nodes.

Moreover, the proposed design approach is fully compliant with the well-consolidated digital design methodology: the genetic algorithm produced the synthesizable VHDL code, while all the design tools actually used in electronic design flow are still used to translate the design in structural and in physical domains.

4 The Evolutionary Algorithm

In order to make filters evolve, a variable population size island distributed evolutionary algorithm has been implemented. Details on the genetic operators can be found in [6].

It is well known that a genetic algorithm may become very inefficient when the genome size increases: this phenomenon, known as bloat, must be limited as much as possible [18]. For this purpose, the algorithm uses an intelligent mutation operator, which removes useless portions of the genome.

814

There is very strong evidence that neutral mutations should occur, as they improve the evolvability [19]. Although, this subject is still undere debate. Our results appear to contribute additional evidence to this view. Since neutral mutations conflict with bloat limitation, a suitable trade-off has to be found. Moreover, to avoid tuning of parameters for different specifications, crossover, mutation and selection parameters are dynamically adapted [20].

4.1 Fitness Function

The evolutionary algorithm must find a design that satisfies user specifications and minimises transition activity of digital gates. These two objectives, however, are hierarchical: specification fulfilment is mandatory, while reduction of digital transitions is just an additional quality, which is meaningful only when the circuit is fully compliant with requests.

Therefore, a two-step fitness function has been devised. First of all, we consider filter specifications (given as a frequency mask): the frequency range is sampled at N equally spaced frequencies ω_i, and the filter response $H(\omega_i)$ is calculated for every ω_i in the pass-band and in the stop-band. The partial error ε_i is set to zero, if $|H(\omega_i)|$ lies within the specifications; otherwise it is proportional to the overshoot or undershoot with respect to the given tolerance. The total error E_{tot} is simply the sum of all partial errors: it is zero when the frequency response of the filter lies within the mask; otherwise, it has a non-zero value. Mask fitness is defined as:

$$f_M = \frac{1}{1 + E_{\text{tot}}} \tag{6}$$

It measures the extent to which the filter complies with frequency specifications and $f_M = 1$ when specifications are completely met.

The second step is the evaluation of the activity fitness, defined as:

$$f_A = 1 + \frac{a}{N_T} \tag{7}$$

where N_T is the number of weighted digital transitions per input sample and a is a constant; its contribution is higher for filters with low transition activity, which is responsible for power consumption.

Finally, the overall fitness is defined as:

$$f = \begin{cases} f_M & \text{if } E_{\text{tot}} > 0 \\ f_M + f_A & \text{otherwise} \end{cases} \tag{8}$$

Such a definition guarantees that circuits complying with specifications have a larger fitness than circuits with a response outside the frequency mask.

4.2 Selection

Different selection strategies have been implemented:

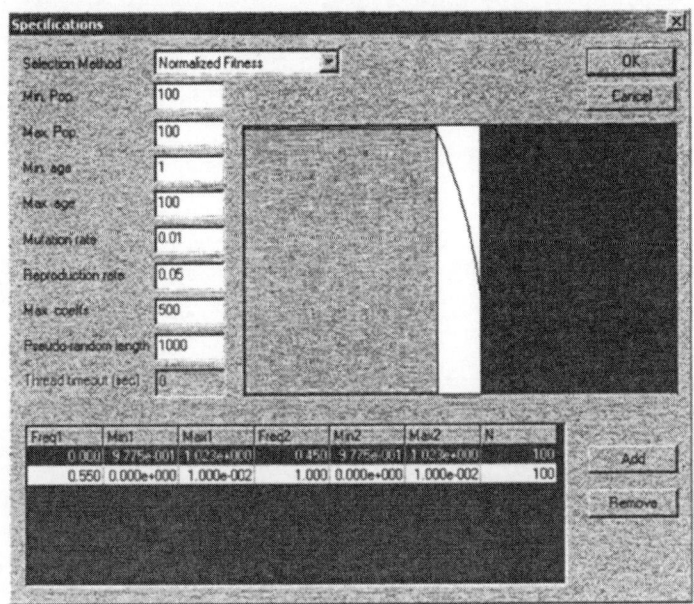

Fig. 5. Client graphical interface with frequency mask

1. an ìagingî rule, which assigns a life expectation to every individual at its birth, according to its ranking: life expectation linearly decreases from 100 to 1 generation as the ranking goes from best to worst;
2. Ýtness-proportionate selection, whereby each individual has a survival probabaility equal to its normalised Ýtness;
3. linear ranking selection with elitism, whereby the best two individuals survive with probability $P_s = 1$, and the individual of rank i has a probability $P_s = 1 - \frac{i}{2Ø}$ of surviving, where Øis a parameter of the algorithm affecting the average population size;
4. selection based on normalised Ýtness and ranking, whereby each individual has a survival probability $P_s = \sqrt{f_i' r_i'}$ where f_i' and r_i' are the normalised Ýtness and rank;
5. histogram-based selection, whereby the selection aims at a uniform distribution of the Ýtness.

Experimental evidence shows that each selection strategy performs better in a particular stage of the evolution. For instance, we have seen that the ìagingî selection favours a faster evolution at early stages, while normalised Ýtness and ranking selection give better results later on. However, a detailed discussion on selection strategies is beyond the scope of this paper. In the examples presented below, all strategies have been used in parallel.

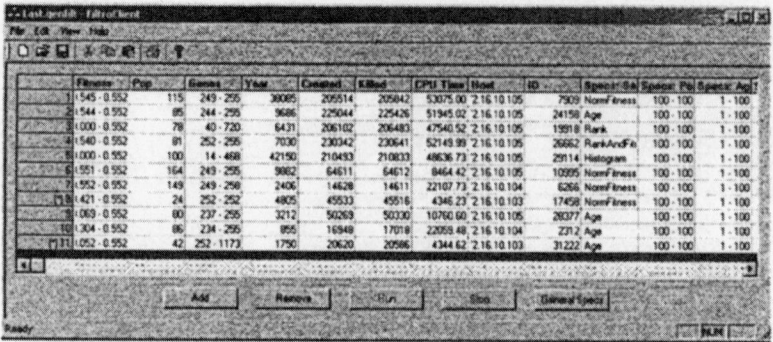

Fig. 6. Client interface with thread list

5 Client-Server Implementation

The algorithm has been implemented with a client-server approach. Communication between the two parts occurs through a TCP/IP socket. The server part may run on several computers, and accepts connections from a client. The client has a graphical user interface, that allows us to input the design parameters. As an example, Fig. 5 illustrates the mask speciÝcations of the Ýlter (i.e., the minimum and maximum attenuation as a function of the normalised frequency).

Using the graphical client interface, a connection to the servers is established and simulated evolutions are run in parallel. Each server starts a new thread for every request from the client. Different threads may have different parameters. The client window is shown in Fig. 6.

Individuals are randomly sent from one thread to another thread, to maximise the success probability of the algorithm. Moreover, when an individual is copied to a different thread, its neutral genes are removed from the genome. This intelligent mutation greatly reduces the probability of bloat.

6 Design Examples

The proposed design method has been applied to the design of two digital Ýlters to be used as decimation stages in a $\Sigma\Delta$ analog-to-digital converter. Filter speciÝcations are listed in Table 2.

Both Ýlters were previously designed using the `remez` function available in the Matlab Signal Processing Toolbox [21], and by approximating ideal coefÝcients with Ýnite precision numbers.

Using the evolutionary algorithm, we were able to obtain designs with better characteristics in terms of both area and power consumption.

For design #1, the genome of the evolved Ýlter has 46 primitives and the resulting frequency response is shown in Fig. 7. It is apparent that the Ýlter meets all design speciÝcations.

817

Table 2. Filter speciÝcations

	Ýlter #1	Ýlter #2
clock frequency	64 MHz	64 MHz
data input rate	16 MHz	8 MHz
data output rate	8 MHz	4 MHz
normalised pass-band	0 ... 0.25	0 ... 0.45
maximum pass-band ripple	0.2 dB	0.2 dB
normalised stop-band	0.75 ... 1	0.55 ... 1
minimum stop-band attenuation	60 dB	40 dB
normalised idonít careî band	0.25 ... 0.75	0.45 ... 0.55

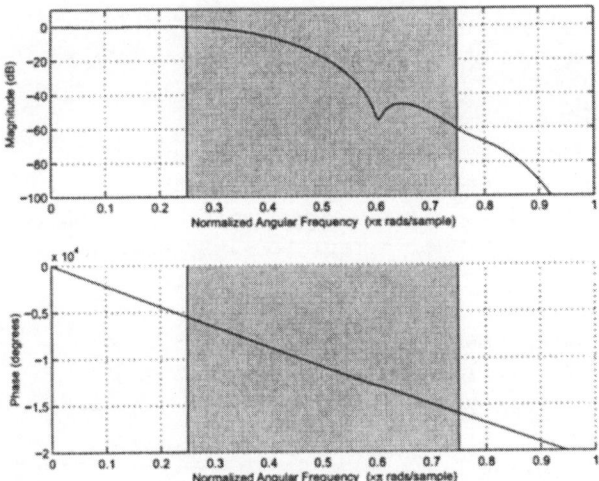

Fig. 7. Frequency response of the evolved Ýlter #1; the grey area represents the ìdonít careî band

Fig. 8 illustrates the VHDL code generated by the algorithm. The RTL (register-transfer logic) architecture [22] is translated by Synopsys into a circuit contining 10,000 equivalent logic gates. The conventional Ýlter implementation required about 40,000 gates to meet the same speciÝcations. By comparing these two Ýgures, we can conclude that the effort of the evolutionary algorithm towards optimisation of transition activity also led to a dramatic reduction of the required hardware, thus reducing the silicon area (and hence the cost) by a factor of 4. From simulation with a pseudorandom input pattern, the evolved Ýlter has a power consumption of 14 mW, while the previously designed Ýlter dissipates as much as 40 mW.

The Ýlter #2 is more demanding, due to its steeper transition band. Its evolutionary design required several days of CPU time on Ýve computers running in parallel. Fig. 9 shows its frequency response.

```
entity filter is
port(
  y:out std_logic_vector(29 downto 0);
  clk:in std_logic;
  rst:in std_logic;
  x:in std_logic_vector(15 downto 0)
);
end entity;

architecture RTL of filter is
--filter signals
signal y0: std_logic_vector(16 downto 0);
signal y1: std_logic_vector(16 downto 0);
...
signal y45: std_logic_vector(29 downto 0);

begin
y <= y45;
gene0: adder generic map(Bits_x1=>x'length,
    Bits_x2=>x'length, Bits_y=>y0'length)
    port map(x1=>x, x2=>x, y=>y0, clk=>clk, rst=>rst);
gene1: change_s generic map(Bits=>y2'length)
    port map(x=>x, y=>y1, clk=>clk, rst=>rst);
...
gene45: adder generic map(Bits_x1=>y43'length,
    Bits_x2=>y42'length, Bits_y=>y45'length)
    port map(x1=>y43, x2=>y42, y=>y45, clk=>clk, rst=>rst);
end RTL;
```

Fig. 8. VHDL synthesisable code generated by the algorithm

Table 3. Characteristics of the Ýlters (after synthesis)

	Ýlter #1		Ýlter #2	
	evolutionary	conventional	evolutionary	conventional
No. of coefÝcients	17	13	104	101
No. of primitives	32	60	220	309
No. of logic gates	10,000	40,000	69,000	105,000
Power dissipation	14 mW	40 mW	37 mW	59 mW

Finally, Table 3 compares the characteristics of evolutionary designs with conventional ones.

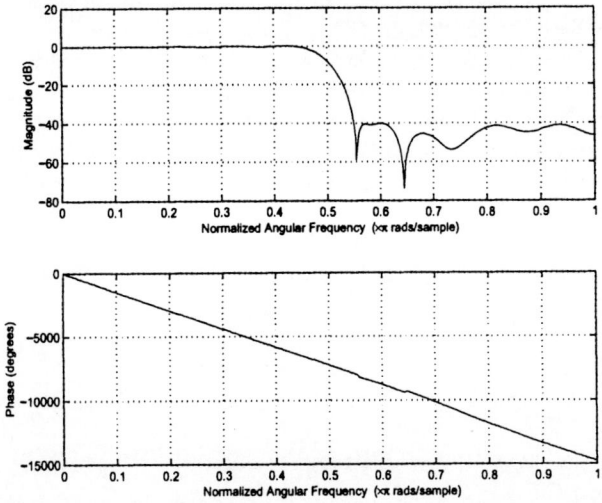

Fig. 9. Frequency response of the evolved Ýlter #2

7 Conclusion

This paper has described an evolutionary approach to the design of digital Ýlters. Genetic encoding of Ýlter primitives has a Ýne granularity which is exploited by the evolutionary algorithm during its random search. The encoding is also a straightforward representation of the time-domain response of the Ýlter, thus allowing a direct evaluation of cost and transition activity of digital blocks. A Ýtness function has been devised to allow multi-objective evolution. The ìbestî result produced by the algorithm is automatically translated into VHDL code, which can be synthesised into a circuit without any additional operation, according to the standard digital design methodology. Client-server implementation of the evolutionary algorithm allows to run multiple threads in parallel.

The results obtained with the simulated evolution show that minimisation of transition activity leads to a dramatic reduction of the hardware with respect to the conventional design methodology, while maintaining the same Ýlter performance.

References

1. Darwin, C.: On the Origin of Species by Means of Natural Selection. John Murray, London, UK (1859)
2. Bäk, T.: Evolutionary Algorithms in Theory and Practice. Oxford University Press, Oxford, UK (1996)
3. Drechsler, R.: Evolutionary Algorithms for VLSI CAD. Kluwer Academic Publishers, Dordrecht, The Netherlands (1998)
4. Sipper, M., Mange, D., Sanchez, E.: Quo vadis evolvable hardware? Communications of the ACM **42** (1999) 50ñ56

5. Thompson, A., Layzell, P.: Analysis of unconventional evolved electronics. Communications of the ACM **42** (1999) 71ñ79

6. Erba, M., Rossi, R., Liberali, V., Tettamanzi, A.G.B.: An evolutionary approach to automatic generation of VHDL code for low-power digital Ýlters. In: Genetic Programming ñ Proc. 4th European Conference (EuroGP2001), Como, Italy (2001) 36ñ50

7. Wakabayashi, K., Okamoto, T.: C-based SoC design Ðow and EDA tools: An ASIC and system vendor perspective. IEEE Trans. Computer-Aided Design of Integr. Circ. and Syst. **19** (2000) 1507ñ1522

8. Tsui, C.Y., Monteiro, J., Pedram, M., Devadas, S., Despain, A.M., Lin, B.: Power estimation methods for sequential logic circuits. IEEE Trans. VLSI Systems **3** (1995) 404ñ416

9. Pedram, M.: Power minimization in IC design: Principles and applications. ACM Trans. on Design Automation of Electronic Systems **1** (1996) 3ñ56

10. Jackson, L.B.: Signals, Systems, and Transforms. Addison-Wesley, Reading, MA, USA (1991)

11. Zhao, Q., Tadokoro, Y.: A simple design of FIR Ýlters with power-of-two coefÝcients. IEEE Trans. Circ. and Syst. **35** (1988) 556ñ570

12. Pirsch, P.: Architectures for Digital Signal Processing. John Wiley & Sons, Chichester, UK (1998)

13. Koza, J.R.: Genetic Programming: on the Programming of Computers by Means of Natural Selection. The MIT Press, Cambridge, MA, USA (1993)

14. Miller, J.F., Thomson, P.: Cartesian genetic programming. In: Genetic Programming ñ Proc. 3rd European Conference (EuroGP 2000), Edinburgh, UK (2000) 121ñ132

15. Miller, J.F., Job, D., Vassilev, V.K.: Principles in the evolutionary design of digital circuits ñ Part I. Genetic Programming and Evolvable Machines **1** (2000) 7ñ35

16. Harris, S.P., Ifeachor, E.C.: Automatic design of frequency sampling Ýlters by hybrid genetic algorithm techniques. IEEE Trans. Signal Processing **46** (1998) 3304ñ3314

17. Lee, A., Ahmadi, M., Jullien, G.A., Lashkari, R.S., Miller, W.C.: Design of 1-D FIR Ýlters with genetic algorithm. In: Proc. IEEE Int. Symp. on Circ. and Syst. Volume III., Orlando, FL, USA (1999) 295ñ298

18. McPhee, N.F., Poli, R.: A schema theory analysis of the evolution of size in genetic programming with linear representation. In: Genetic Programming ñ Proc. 4th European Conference (EuroGP 2001), Como, Italy (2001) 108ñ125

19. Yu, T., Miller, J.: Neutrality and evolvability of boolean function landscape. In: Genetic Programming ñ Proc. 4th European Conference (EuroGP 2001), Como, Italy (2001) 204ñ217

20. Niehaus, J., Banzhaf, W.: Adaption of operator probabilities in genetic programming. In: Genetic Programming ñ Proc. 4th European Conference (EuroGP 2001), Como, Italy (2001) 325ñ336

21. The Mathworks, Inc.: Signal Processing Toolbox, Natick, MA, USA. (1983)

22. Weste, N.H.E., Eshraghian, K.: Principles of CMOS VLSI Design: a System Perspective (2nd edition). Addison-Wesley, Reading, MA, USA (1993)

A Comparision of Soft Computing Methods for Reservoir Simulation

Guadalupe Janoski, Srinivas Mukkamala, Andrew H. Sung

silfalco@cs.nmt.edu, srinivas@cs.nmt.edu, sung@cs.nmt.edu
Department of Computer Science
New Mexico Institute of Mining and technology
Socorro NM, 87801

Abstract. As time progresses, more and more oil reservoirs reach maturity; consequently, secondary and tertiary methods of oil recovery have become increasingly important in the petroleum industry. This reality has increased the industry's interest in using simulation as a tool for reservoir evaluation and management to minimize costs and increase efficiency. This paper presents and compares several control methods in regards to the well-known reservoir simulation task of history matching that is performed to calibrate simulators.

1   Introduction

As we enter the new century, the petroleum industry is becoming increasingly dependent on secondary and tertiary methods of oil recovery. This necessity has added to the industry's interest in using simulation as a tool for reservoir evaluation and management to minimize costs and increase efficiency. This paper presents that a combination of soft computing algorithms and cluster computing techniques provides realistic hope for obtaining accurate simulation results in a cost-effective fashion. In particular, we show that parallelized genetic algorithms can be used to perform reservoir production history matching and obtain solutions efficiently.

An important step in calibrating a petroleum reservoir simulator (in our case MASTER, developed by the U.S. Department of Energy), is to perform history matching on a particular reservoir, or field[1,2]. History matching predicts the production of a petroleum reservoir based on its past history. Initial calibration of the simulator is achieved by matching simulator predicted production curves (consisting of the output of oil, gas, water in our experiment) to the reservoir's historical production (in our problem a set of data spanning 1960-1991). This attempted curve matching is named *history matching*. While appearing simple, it is an extremely computationally complex problem. For example dealing with only a small 8 production well section, which we wish to match, we must deal with a search space of over 2^{12954} different possible solutions.

Part of the problem overhead is that fact that we must include 17 wells in the surrounding area for environmental data, and use a multi-layer grided cube consisting of 7 layers, 25 simplified grid regions, with each grid area having over

32 parameters with real number ranges. This is easily visible by using the single level map in Fig. 1 of the well layout [3].

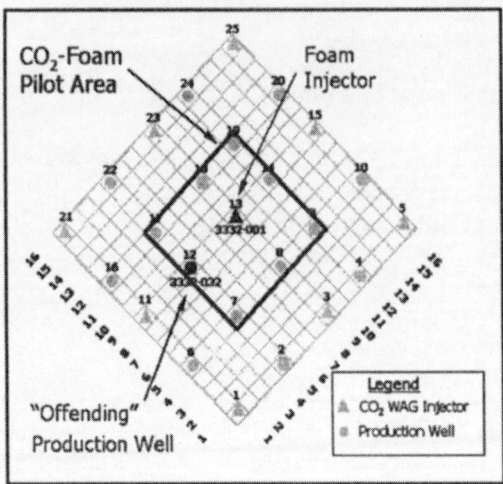

Fig. 1. Well Map. The above image is a well map of the area that was chosen to be used for simulation. The interior 8 production wells {7, 8, 9, 12, 14, 17, 18, 19} were the wells that were to be matched, while the remaining wells were provided for environmental data.

This leads us to three problems. Firstly, we have an enormous data space in which we must locate the best solution, since by nature of the simulator and the data tracked we may never reach an exact solution.

Secondly, we would never be able to solve the problem using traditional methods (trail and error with manual adjustment of parameters, using a single computer) without exponentially increased computational power. Therefore we must hunt for solutions smartly through quick, parallelized searches-thus the ARIA controller (described later), and usage of the cluster.

Thirdly there exists a need for the reduction or elimination of human intervention in the simulation process. The need of a human simulation engineer to manually perform history matching to calibrate a simulator often becomes the most expensive part of the task. Minimizing the necessity for such intervention is thus of high importance, in terms of both cost and efficiency. This paper address this by using a fuzzy/genetic approach for a parameter control to obtain a history match.

In dealing with these problems, we know that the size of the problem will never be reduced. In fact it will only increase with time, since more accurate simulation results depend on finer grids. The other two obstacles, however, can be overcome by implementing a controller for automatic parameter adjustment to minimize human intervention, and executing the integrated simulator-controller on a cluster of computers.

This paper describes our initial investigation pf performing history matching for a reservoir in southern New Mexico that was the site of a CO_2 injection project. Promising preliminary results that show the potential of our approach are presented. In the following sections, we describe the basics of the simulator and the history match model, and several methods for parameter adjustment.

2 Reservoir & Simulator Background

To advance the CO_2-foam technology for improved oil recovery, a pilot area in EVGSAU (covering 7025 acres, in Lea County, New Mexico) was selected in 1990 as a site for a foam field trial to comprehensively evaluate the use of foam for improving the effectiveness of CO_2 injection projects. Specifically, the prime directive of the foam field trail was to prove that a foam could be generated and that it could aid in suppressing the rapid CO_2 breakthrough by reducing the mobility of CO_2 in the reservoir. Operation of the foam field trial began in 1991 and ended in 1993. The response from the foam field trial was very positive, it successfully demonstrated that a strong foam could be formed *in situ* at reservoir conditions and that the diversion of CO_2 to previously bypassed zones/areas due to foam resulted in increased oil production and dramatically decreased CO_2 production.

As part of the CO_2 project, the multi-component pseudo-miscible simulator MASTER (Miscible Applied Simulation Techniques for Energy Recovery), which was developed by the U.S. Department of Energy, was modified by incorporating a foam model and used to conduct a history match study on the pilot area at EVGSAU to understand the process mechanisms and sensitive parameters[1]. The ultimate purpose was to establish a foam predictive model for CO_2-foam processes.[3,4].

3 Expert & Fuzzy Control

Our initial study of the problem was done by a 200 case parameter value study so as to tell the effects of differing parameters. This was extremely complex in deciphering the results since increasing the permeability in one location may have the opposite effect as an increase in another location when both production and injection wells (and their placement times) must be taken into account. To deal with the results and collate the data an initial simple expert system proved to be ideal in that the rapid prototyping and quick modification ability allowed a controller to be built. This initial expert system also proved to be extremely beneficial as a basis for the later fuzzy, and genetic algorithm controllers.

3.1 Expert System

Our initial studies began with the creation of a simple expert system (ES) based controller, which would later form the basis of several other controllers. The rules of the ES controller were formulated empirically from the initial parameter study. The ES was composed of 25 IF-THEN rule groups; one rule per well. These rules used a combination of actual well error values (current parameter set history vs. case), and predicted well error values. See Fig. 2. For ease of use the error values were divided into one of nine ranges that described the degree of error within the match and type: below or above the actual history. Each rule set was rune in sequence and the resulting predicted set passed to the next rule to be used as an actual set.

While the ES controller method proved invaluable as it allowed for rapid proto-typing, and quick reduction of error in the match, it was not without its faults. The primary problem was the granularity induced by having only 9 error ranges. Standardized parameter alteration values, tended to cause oscillation as error ranges would tend to bounce between two opposing sets (such as H to L, and L to H), in later runs. See Figure 3. This was primarily due to the fact that wells 8 and 12 tended to work inversely of each other within the primary depletion period this led to oscillation which tended to occur as the match on one well would be improved, the other well's match would worsen. Despite this reductions in error by over 800% by the fourth or fifth iteration of the ES were not uncommon.

```
If well error value for well 8 is SH Slightly High
and well error value for well 12 is SL Slightly Low
then decrease parameter 3 by 30. Change predicted
set for well 8 and 12 to K.
```

Fig. 2. The above is a partial IF-THEN rule for parameter 3, which is located at well 3 on the well map. The underlined text denotes well error ranges.

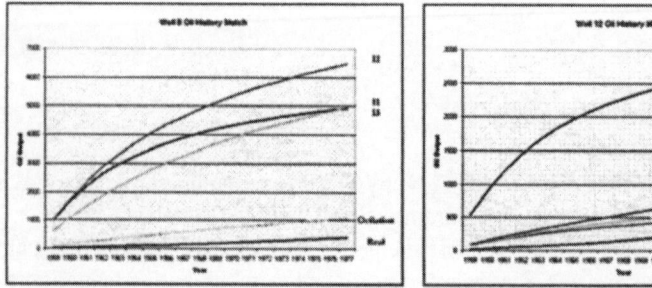

Fig. 3. The above two graphs, are a the result, of a match on the primary depletion period, for wells 8 and 12 (left and right respectively). As can be seen, initial error of the match decreases rapidly, towards the case being matched (actual historical data in this simulation).

3.2 Fuzzy Control

The granularity problem, along with the fact that as we would need to later increase the number of parameters, and increasing rule set size and complexity , along with dealing with the secondary depletion period required that we find a better control method than the ES. As our next step we decided on a fuzzy logic based control system. This system would allow us to go from a twenty plus page rule sets, to a few simple tables, several of which could be reduced to a few simple equations.

The fuzzy controller (FC) also proved beneficial in that this system could be easily automated as it when through the simulation cycle. It was this controller that truly began dealing with the problem of reducing human intervention.

The purpose of the fuzzy controller was to do the parameter adjustment automatically and eliminate human intervention. The benefits of fuzzy control in this application are the ability to get around the problems of complexity in formulating exact rules and to deal with situations where there are multiple meta rules that may be applicable under similar circumstances. For example, expert opinion about permeability adjustment leads to the development of three different meta-rules:

1. If both wells' outputs are too high, then choose those blocks whose reduction in permeability leads to low outputs.
2. If wells' outputs are too low, then choose those blocks whose increase in permeability leads to high outputs.
3. If one well's output is too high and the other's is too low, then choose those blocks whose alteration in permeability leads to proportional, corrective shifts of outputs.

Rules of the third type were the most difficult to obtain, since many factors need to be considered before a decision was made regarding which blocks' permeability to increase and which blocks to decrease, thus the need for developing the rules empirically. If it was not for the ES that formed a basis for the fuzzy controller, and the initial parameter study creating these rules would have been impossible as some of the wells interactions where highly complex.

The fuzzy controller consists of sections:

1. Fuzzification Module: Accepts condition/Input and calculated membership grades to express measurement uncertainties.
2. Fuzzy Inference Engine: Uses the fuzzified measurements and the rules in the rule base to evaluate the measurements.
3. Fuzzy Rule Base: contains the list of fuzzy rules.
4. Defuzzification Module: converts the conclusion reached by the inference engine, into a single real number answer

The primary benefits of using fuzzy control is that it is easy to design and tune, and it avoids the difficulty of formulating exact rules for control actions. The fuzzy controller's rules are empirically obtained, based on a parameter study in which a single well's permeability value was altered while the rest of the 24 permeability values were held constant. The fuzzy controller implemented for permeability adjustment is of the simplest kind in that percentage errors and control actions are fuzzified, but only rarely will more than one rule fire[5,6]. The control action applied is thus usually only scaled by the membership grade of the percentage error in the error fuzzy set. The adaptive controller works as follows.

1. Fuzzification is accomplished by usage of membership functions. After a simulation is run, an error calculation is made from the simulated and the synthetic case or historical data based on a percent error formula. This value is then used to determine error values membership in each fuzzy set: {EL Extremely Low, VL Very Low, L Low, SL Slightly Low, K within tolerance, SH Slightly High, H High, VH Very High, EH Extremely High}. The corresponding fuzzy set values are -4, -3, -2, -1, 0, 1, 2, 3, 4, respectively.

2. Inference begins once the membership grades are calculated. It assigns the fuzzy set with the highest membership value for each well. If an equilibrium condition is reached between two sets, the set value closest to K is chosen.

3. Rule Firing is our next step. Within the fuzzy rule base there are 3 types of rules: I (increase production rules), D (decrease production rules), and P (shift production from one well to another). Based on the fuzzy set assigned to each well, we can decide the rule type that needs to be applied.. Based on the fuzzy set value assigned to each well, we can calculate the average set distance from K and decide the change degree (firing strength) of a rule that needs to be applied, that needs to be applied.

4. The final step is application of the control action. The action taken depends on the chosen rule type and the degree change needed. The parameters for the next simulation run are now altered.

Many experiments have been conducted [7]. The fuzzy controller's performance depends, naturally, on the definition of fuzzy sets for error and the definition of the fuzzy sets for control actions; therefore, the rule base needs to be fine tuned for optimal performance. Since the rules must be based on empirical observations, other factors, such as scaling factors of the controller [7], may not be quite as critical. The basic idea of using a fuzzy controller for automatic parameter adjustment in history matching, however, has been validated by using a specific controller with crisp control actions. In this case we were able to obtain very good matches within 5 iterations for the two wells over their primary production period of 18 years. Previously, with manual adjustment, such close matches would easily take several weeks to a few months to achieve.

4 Genetic Algorithms

Initial genetic algorithm (GA) trials were run using differing crossover methods. These studies proved interesting in that little information was needed in creating the GA system, but at the same time proved to have a huge drawback. As simulation times could range up to 45 minutes on even a 600MHz, creating initial populations, and simulating future generations, became extremely costly providing a large negative to this method. As a result in our study smaller populations for initial testing were used, thus limiting the GA, as large degrees of similarity occurred between population members in succeeding generations.

This method also proved to be interesting as an initial study would not require large amounts of time to study the problem and adapt a GA to solve it. In contrast, while results were promising, much fine-tuning based on previously acquired knowledge was needed to take this method to the next level.

In doing using multipoint crossover. Population improvement tended toward only a 1-3% change in the initial few generations. In using standard crossover the best results were found using a complete generational replacement scheme with small random number of crossover points for each new child.

5 Hybrid Systems: A Classifier System

In dealing with the previous systems one obvious question is how could we get a system to learn its own rules. The following hybrid system is the result.

ARIA (Automatic Recombinant Input Approach) uses a nonstandard genetic based learning machine (GBML) to create a set of rules that may be used for atom rule creation for control in the history matching.

Each rule consists of a set of atom like subsections. For example in rule 1 below each line is an atom section that we work with:

Rule 1:
Error Environment Match:
Error Well X=range
Actions:
Change Parameter N by X
Statistics:
Age, Uses, Accuracy

Classifier 1:
Error Calculations:
    Error for well N= Error
Parameter List:
    Parameter N=pN

Fig. 4. ARIA example rule and classifier

ARIA consists of rule populations that are tested using actual application data, and are tracked based on their effectiveness, in altering a well parameter set. It uses a standard genetic algorithm classifier messaging system.

The system consists of four parts:

1. Part one is the error reception part (environmental interface) in which a parameter set to be adjusted is received.
2. Part two is the rule base of previously composed rules.
3. Part three is a genetic algorithm with fuzzy control section that creates new rules when a rule is not available from the database.
4. The final part is the messaging system that tracks a rule's effectiveness, matches a rule to a parameter set and takes relevant action.

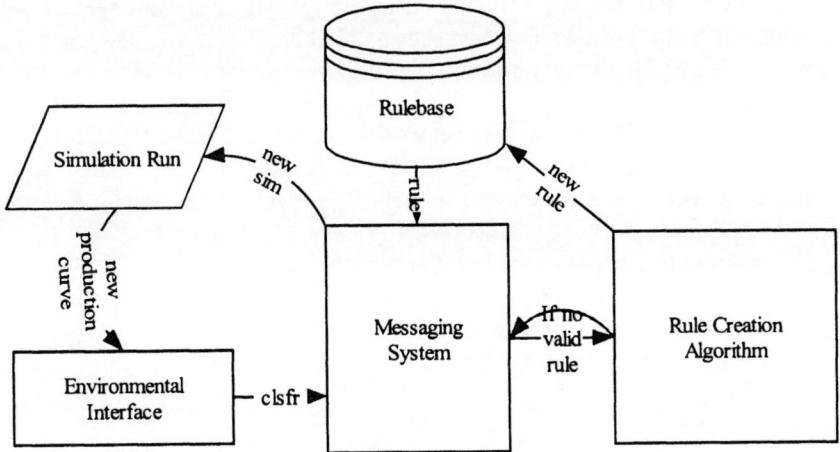

Fig. 5. ARIA pictorial overview

The first part of the ARIA process just accepts parameter simulation data, and creates error estimates from the actual historical production data creating the first part of our environmental awareness. This error calculation is done by using a Sum of Squared Errors calculation between the predicted output of each well and its historical values. Once the SSE has been calculated for each well, these 8 values become the environmental pattern (classifier) which will be used in the messaging system for rule matching. Figure 6. is an example of an environmental classifier. It has two parts the Error calculations, and the list of parameters that belonged to the simulation data.

These error values in the classifier are then matched based on a set of fuzzy rules in the messaging system to appropriate action rules. The fuzzy control in this section is very basic and consists of simplistic rules that determine if a rule exists within range, a tolerance factor, and rates each rule by its statistical information, and then determines which is the most appropriate rule. This is done by attempting to find a rule whose Error Environmental Match ranges for each well bracket the classifiers error calculations. Since there do exist 8 error calculations the idea of a tolerance factor was introduced, in which not all 8 error calculations must be in range. This calculation is done by using an averaging function to calculate how out of range the error set is. If an adequate rule is found, it is then used and statistical success or failure data of its use on the simulation parameters is average together. On the other hand, if an appropriate rule cannot be located the ARIA system invokes the genetic fuzzy logic creation algorithm to create a new rule, which is then used.

This method has shown some promise in application, and is merely an extension off of the previous work for the MASTER WEB project in which fuzzy control was applied resulting in convergence and error control within ten generation, and 200% error ranges, proving to be a very quick and accurate system. Currently the system has been running small numbers of iterations, as tests are being run to determine the best initial rule population. Currently small numbers of changes, that rely on being able to affect parameters within the lower third of their value ranges, without causing parameters to go out side of their allowed values, have shown the most successful ability to converge to a solution. They have been able to come within a 30 to 70% error within approximately 15 iterations.

The size of the rule base has also been shown to have a significant effect on the number of iterations, as the larger the size the more likely an appropriate rule will be found. Created rules have are extremely dependent on the genetic algorithm used, as wells have complex interactions.

This method had several benefits. For example, as we started to deal with the secondary depletion period and the increasingly complex well interactions, this system could actually discover them. This provided surprising knowledge, and a system capable to adaptation, even if it started with false premises.

6 Neural Networks

Using a neural network for modeling has produced interesting results. In this section we present the results of two different networks. The first section will display results obtaining permeability values using period I data with attempting to history match to a synthetic case. In section 6.2 we present the results of a second network that history matches using an increased number of parameters (i.e. an addition 7 relative permeability parameters in addition to the 25 permeability parameters).

6.1 Synthetic Case

The neural net method we chosen for the problem takes a group of predicted oil production output as training data. The data was acquired by running the MASTER simulator with a set of ranged values so as to cover the hypothetical case's production output curve. This resulted in a set of curves, which bracketed the synthetic case in the first neural network, and the actual history in the second. It was necessary for the training data to cover the synthetic case history so to restrict the range of the problems. The figure below shows a very small set of cases that cover the history. The solid line in is the synthetic case history.

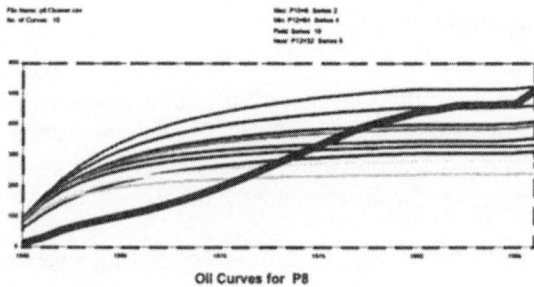

Oil Curves for P8

Fig. 6. The figure demonstrates the bracketing curves for the synthetic case (solid line).

Once the network is well trained, we feed the network with the historical data to get 25 permeability parameters and 7 relative permeability parameters. We then feed these parameters into the MASTER simulator to check if these parameters are acceptable and to create an estimate of the errors for these parameters.

The network we built for this study is a three-layer feed forward network, with 27 input units (historical data), 30 hidden units and 32 outputs (permeabilities). The scaled conjugate gradient descent algorithm is used in training. Figure 5 shows the comparisons between the desired output and the output from a trained network. We can see good matches between predicted value and desired value except for one pair. This mismatch can likely be attributed to the fact that certain permeability values have little effect on the output history. (For example during the first 20 years the 25th permeability value causes less than a 1% change across its complete value range.) Furthermore, Fig. 7 above shows an experimental result in using the neural network to match the chosen hypothetical case, displaying a very close match. Currently the training data and testing data are all simulation results from MASTER simulator. The next section will demonstrate results obtained using real historical data with an increased number of parameters.

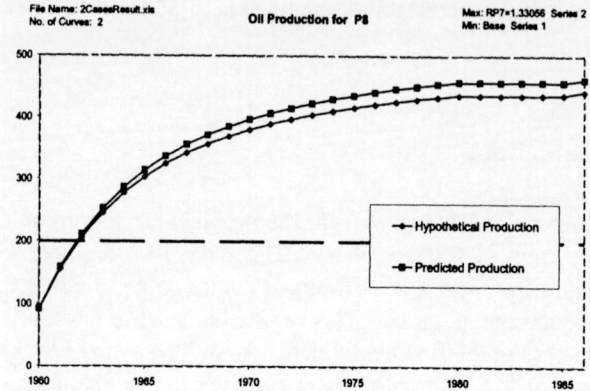

Fig. 7. Synthetic case match using the neural network

6.2 Extended Synthetic Case

The second network is a three-layer feed forward network, with 40 input units (historical data), 20 hidden units, and 32 outputs (7 relative permeability parameters in addition to 25 permeability parameters) which uses gradient decent with momentum algorithm for faster training. In this network we use data from periods I-III.

Using the Mat lab Neural Network Toolbox, we created a network and trained it on 624 pre-chosen cases using the "Gradient descent w/momentum & adaptive linear back propagation" method. We ran the training for 500 epochs, although the system tended to stabilize by 300 epochs.

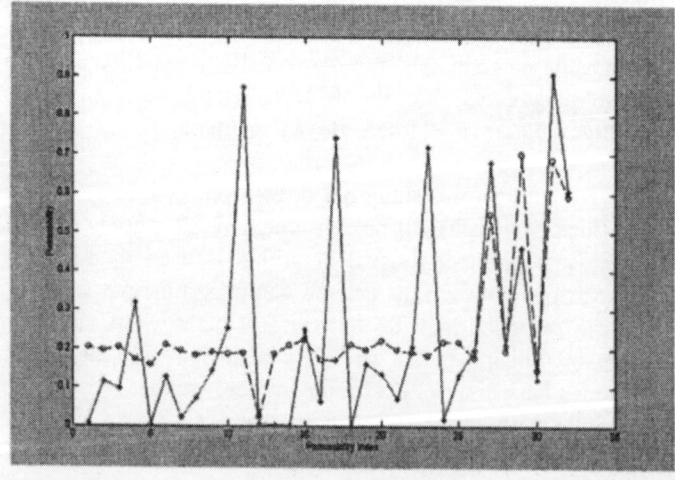

Fig 8. Network Results

The graph of the progress of the training can be seen here below. After training, we plotted one data set used to train the network against the output of the network. As you can see in the graph in Fig. 8, the network seemed unable to match the permeabilities accurately, but worked rather well on the relative permeabilities. You will note that all permeabilities are between 28 and 60. This is consistent with Dr. Chang's suggestion that 50 are a sort of "expected value". Finally, we passed these values to the simulator to see how closely the inputs drive the simulator to the actual production.

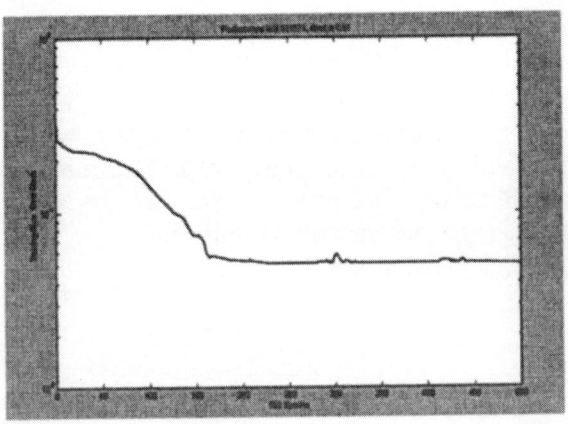

Fig 9. Network Training

As expected, our results were not surprising, using the extended data did not provide us with as perfect of a match as the first network. This can be partially attributed to accumulative data corruption due to such things as well workovers, stock tank oil reserves, etc.

6.3 Neural Network Results

This method has a big advantage to it. Training may be initially costly, but the method has the benefit that once the neural net is trained, the solution can be obtained rapidly-unlike the ES and FC methods which usually take many simulation cycles to minimize error and thus take longer time to find solution.

7 Conclusion

While simulation experts have traditionally done history matching semi-manually, soft computing algorithms offer a great deal of promise for reservoir simulation in general, and history matching in particular. While these algorithms have produced great experimental results in history matching, one must realize that there may be high costs within their use and design. For example the neural network may be costly (primarily in simulation time to create the training set), though actual run

832

time after it has been trained is quick; while the expert and fuzzy controllers run several simulation iterations minimizing error each time.

Preliminary results of applying the soft computing algorithms with the MASTER simulator on the EVGSAU reservoir have shown good matches within hours. These results would have taken weeks to achieve using conventional methods. The algorithms' applicability is also sufficiently general, clearly demonstrating the potential of this approach.

Acknowledgement

We would like to gratefully acknowledge that support for this research was received from Sandia National Laboratories and the State of New Mexico. Dr. Eric Chang (previously with the Computer Science Department of New Mexico Tech) and Dr. Reid Grigg (of the Petroleum Recovery Research Center of New Mexico Tech) initiated our interest in this project and made invaluable contributions during its development.

References

1. Ammer, J.R., Brummert, A.C., and Sams, W.N. (1991) "Miscible Applied Simulation Techniques for Energy Recovery – Version 2.0." Report to U.S. Department of Energy, Contract No. DOE/BC–91/2/SP.
2. Aziz, K., and A. Settari. (1979) Petroleum Reservoir Simulation. London: Applied Science.
3. Chang, S. -H. and Grigg, R. B. "History Matching and Modeling the CO2-Foam Pilot Test at EVGSAU." Paper SPE 39793 presented at the 1998 SPE Permian Basin Oil and Gas Recovery Conference, Midland, Texas.
4. Martin, F.D., Stevens, J.E., and Harpole, K.J. (Nov. 1995) "CO2-Foam Field Test at the East Vacuum Grayburg/San Andres Unit." SPERE 266.
5. Klir, G.J. and Yuan, B. (1995) Fuzzy Sets and Fuzzy Logic: Theory and Applications, Prentice-Hall.
6. Jang, J.-S. R., Sun, C.-T., and Mizutani, E. (1997) Neural-Fuzzy and Soft Computing, Prentice-Hall
7. Janoski, G. (1999) "MASTER Web Reservoir Simulation," Technical Report, Department of Computer Science, New Mexico Tech.
8. Palm R. (1995) "Scaling of Fuzzy Controllers Using the Cross-Correlation," IEEE Tran. Fuzzy Systems, Vol. 3, no. 1, pp.116-123.

Multistage Decision Making for a Fuzzy Automaton by Simulated Annealing

Jiri Pospichal, Vladimir Kvasnicka

Slovak Technical University, Dept. of Mathematics, 812 37 Bratislava, Slovakia

Abstract. A new optimization method, which stochastically builds up a solution step by step in combination with simulated annealing, is used for a multistage decision making of a finite-state automaton. The quality of the new algorithm for larger scale problems was tested by two tasks: (1) maximizing the probability of goal satisfaction with fuzzy goals subject to fuzzy constraints and (2) minimizing the length of a control sequence leading to a specified termination state. The new method required orders of magnitude fewer solution evaluations in comparison with a "classical" genetic algorithm.

1 Introduction

The presented approach presumes, that the model of a studied system under control is known together with an explicit performance function. This restriction for multistage control under fuzziness was used already by Bellman and Zadeh in 1970 [2]. One of the most simple tasks deals with a fuzzy constraint and a fuzzy goal for each decision step, while each state of the controlled system as well as each decision step may be crisp and a termination time is specified. The typical task of the control algorithm is to attain goals and satisfy constraints.

Multistage decision making is usually solved by techniques like the dynamic programming or the branch and bound method. When control steps are crisp and goals and constraints are fuzzy, these methods need defuzzification. The maximizing decision is typically used for its simplicity, though other defuzzification approaches like a center-of-gravity would be better.

When the control steps and states are fuzzy, an efficient application of the dynamic programming and the branch and bound methods also implies restrictions on the used aggregation operator. The min-type fuzzy decision is mostly used, though dynamic programming also works well for a product-type, weighted-sum-type and max-type fuzzy decisions. Various constraints added to a problem cause difficulties to these techniques. They are also plagued by dimensionality problem, which makes it difficult to obtain a solution for practical tasks of a non-trivial size.

Evidently, deterministic methods restrain our efforts only to tasks, which we know how to solve, and cause snipping and mutilating of other problems to fit the requirements of these available techniques.

Kacsprzyk [7,8], who applied genetic algorithms and neural networks to multistage fuzzy control, was the first one who answered the need for new methods. Especially genetic algorithm as an evolutionary optimization technique suitable for multimodal high-dimensional problems is a good choice for this type of tasks, since it does not impose any requirements on a defuzzification procedure or aggregation operator. However, this is not the only robust optimization technique available. Typically a huge number of potential solutions must be evaluated. It is inefficient, when a change of one control step can substantially decrease performance of further control steps, which shows after its basic operators crossover and mutation change subsequence(s) of control steps. The interdependence problem can be illustrated by instructions how to get in a city from one place to another. If we change one instruction (e.g., replace a control step "turn to the left" by a step "turn to the right"), the sequence of instructions would most probably get us very far from our desired destination.

In this case a method without crossover should be preferable. A control sequence should be built step by step, decision about each step should use an information about the part of the sequence already created in previous steps. These requirements are satisfied by the further described modification of a simulated annealing approach. An operation generating new feasible solutions from a currently accepted solution is tailored for given tasks to enhance the performance.

An automaton $A = \langle U, X, f \rangle$ is studied, where U is the set of input states and X is the set of internal states (equated with output states) of A and $f\colon X \times U \to X$ is the state-transition function of A. For a discrete time t ($t \in N$, when N corresponds to a termination time) the next internal state $x_{t+1} \in X$ is a function of its present internal state x_t and the input state $u_t \in U$, $x_{t+1} = f(x_t, u_t)$. Fuzzy constraints $C^t(u_t)$ may be imposed on u_t and fuzzy goals $G^{t+1}(x_{t+1})$ on each x_{t+1} for each $t=0,1,...,N-1$. Optimization of this deterministic system consists in a search for a best sequence of input states (controls) $u_0,...,u_{N-1}$ for a given initial internal state x_0. Instead of the usual random replacement of a control u, a new "tail replacement" method is used, where the first part $u_0,...,u_k$ (of a randomly chosen length k) of the sequence $u_0,...,u_{N-1}$ is copied and the rest of the sequence $u_{k+1},...,u_{N-1}$ is constructed step by step. (When in the previous example of directions to some destination in a city a value of entry u_i "turn to the left" is used instead of "turn to the right", we are able afterwards to get to our destination, because we iteratively change also all the following instructions.) In each step the next value of an entry u_i (from $i=k+1$ up to $N-1$) is pseudorandomly selected by a roulette wheel adopted from genetic algorithm approach. This approach requires a heuristic evaluation of each admissible value of the entry u_i by a "fitness" function. (If we have to choose from instructions "turn to the left", "turn to the right" or "go straight" and we know, in which geographic direction our destination is – e.g. it is a high tower which can be seen from far away, the greatest fitness would logically go to the control, which would turn us toward it.)

Pseudorandom selection means, that an entry u_i with a greater value of a "fitness" function would be selected with a greater probability. This "fitness" function can be computed using a fuzzy goal and/or a fuzzy constraint for a given time step. A value of the entry u_i with a better "fitness" should better satisfy constraints C^t, and at the same time, when applied to a current state x_i, it should produce a new state x_{i+1} satisfying better a goal G^{i+1}. (In the city directions example, a better goal satisfaction might mean a position x_{i+1} in a city closer to our desired destination, and constrains could mean, that the instruction u_i "turn to the left" would lead us to a steep hill, and we do not want to exhaust ourselves.)

The desire to attain a goal G and satisfy a constraint C can be translated into a maximization of a fuzzy expression $\mu_{C^i}(u_i)*\mu_{G^{i+1}}(x_{i+1})$, where $\mu_{C^i}(u_i)$ is a membership function of constraint for a given time i, which maps U into the unit interval [0,1], similarly $\mu_{G^{i+1}}(x_{i+1})$ is a membership function of goal for a time $i+1$, and "*" is an aggregation operator (mostly minimum or product).

The newly proposed algorithm is quite general. When the goals and the constraints are given for each time step, it is easy to construct a function, which will evaluate a suitability of a decision for the current time step. This function would be used as a heuristic assisting in construction of a sequence of decisions, which would be best from a global perspective (while the "global perspective function" is related in some way to the goals and constraints for each time step, like global optimization goal to local optimization heuristics).

Unfortunately, no library or a test set of examples yet exist for this type of multistage decision making. To prove the efficiency of the proposed algorithm, it is compared with a genetic algorithm applied to artificial cases of multistage control; a description of real practical tasks would be too lengthy.

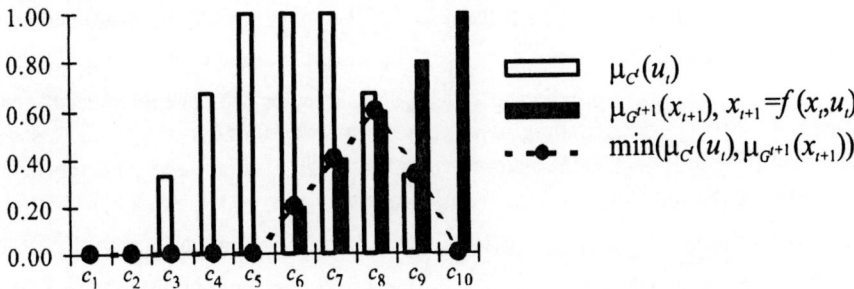

Fig. 1. Hollow rectangles show grades of a membership function for the fuzzy constraint C^t for each possible control step for a given time stage t. Filled rectangles show grades of a membership functions for the fuzzy goal G^{t+1} in the next time stage $t+1$, for a state resulting from the application of the adjacent control step $u_t \in U = \{c_1,...,c_{10}\}$ on a state x_t (the same state x_t for applications of all u_t). The decision making problem in fuzzy environment "attain goal G and satisfy constraint C" can be translated into maximization of $\mu_{C^t}(u_t)*\mu_{G^{t+1}}(x_{t+1})$ where * may be min aggregation operator as in the figure; its result is shown as black circles connected by dotted line, the best control step in the figure is clearly $u_t = c_8$.

The first task is based on a textbook example from Kacprzyk [8], where different fuzzy constraints C and fuzzy goals G are imposed at each time stage t on the control step u_t and the resulting state x_t, and the termination time is given. Since the fuzzy constraints do not depend on a current state x_i, this task is simpler than previously mentioned city directions. The optimization task is to maximize the term $\mu_{C^0}(u_0)*\mu_{G^1}(x_1)*\mu_{C^1}(u_1)*\mu_{G^2}(x_2)*...*\mu_{C^{N-1}}(u_{N-1})*\mu_{G^N}(x_N)$ for a specified aggregation operator * (see Fig. 1).

The other task is inspired by a knight path problem. Consider a knight, starting in the left lower corner of the chessboard, which must visit all the positions in the upper row and return to the right lower corner of the chessboard using the smallest number of moves. A method solving this recreational problem can be used for serious applications, where several goals must be all achieved in an arbitrary (possibly interrupted) succession, without dependence on the order, in which they were achieved, except for the last goal. No fuzzy constraints will be considered, but the performance of the algorithm will be supported by an introduction of a fuzzy variable "closeness to a goal", used for a construction of a control sequence.

2 Multistage control with fuzzy constraints and goals

2.1 Illustrative task description

Let us have an automaton $A = \langle U, X, f \rangle$ with the state space – the set of internal states (equated with output states) $X = \{s_1, \dots, s_{20}\}$, the control space – the set of input states (controls) $U = \{c_1, \dots, c_{32}\}$, the planning horizon $N = 10$ (length of a sequence of controls, after which the result of the sequence will be evaluated as a whole) and the initial state $x_0 = s_1$. The controls are "evenly spaced" real numbers in $[0,1]$ corresponding to $c_1, \dots, c_{20}$, while states are "evenly spaced" real numbers in $[0,1]$ corresponding to $s_1, \dots, s_{32}$, with $s_1 = c_1 = 0$ and $s_{20} = c_{32} = 1$. The state transition $f: X \times U \to X$ is the state-transition function of A (see Fig. 2) determined by an equation $x_{t+1} = c_a$, $a = \text{round}[\min[\max[0, x_t -0.05 +u_t/2.8)],1] \times 19]$, expressed formally as $x_{t+1} = f(x_t, u_t)$, where round[x] gives the integer closest to x. When the automaton is in a discrete time t, the next internal state $x_{t+1} \in X$ is a function of its present internal state x_t and input state $u_t \in U$, $x_{t+1} = f(x_t, u_t)$. Any of $c_1, \dots, c_{32}$ may be substituted for u_t.

Fuzzy constraint for a time t is now defined in the set of these options as a fuzzy set C^t (a separate constraint is specified for each time t) characterized by its membership function $\mu_{C^t} : U \to [0,1]$, so that $\mu_{C^t}(u_t) \in [0,1]$.

The fuzzy goal for a time stage $t+1$, which is imposed on an inner state x_{t+1} obtained from $x_{t+1} = f(x_t, u_t)$, is defined as a fuzzy set G^{t+1} in the set of options X, characterized by its membership function $\mu_{G^{t+1}} : X \to [0,1]$, so that $\mu_{G^{t+1}}(x_{t+1}) \in [0,1]$. The fuzzy constraints and fuzzy goals at the consecutive control stages are given as trapezoid fuzzy numbers in $[0,1]$, that are equated with the 4-touples (a,b,c,d), where a and d determine the interval, where the fuzzy number has a nonzero membership function, while b and c determine the interval with a membership function equal to 1. The fuzzy constraints and goals (see Fig. 2) are:

$$C^0 = (c_1, c_1, c_4, c_{32}) \quad C^5 = (c_1, c_1, c_{13}, c_{32}) \quad G^1 = (s_1, s_2, s_7, s_9) \quad G^6 = (s_6, s_{10}, s_{16}, s_{18})$$
$$C^1 = (c_1, c_1, c_7, c_{32}) \quad C^6 = (c_1, c_1, c_{15}, c_{32}) \quad G^2 = (s_2, s_3, s_9, s_{11}) \quad G^7 = (s_7, s_{11}, s_{16}, s_{18})$$
$$C^2 = (c_1, c_1, c_9, c_{32}) \quad C^7 = (c_1, c_1, c_{17}, c_{32}) \quad G^3 = (s_3, s_5, s_9, s_{11}) \quad G^8 = (s_9, s_{14}, s_{18}, s_{20})$$
$$C^3 = (c_1, c_1, c_{10}, c_{32}) \quad C^8 = (c_1, c_1, c_{18}, c_{32}) \quad G^4 = (s_4, s_7, s_{12}, s_{14}) \quad G^9 = (s_{11}, s_{16}, s_{20}, s_{20})$$
$$C^4 = (c_1, c_1, c_{12}, c_{32}) \quad C^9 = (c_1, c_1, c_{20}, c_{32}) \quad G^5 = (s_5, s_8, s_{14}, s_{16}) \quad G^{10} = (s_{14}, s_{18}, s_{20}, s_{20})$$

The upper indices determine discrete time, when the constraints to the control steps and goals relating to the resulting states should be used. The task is derived from a similar task by Kacprzyk [8], who unfortunately did not describe the whole state transition equation and therefore his results are not entirely reproducible.

The function used for evaluation of a control sequence is

$$\mu_D(D,S \mid x_0) = \mu_{C^0}(u_0) * \mu_{G^1}(x_1) * \mu_{C^1}(u_1) * \mu_{G^2}(x_2) * \ldots * \mu_{C^9}(u_9) * \mu_{G^{10}}(x_{10}) \quad (1)$$

where the symbol '*' is interpreted as an aggregation operator and D is a sequence of control steps $u_0,\ldots, u_9$, while S is a sequence of states $x_1,\ldots, x_{10}$. The min-type and the product-type aggregation operator was used in this task, but genetic algorithm as well as simulated annealing can easily use also other aggregation operators.

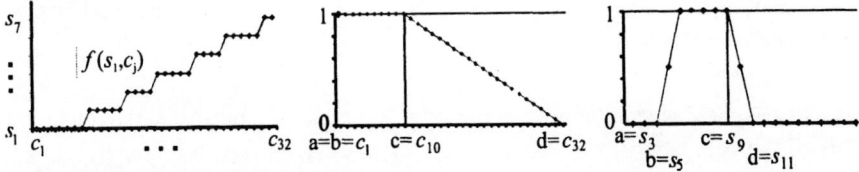

Fig. 2. Graphical representation of a state transition function, when applied on the state s_1. For $c_1 \ldots c_7$ the result is still s_1, for $c_8,\ldots,c_{12}$ the result is s_2, etc. To represent the whole function for 20 states would require 20 such figures, one for a transition function applied to a particular state. The second and third plot represent constraints $C^3 = \{c_1, c_1, c_{10}, c_{32}\}$ and goals $G^3 = \{s_3, s_5, s_9, s_{11}\}$ shown as trapezoid fuzzy numbers.

2.2 Solution by a genetic algorithm

The genetic algorithm used for comparison with our adapted simulated annealing algorithm was based on general description of Kacprzyk [8]. In our comparison we used discrete alleles in a chromosome, which consisted of a sequence of 10 control integer values ranging from 1 to 32. Mutation randomly changed a value of an allele to one of the 32 possible values with a probability 0.1. One-point crossover was used with a probability 0.6 for a couple of chromosomes selected by a roulette wheel [1], otherwise the chromosomes were copied into a new generation without a change. Crossover operated on integer sequences, not on their binary codes. The population size was 250; the results in Fig. 3 correspond to 1000 generations with the min-type aggregation operator. The best value for the presented run was achieved after 126 500 evaluations of chromosomes (if we do not recalculate the fitness of chromosomes, which were not changed by a crossover, we get 126 500×0.6=75 900 evaluations), which corresponds to 506 generations. To get similar result as Kacprzyk, parameters of the genetic algorithm had to be changed, particularly the mutation rate was increased. This slight discrepancy can be explained by probable use of binary coded chromosomes by Kacprzyk.

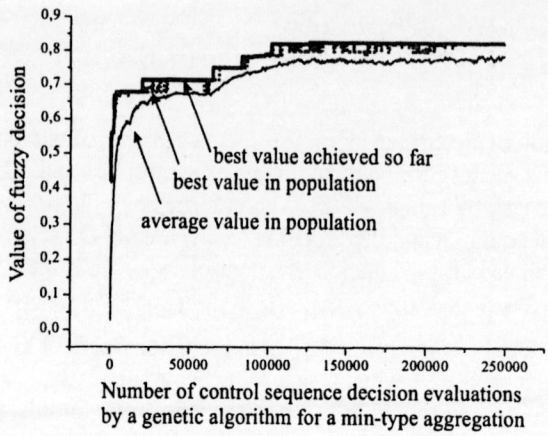

Number of control sequence decision evaluations
by a genetic algorithm for a min-type aggregation

Fig. 3. An example of results of one run of the genetic algorithm for a min type aggregation operator. Although the number of control sequence decision (i.e. chromosomes) evaluation seems to be high, it corresponds only to 1000 generations, and the best known solution was found in half of the run. This corresponds to a similar example by Kacprzyk [8].

The best achieved value for the min-type aggregation operator is 0.826087 with one of the best control sequences $D^*=c_8, c_9, c_{13}, c_{13}, c_{10}, c_{15}, c_{13}, c_{16}, c_{16}, c_{18}$ with corresponding state sequence $S^*=s_1, s_2, s_3, s_5, s_7, s_8, s_{10}, s_{12}, s_{15}, s_{18}, s_{20}$. The same value was achieved only in 4 runs from 10. These 4 runs needed in average 159 500 evaluations to achieve this value, and the average best results of the rest 6 runs after 250 000 evaluations was 0.784212.

The best achieved value for the product-type aggregation operator is 0.582644 with one of the best control sequences $D^*=c_8, c_8, c_{13}, c_{13}, c_8, c_{13}, c_{15}, c_{14}, c_{13}, c_{18}$ with corresponding state sequence $S^*=s_1, s_2, s_3, s_5, s_7, s_8, s_{10}, s_{12}, s_{14}, s_{16}, s_{19}$. From 10 runs, only in 3 runs was achieved solution with the same value. These 3 runs needed in average 53 250 evaluations, and the average best results of the rest 7 runs after 250 000 evaluations was 0.545596.

2.3 Simulated annealing algorithm adapted for multistage control

The simulated annealing algorithm [11,13,16,9] is based on an analogy between annealing of solids and large scale optimization problems. Annealing means heating up solids to a high temperature T, when their particles are randomly positioned, and slow cooling, when the particles arrange themselves into a state with a low energy. On the base of Boltzmann distribution in 1953 Metropolis designed the Monte Carlo Method [11], in which a small random perturbation of a state (or provisional solution of an optimization) is generated, and its energy $E_{perturbed}$ (value) is compared with an energy $E_{current}$ (value) of a current state (solution). If the perturbed state is better than the current one (i.e. it has a lower energy or a better value), the current state is replaced by the perturbed one. However, the current state can be replaced by the perturbed one even if the perturbed state has a higher energy (or lower value). The probability of acceptance is calculated by a formula

$$Pr(perturbed \leftarrow current) = \min(1, exp(-(E_{perturbed} - E_{current})/T) \tag{2}$$

```
procedure Metropolis_algorithm
           (input:D_ini,S_ini,k_max,T;output:D_out,S_out);
begin k:=0; D:=D_ini; S:=S_ini;
      while k<k_max do
      begin k:=k+1; perturbation(D,S,D',S',0);
            Pr:=min(1,exp(-(μ_D(D',S'|x_0)- μ_D(D,S|x_0))/T));
            if random<Pr then begin D:=D'; S:=S'; end;
      end;
      D_out:=D; S_out:=S;
end;
```

Algorithm 1. An implementation of Metropolis algorithm. Procedure is initialized by a control sequence D_{ini} which together with an initial state x_0 determines a corresponding sequence of states S_{ini}. The inner loop is repeated k_{max} times (this number should be sufficiently high to achieve "thermal equilibrium"). Perturbation modifies the control sequence together with the state sequence. Acceptance of a new control and state sequences is decided by Metropolis criterion realized for a "temperature" T. After termination of the Metropolis algorithm, the output D_{out}, S_{out} corresponds to the last accepted solution.

Of course, if the optimization seeks a maximum value, as it is in the present problem, the energy in the formula (2) is replaced by a negative value. In order to achieve an equivalent of a thermal equilibrium for a given temperature T, this algorithm is repeated k_{max} times, as it is shown in Algorithm 1. Since the algorithm worked quite well, the k_{max} was set to 10, which is quite a low value. This number should be substantially higher for more difficult problems.

Simulated annealing method is a sequence of Metropolis algorithms performed for a sequence of decreasing values of temperature. In the present example we have used quite a primitive method of decreasing of a temperature $T:=\alpha*T$, where the decreasing of a temperature was quite steep with parameter α equal to 0.95. The initial temperature T_{max} was set to a high value, typically this value should be such, that about a half of perturbed states would be accepted at the beginning, and the minimum temperature T_{min} set here to 0.01 would be probably lower. However, the results of the simulated annealing algorithm were very good in comparison with the genetic algorithm. Since the parameters of genetic algorithm used in the present paper were not optimized since they were adopted from Kacprzyk [8], in order to get an unbiased comparison it would not have been fair to finely tune all the parameters of the simulated annealing.

```
procedure Simulated_annealing
           (input:T_min,T_max,k_max,α;output:D_opt,S_opt);
begin D:=empty sequence; S:=empty sequence;
      perturbation(D,S,D_ini,S_ini,1); T:=T_max;
      while T>T_min do
```

```
        begin Metropolis_algorithm
              (input:D_ini,S_ini,k_max,T;output:D_out,S_out);
              D_ini:=D_out; S_ini:=S_out; T:=α*T;
        end;
        D_opt:=D_out; S_opt:=S_out;
end;
```

Algorithm 2. An implementation of simulated annealing, input parameters are T_{min}, T_{max}, k_{max}, α, output parameters are D_{opt}, S_{opt}. Algorithm is initialized by a pseudorandomly generated initial solution D_{ini}, S_{ini} and by a maximal temperature T_{max}. While-loop is repeated for $T > T_{min}$, decrease of temperature T is controlled by a formula $T:=\alpha*T$. After the termination of the while-loop the result D_{opt}, S_{opt} is considered as the final solution.

The most substantial extension of the presented approach in comparison with a regular simulated annealing method is the perturbation algorithm. A perturbation is typically a small change of a current solution, similar to a mutation in genetic algorithm. However, here the perturbation used two substantial improvements adjusted to the needs of a multistage control. Firstly, since one change of one control step may substantially affect efficiency of all the following control steps, not only one randomly chosen control step was changed during a perturbation, but all the following steps were changed as well. Secondly, since it is easy to obtain a partial information about quality of each possible control step at a given time level, it would be a waste not to use it. Starting from a given state at a given discrete time level, each control step can be evaluated according to its eligibility. This includes also a suitability of the next state, which results from application of the considered control step to a given state. This evaluation can be obtained from the constraints imposed on the considered control step and from the measure, to which the resulting state satisfies the goal for the next time stage. Of course, one can not estimate correctly an influence of the considered control step on control steps in further time stages. Nevertheless, one can correctly estimate, that if for example the constraints applied for a considered control step are satisfied to a zero level, it is a wrong control step, no matter how good the further steps could be (at least for min- or product-type aggregation).

Given a state x_i in the ith time level, a considered control step c_j is evaluated by a value $\mu_{C^i}(c_j)*\mu_{G^{i+1}}(f(x_i,c_j))$. The probability of the choice of a given control step is then determined by a formula

$$Pr(u_i \leftarrow c_j) = \frac{\left(\mu_{C^i}(c_j)*\mu_{G^{i+1}}(f(x_i,c_j))\right)^{power}}{\sum_{j=1}^{m}\left(\mu_{C^i}(c_j)*\mu_{G^{i+1}}(f(x_i,c_j))\right)^{power}} \tag{3}$$

where the summation goes through all possible control steps. The parameter *power*, an exponent to which the values are raised, is used to enhance the differences between evaluation of considered control steps. In the present example this exponent is set to 20. When all the control steps would be evaluated by zero, all probabilities would be set to $1/m$, where $m = |U|$.

```
procedure perturbation(D,S,D',S',ini);
begin if ini=1 then i:=1; else i:=random(planning horizon);
```

```
      j=1;
      while j<i do  begin  u'_j:=u_j;  x'_{j+1}:=x_{j+1};  j:=j+1;  end;
      for j=i to planning horizon
      begin evaluate all feasible control steps according
            to current state x'_j, constraints_j and goals_{j+1};
            select by a roulette wheel the next control
            step u*_j;  u'_j:=u*_j;  x'_{j+1}:=f(x'_j,u'_j)
      end;
end;
```

Algorithm 3. A perturbation is the core contribution of the presented approach to the simulated annealing multistage decision process. It does not change randomly just one element of a control sequence like in a typical application of simulated annealing. Since suitability of control steps depends heavily on the state at the corresponding level (which is actually result of previous steps), all the next steps after a randomly chosen time are pseudorandomly selected anew. The function random(x) gives integer $\in[0,x-1]$.

The actual selection of a control step is based on the so called roulette wheel adopted from genetic algorithm [1]. Firstly a uniformly distributed random number is drawn from the range $r \in (0,1)$. Then a control step $c_k \in U = \{c_1,...,c_m\}$ is used in a control sequence u_i with an index k determined by

$$k(r) = \max\left\{k \in \{1,m\}: \ r < \sum_{j=1}^{k} Pr(u_i \leftarrow c_j)\right\}. \tag{4}$$

The algorithm used parameters $\alpha=0.95$, $k_{max}=10$, $T_{max}=3.0$ and $T_{min}=0.01$. From 100 runs, algorithm found the best solution in all runs, with average number of trials (control sequence evaluations) equal to 15.9. A solution with the same value was found by a genetic algorithm in the best case after 75 900 trials. For the product-type aggregation operator the algorithm found the best solution in all 100 runs, with average number of trials 38.7. These results are slightly flawed by the fact, that during the buildup of solutions the simulated annealing algorithm had to evaluate all feasible control steps at each level of the control sequence design. This was not necessary in the genetic algorithm. Number of evaluations of simple control steps, that for a simulated annealing approach is number of trial solutions multiplied by 32 (number of possible control steps) and by 5 – average number of control steps that are replaced by a perturbation, we get 15.9×32×5=2544 for min-type aggregation, respectively 38.7×32×5=6192 for product-type aggregation. In the genetic algorithm, the number of trial solutions is multiplied only by 10, and by 0.6 (crossover probability 0.6), resulting in 954 000 (respectively 319 500) control step evaluations. This shows, that even if we do not count the number of evaluated solutions, but the number of evaluations of simple control steps, modified simulated annealing is still better by two orders of magnitude. In that result we even did not consider, that the genetic algorithm in most of cases failed to find a solution of the same quality as the simulated annealing algorithm.

The results are impressive, even though the modeled task is not so extremely difficult. The temperature decreasing scheme was very primitive and the described method could be still improved by using e.g. fast simulated annealing [6].

3 Multistage control with auxiliary fuzzy goals

The second illustrative task involves a deterministic system which terminates after achieving all its goals in whatever order, except for the last one. The number of control steps should be minimized. The suitability of each control step at a time level t depends not only on a current state x_t but also on the already achieved goals, i.e. on the past trajectory. The value of a proper control sequence that achieved all the goals including the final one is evaluated inversely to its length.

Each of the currently admissible control steps is always evaluated by a function calculating the shortest distance of the new state from the nearest of eligible goals.

To visualize more clearly the task, an example is used from chess (similar tasks were studied by another version of simulated annealing algorithm [14]).

3.1 Illustrative task description – knight path on a chessboard

An 8×8 chessboard is given, its squares indexed sequentially from left to right, and bottom to top from 1 to 64. A knight piece is placed on the prescribed initial square 1 of an empty chessboard. It must visit eight prescribed squares – goals (57,58,59,60,61,62,63,64) in an arbitrary succession and after that go to the prescribed terminal square 8. The length of the path should be as short as possible.

The state space is determined by a position $X = \{s_1,\ldots,s_{64}\} = \{1,\ldots,64\}$ of the knight, the control space $U = \{c_1,\ldots,c_8\} = \{-17,-15,-10,-6,6,10,15,17\}$ represents possible moves of the knight. Certain control steps would be forbidden in some positions, since the knight must not go out of the chessboard. The maximum planning horizon would be set to $N = 64$ and the initial state is $x_0 = s_1$. The state transition equation is given as $x_{t+1} = f(x_t, u_t)$, where $f(x_t,u_t) = x_t + u_t$, for $(x_t + u_t) \in X$ and ($|(x_t - 1) \bmod 8 - (x_t + u_t - 1) \bmod 8| + |\lfloor (x_t - 1)/8 \rfloor - \lfloor (x_t + u_t - 1)/8 \rfloor|$)=3, and application of u_t is forbidden otherwise. The function $|z|$ means an absolute value of z, and $\lfloor z \rfloor$ gives the greatest integer less than or equal to z.

3.2 Simulated annealing algorithm adapted for multistage control

The fuzzy goal membership values are calculated as real numbers from the interval [0,1]. For each square at each time stage for a given control sequence a fuzzy goal value can be calculated from the number of moves that are necessary to achieve a nearest eligible goal. A goal square position is eligible if it was not yet visited, terminal position is eligible only after all other goals were visited. The fuzzy goal value of an eligible goal square is set to 1, for any other position it is calculated as 1/(no. of moves to a nearest eligible goal+1). It means, that fuzzy goal values for some position calculated with respect to each eligible goal are then aggregated by a max-type operator. Since it might be computationally demanding to calculate the number of moves to the nearest eligible goal from each square, this number is calculated only for eligible goals achievable in three or less moves,

843

positions farther than three moves from an eligible goal are evaluated by zero. The fuzzy goal values are evaluated at the beginning of the control sequence construction and reevaluated only when a goal is reached. This goal position is then removed from the set of eligible goals and all square positions are reevaluated with respect to their distance to currently eligible goals. After all other goals are visited, the final position (desired terminal square of the knight) is "turned on" as a goal, so that fuzzy goal values of each position are calculated similarly as for the previous goals.

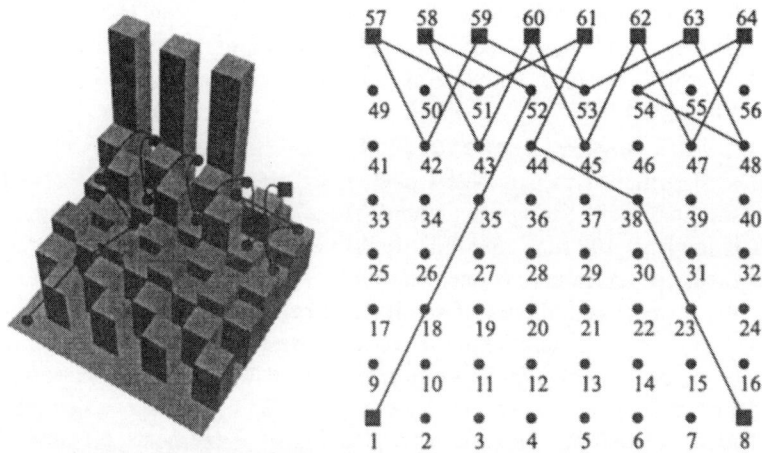

Fig. 4. The graphic on the left represents fuzzy goal values for an unfinished path of a knight on a chessboard, when the knight starts at the left bottom corner (positions of squares on the knight's path are shown by black circles, moves by curved lines) and it should visit all the squares in the last row (goals) and return to the right bottom corner. The current position of a knight is shown by a black square, height of each bar shows a fuzzy goal value for that square position (i.e. represent inverse distance from the nearest eligible goal). The result of such a multistage decision process calculated by adapted simulated annealing is on the right - a shortest path of a knight on a chessboard (with a length 23), with required start from a position 1, termination at position 8 and visiting positions 57,...,64 (marked by black squares).

Given a state x_i at the ith time level, a considered control step c_j is evaluated by a value $\mu_{G^{i+1}}(f(x_i, c_j))$, where the function $\mu_{G^{i+1}}$ is not defined in an assignment, but its evaluation takes into account the previous part of the control sequence. The probability of the choice of a given control step is then determined by a formula (5) for all control steps, which are not forbidden.

$$Pr(u_i \leftarrow c_j) = \frac{\left(\mu_{G^{i+1}}\left(f\left(x_i, c_j\right)\right)\right)^{power}}{\sum_{j=1}^{m}\left(\mu_{G^{i+1}}\left(f\left(x_i, c_j\right)\right)\right)^{power}} \quad (5)$$

The summation goes through all possible control steps and the parameter *power*, an exponent to which the values are raised, is used to enhance the differences between evaluation of considered control steps. In the present example this exponent is set to a number 10. In case that all the control steps would be evaluated by zero, all probabilities are set to $1/m$, where m is the number of eligible control steps. The actual selection of the control step is based on the so called roulette wheel adopted from genetic algorithm [1], as described in the previous example.

When all the goals were visited including the terminal one, or when a number of control steps reaches a limit, the value of a control sequence is calculated as fitness=length_of_control_sequence + 64 × number_of_missed_subgoals.

The probability of acceptance in the Algorithm 1 is for the present example modified as follows:

```
Pr:=min(1,exp(-(fitness(D',S'|xₒ)-fitness(D,S|xₒ))/T));
```

The algorithm used parameters α=0.95, k_{max}=10, T_{max}=3.0 and T_{min}=0.01. The best solution found during calculations had a length 23 and the algorithm needed in average from 100 runs only 6.02 trials to achieve a solution of the same length. The simulated annealing here actually did not have much significance, the special way of consecutive building of a control sequence with help of fuzzy goals was powerful enough to achieve these excellent results. The results achieved by a genetic algorithm solving the same task are not reported here, because genetic algorithm was not able to find such a good solution in any reasonable time.

The Fig. 4 shows a graphic representation of fuzzy goals for an unfinished control sequence and a best control sequence achieved by the algorithm.

4 Conclusions

The adapted simulated annealing approach is compared with a genetic algorithm. Simulated annealing was merged with a consecutive step building, with each step chosen by a roulette wheel approach, when evaluation of a possible step is based on its local suitability calculated from fuzzy constraint and fuzzy goal satisfaction. Perturbation then replaces a whole "tail" of a control sequence instead of replacing only one entry.

The approach in the section 2 can be easily applied for socioeconomic regional development, research and development planning, scheduling and resource allocation, described in [4,8,10,17]. The only adjustments would be necessary in evaluation of the global fitness function in Metropolis algorithm and evaluation of single control steps in the perturbation procedure. The section 3 can be easily adjusted for routing problems described in [3,5,12,15]. The algorithm proved to be for the presented tasks exceptionally effective, with number of evaluated solutions lower by orders of magnitude in comparison with a genetic algorithm. The effectiveness of the presented algorithm for a multistage decision making is however subject to the ability to evaluate to some extent the next control steps for each state, even though the step evaluation may not reflect exactly the influence of the step on the global task evaluation.

Acknowledgments. This work was supported by the grants # 1/7336/20 and # 1/8107/01 of the Scientific Grant Agency of Slovak Republic.

References

1. Baeck T, Fogel DB, Michalewicz Z (eds) (1997) Handbook of evolutionary computation. IOP Publishing, Bristol
2. Bellman RE, Zadeh LA (1970) Decision–making in a fuzzy environment. Management Science, **17(4)**: 141-165
3. Braysy O (1999) A new algorithm for the vehicle routing problem with time windows based on the hybridization of a genetic algorithm and route construction heuristics. Proceedings of the University of Vaasa, Research papers 227, Vaasa, Finland
4. Esogbue AO, Liu B (1997) On Stochastic Multistage Decision Making Under Fuzzy Criteria. In: Proceedings of FUZZY '97: The International Conference on Fuzzy Logic and Applications, Zichron Yachov, Israel , May 18-21, 1997, pp 307-315
5. Cheng R, Gen M (1996) Fuzzy Vehicle Routing and Scheduling Problem Using Genetic Algorithms. In: Herrera F, Verdegay JL (eds) Genetic Algorithms and Soft Computing, Physica-Verlag, pp 683-709
6. Ingber AL (1989) Very fast simulated re-annealing. J Mathematical Computer Modelling **12 (8)**: 967-973
7. Kaczprzyk J (1997) Multistage Evolutionary Optimization of Fuzzy Systems – Application to Optimal Fuzzy Control. In: Pedrycz W (ed) Fuzzy evolutionary computation. Kluwer Academic Publishers, Boston
8. Kaczprzyk J (1997) Multistage Fuzzy Control: A model-based approach to control and decision making. Wiley, Chichester, UK
9. Kvasnicka V, Pospichal J (1996) Simulated Annealing. Communications in Mathematical Chemistry (MATCH) **34**: 7-49
10. Li L, Lai KK (2000) Fuzzy dynamic programming approach to hybrid multiobjective multistage decision-making problems. Fuzzy Sets and Systems,**117(1)**: 13-25
11. Metropolis N, Rosenbluth AW, Rosenbluth MN, Teller AH, Teller E (1953) Equation of State Calculations for Fast Computing Machines. J. Chem. Phys. **21**: 1087-1092
12. Mole RH, Jameson SR (1976) A sequential route-building algorithm employing a generalised savings criterion. Operational Research Quarterly **27(2)**:503-511
13. Otten RHJM, Van Ginneken LPPP (1989) The Annealing Algorithm. Kluwer, Boston
14. Pospichal J, Kvasnicka V (2000) Optimization as a Multistage Decision Making. In: Sincak P, Vascak J, Kvasnicka V, Mesiar R (eds) The State of the Art in Computational Intelligence. Physica Verlag, Heidelberg, pp 175-181
15. Van Breedam A (1995) Improvement heuristics for the vehicle routing problem based on simulated annealing. European Journal of Operational Research **86**: 480-490.
16. Van Laarhoven PMJ, Aarts EHL (1987) Simulated Annealing: Theory and Applications. Reidel, Dordrecht (The Netherlands)
17. Weber K, Sun Z (2000) Fuzzy Stochastic Multistage Decision Process With Implicitly Given Termination Time. In: Hampel R, Wagenknecht M, Chaker N (eds) Fuzzy Control Application - Theory and Practice. Physica-Verlag, Heidelberg

Index of Contributors

Subject Index